Lecture Notes in Mechanical Engineering

Lecture Notes in Mechanical Engineering (LNME) publishes the latest developments in Mechanical Engineering—quickly, informally and with high quality. Original research or contributions reported in proceedings and post-proceedings represents the core of LNME. Volumes published in LNME embrace all aspects, subfields and new challenges of mechanical engineering.

To submit a proposal or request further information, please contact the Springer Editor of your location:

Europe, USA, Africa: Leontina Di Cecco at Leontina.dicecco@springer.com
China: Ella Zhang at ella.zhang@cn.springernature.com
India, Rest of Asia, Australia, New Zealand: Swati Meherishi at swati.meherishi@springer.com

Topics in the series include:

- Engineering Design
- Machinery and Machine Elements
- Mechanical Structures and Stress Analysis
- Automotive Engineering
- Engine Technology
- Aerospace Technology and Astronautics
- Nanotechnology and Microengineering
- Control, Robotics, Mechatronics
- MEMS
- Theoretical and Applied Mechanics
- Dynamical Systems, Control
- Fluid Mechanics
- Engineering Thermodynamics, Heat and Mass Transfer
- Manufacturing Engineering and Smart Manufacturing
- Precision Engineering, Instrumentation, Measurement
- Materials Engineering
- Tribology and Surface Technology

Indexed by SCOPUS, EI Compendex, and INSPEC.

All books published in the series are evaluated by Web of Science for the Conference Proceedings Citation Index (CPCI).

To submit a proposal for a monograph, please check our Springer Tracts in Mechanical Engineering at https://link.springer.com/bookseries/11693.

Justyna Trojanowska · Agnieszka Kujawińska ·
Naiming Xie · Jozef Husár · Thanh T. Tran
Editors

Advances in Manufacturing V

Volume 2 - Production Engineering: Factory of the Future

Springer

Editors
Justyna Trojanowska
Faculty of Mechanical Engineering
Poznan University of Technology
Poznań, Poland

Agnieszka Kujawińska
Faculty of Mechanical Engineering
Poznań University of Technology
Poznań, Poland

Naiming Xie
College of Economics and Management
Nanjing University of Aeronautics
and Astronautics
Nanjing, China

Jozef Husár
Faculty of Manufacturing Technologies
Technical University of Košice
Prešov, Slovakia

Thanh T. Tran
RMIT Vietnam University
Ho Chi Minh city, Vietnam

ISSN 2195-4356 ISSN 2195-4364 (electronic)
Lecture Notes in Mechanical Engineering
ISBN 978-3-032-21653-3 ISBN 978-3-032-21654-0 (eBook)
https://doi.org/10.1007/978-3-032-21654-0

This Springer imprint is published by the registered company Springer Nature Switzerland AG
The registered company address is: Gewerbestrasse 11, 6330 Cham, Switzerland

Preface

This volume of Lecture Notes in Mechanical Engineering gathers accepted papers presented at the 9th International Scientific-Technical Conference MANUFACTURING 2026, held in Poznan, Poland on May 19–21, 2026. The conference was hosted by the Faculty of Mechanical Engineering, Poznan University of Technology, Poland.

The primary objective of the MANUFACTURING 2026 was to showcase cutting-edge advancements across the broad spectrum of mechanical engineering, providing a dynamic forum for the exchange of innovative ideas and expert opinions. Focused on the theme of the Factory of the Future, the conference serves as a vital platform for dialogue on industrial evolution and science-industry synergy. Its global reach continues to expand, as evidenced by the consistently increasing participation from international research centers. Furthermore, the event fosters the integration of academia with the economic sector, encouraging collaboration with industry leaders and management professionals interested in implementing research results into business practice. The scope encompassed the following areas:

- Mechanical Engineering
- Production Engineering
- Quality Engineering
- Measurement and Control Systems
- Biomedical Engineering

The global significance of the conference is reflected in the diversity of the submissions. The Organizing Committee received 219 contributions from 37 countries. To ensure the highest academic standards, all papers underwent a rigorous double-blind peer review process. Each manuscript was evaluated by at least three independent international experts, with a strict limit of five assignments per reviewer to guarantee depth and quality of feedback. Following this comprehensive appraisal, 120 papers were accepted for publication (an acceptance rate of 55%), representing the work of 347 authors from 25 countries. A total of 109 articles were published in the proceedings Advances in Manufacturing V, while 11 selected papers appeared in the journal Management and Production Engineering Review, issued under the auspices of the Committee on Production Engineering of the Polish Academy of Sciences and the Polish Society for Production Management.

The review process focused on originality of paper, significance of presented results, comprehensibility and quality of the presentation, relevance for this conference and overall impression. Furthermore, reviewers meticulously assessed the timeliness of the research, the adequacy of the literature review, methodological rigor, and the clarity of conclusions, as well as the quality of English. Feedback provided to authors included specific recommendations for improving methodology, data presentation, and editorial quality. Acceptance required at least two positive reviews and a full resolution of all reviewers' concerns. This international effort involved experts from 30 countries across Europe, North America, South Asia, and North Africa. Additionally, all manuscripts were checked using anti-plagiarism software, which allowed the similarity index to be less than 15%.

The book **Advances in Manufacturing V** is organized into five volumes that correspond to the main conference disciplines mentioned above.

The **Advances in Manufacturing V - Volume 2 - Production Engineering: Factory of the Future** volume focuses on the paradigm shift toward the Factory of the Future, gathering cutting-edge research and technologies designed to bridge the gap between theoretical excellence and industrial application. The content is structured into five thematic parts, reflecting the multidisciplinary nature of modern production engineering.

- Digital Twins and Virtual Engineering: Covers the evolution of Cognitive and Generative AI-based twins, the use of Cross Reality (XR) for system control, and the role of teamwork in virtual projects.
- AI in Design and Management: Focuses on AI integrability (AIRI), Explainable AI (XAI) in decision-making, and machine learning applications for circular manufacturing and sustainable product design.
- Additive Manufacturing Technologies: Reports on advancements in FDM processes, specifically regarding carbon fiber-reinforced materials, non-contact quality analysis, and intelligent monitoring.
- Monitoring and Control Systems: Highlights TinyML-based vision supervision, production flow control in smart factories (e.g., Digit.io), and scheduling models for discrete production.
- Operational Excellence and Logistics: Addresses Human-Machine-Robot collaboration, cybersecurity, and the evolution of maintenance strategies within sustainable distribution networks.

Comprising 27 chapters prepared by 81 authors from 10 countries, this volume serves as a roadmap for the effective implementation of Industry 4.0 and 5.0 solutions, fostering a vital synergy between science and the global economy.

We wish to express our sincere gratitude to the members of the Scientific and Program Committees for their dedication and expertise during the review process. We also extend our thanks to the authors, sponsors, and all contributors whose commitment was instrumental in making MANUFACTURING 2026 a resounding success.

May 2026

Justyna Trojanowska
Agnieszka Kujawińska
Naiming Xie
Jozef Husár
Thanh T. Tran

Organization

Steering Committee

General Chair

Adam Hamrol	Poznan University of Technology, Poland

Chairs

Bartosz Gapiński	Poznan University of Technology, Poland
Magdalena Hryb	Poznan University of Technology, Poland
Justyna Trojanowska	Poznan University of Technology, Poland

Conference Founders

Zenobia Weiss	Poznan University of Technology, Poland
Adam Hamrol	Poznan University of Technology, Poland

Scientific Committee

Katarzyna Antosz	Rzeszow University of Technology, Poland
Christopher Brown	Worcester Polytechnic Institute, USA
Anna Burduk	Wrocław University of Science and Technology, Poland
Suthep Butdee	Rajamangala University of Technology Krungthep, Thailand
Marcin Butlewski	Poznan University of Technology, Poland
Olaf Ciszak	Poznan University of Technology, Poland
Reggie Davidrajuh	University of Stavanger, Norway
Ewa Dostatni	Poznan University of Technology, Poland
Jan Duda	Cracow University of Technology, Poland
Mosè Gallo	University of Naples Federico II, Italy
Bartosz Gapiński	Poznan University of Technology, Poland
Józef Gawlik	Cracow University of Technology, Poland
Adam Gąska	Cracow University of Technology, Poland
Arkadiusz Gola	Lublin University of Technology, Poland

Filip Górski	Poznan University of Technology, Poland
Marta Grabowska	Poznan University of Technology, Poland
Adam Hamrol	Poznan University of Technology, Poland
Magdalena Hryb	Poznan University of Technology, Poland
Jozef Husár	Technical University of Košice, Slovakia
Aminul Islam	Technical University of Denmark, Denmark
Vitalii Ivanov	Sumy State University, Ukraine
Michał Jakubowicz	Poznan University of Technology, Poland
Andrzej Jardzioch	West Pomeranian University of Technology in Szczecin, Poland
Sabahudin Jasarevic	University of Zenica, Bosnia & Herzegovina
Małgorzata Jasiulewicz-Kaczmarek	Poznan University of Technology, Poland
Anna Karwasz	Poznan University of Technology, Poland
Sławomir Kłos	Poznan University of Technology, Poland
Lucia Knapčíková	Technical University of Košice, Slovakia
Agnieszka Kujawińska	Poznan University of Technology, Poland
Vijaya Kumar Manupati	National Institute of Industrial Engineering, India
Stanisław Legutko	Poznan University of Technology, Poland
Sławomir Luściński	Kielce University of Technology, Poland
José Machado	University of Minho, Portugal
Damjan Maletic	University of Maribor, Slovenia
Dariusz Mikołajewski	Kazimierz Wielki University, Poland
Piotr Moncarz	Stanford University, USA
Magdalena Niemczewska-Wójcik	Cracow University of Technology, Poland
Przemysław Niewiadomski	University of Zielona Góra, Poland
Razvan Ioan Pacurar	Technical University of Cluj-Napoca, Romania
Piotr Paczos	Poznan University of Technology, Poland
Mirosław Pajor	Maritime University of Szczecin, Poland
Justyna Patalas-Maliszewska	University of Zielona Góra, Poland
Ivan Pavlenko	Sumy State University, Ukraine
Erwin Rauch	Free University of Bozen-Bolzano, Italy
Paulina Rewers	Poznan University of Technology, Poland
Izabela Rojek	Kazimierz Wielki University, Poland
Michał Rychlik	Poznan University of Technology, Poland
Krzysztof Santarek	Warsaw University of Technology, Poland
Sebastian Skoczypiec	Cracow University of Technology, Poland
Bożena Skołud	Silesian University of Technology, Poland
Jerzy Andrzej Sładek	Cracow University of Technology, Poland
Beata Starzyńska	Poznan University of Technology, Poland
Liangyan Tao	Nanjing University of Aeronautics and Astronautics, China

Anna Timofiejczuk	Silesian University of Technology, Poland
Justyna Trojanowska	Poznan University of Technology, Poland
Wiesław Urban	Bialystok University of Technology, Poland
Leonilde Varela	University of Minho, Portugal
Nikola Vitković	University of Niš, Serbia
Michał Wieczorowski	Poznan University of Technology, Poland
Ewa Więcek-Janka	Poznan University of Technology, Poland
Szymon Wojciechowski	Poznan University of Technology, Poland
Ralf Woll	BTU Cottbus-Senftenberg, Germany
Naiming Xie	Nanjing University of Aeronautics and Astronautics, China
Jiehan Zhou	Shandong University of Science and Technology, China

Program Committee

Adam Gąska	Cracow University of Technology, Poland
Adam Hamrol	Poznan University of Technology, Poland
Agnieszka Kaczmarek-Pawelska	University of Zielona Góra, Poland
Agnieszka Stachowiak	Poznan University of Technology, Poland
Alejandro Pereira	University of Vigo, Portugal
Aleksandar Miltenovic	University of Nis, Serbia
Alice Chlupová	Czech Academy of Sciences, Czech Republic
Alzbeta Sapietova	University of Žilina, Slovakia
Ancuta Pacurar	Technical University of Cluj-Napoca, Romania
André Guimarães	University of Beira Interior, Portugal
Andrzej Jardzioch	West Pomeranian University of Technology in Szczecin, Poland
Angelika Hofer	Free University of Bozen-Bolzano, Italy
Anna Borucka	Military University of Technology Warsaw, Poland
Anna Burduk	Wroclaw University of Science and Technology, Poland
Anna Karwasz	Poznan University of Technology, Poland
Anna M. Deptuła	Opole University of Technology, Poland
Anna Saniuk	University of Zielona Góra, Poland
Arkadiusz Gola	Lublin University of Technology, Poland
Bartosz Gapinski	Poznan University of Technology, Poland
Bogdan Dybała	Wroclaw University of Science and Technology, Poland
Christopher Brown	Worcester Polytechnic Institute, USA

Cristina Stefana Miron-Borzan	Technical University of Cluj-Napoca, Romania
Dagmara Łapczyńska	Wroclaw University of Science and Technology, Poland
Damian Grajewski	Poznan University of Technology, Poland
Damian Krenczyk	Silesian University of Technology, Poland
Dan-Sorin Comsa	Technical University of Cluj-Napoca, Romania
Daniel Medyński	Witelon Collegium State University, Poland
Dariusz Mikołajewski	Kazimierz Wielki University, Poland
Djahida Belayadi	National School of Cybersecurity, Algeria
Dražan Kozak	University of Slavonski Brod, Croatia
Edward Kozłowski	Lublin University of Technology, Poland
Emilia Campean	Technical University of Cluj-Napoca, Romania
Emilia Sabau	Technical University of Cluj-Napoca, Romania
Eryk Szwarc	Koszalin University of Technology, Poland
Ewa Dostatni	Poznan University of Technology, Poland
Ewa Więcek-Janka	Poznan University of Technology, Poland
Gabriel Stolárik	Technical University of Košice, Slovakia
Grzegorz Pająk	University of Zielona Góra, Poland
Hanna Łosyk	University of Zielona Góra, Poland
Hubert Wojciechowski	Poznan University of Technology, Poland
Izabela Kudelska	Poznan University of Technology, Poland
Izabela Rojek	Kazimierz Wielki University, Poland
Jacek Diakun	Poznan University of Technology, Poland
Jakub Kaščák	Technical University of Košice, Slovakia
Jakub Kopowski	Kazimierz Wielki University, Poland
Jerzy Jozwik	Lublin University of Technology, Poland
Joanna Cyganiuk	University of Zielona Góra, Poland
Joanna Helman	Wroclaw University of Science and Technology, Poland
Joanna Wilczarska	Bydgoszcz University of Science and Technology, Poland
Jozef Husár	Technical University of Košice, Slovakia
Jozef Török	Technical University of Košice, Slovakia
Józef Kuczmaszewski	Lublin University of Technology, Poland
Justyna Patalas-Maliszewska	University of Zielona Góra, Poland
Kamil Krot	Wroclaw University of Science and Technology, Poland
Kamil Leksycki	University of Zielona Góra, Poland
Kamil Musiał	Wroclaw University of Science and Technology, Poland
Karol Dąbrowski	University of Zielona Góra, Poland
Katarzyna Antosz	Rzeszow University of Technology, Poland

Katarzyna Cheba	West Pomeranian University of Technology in Szczecin, Poland
Katarzyna Kazimierska-Drobny	Kazimierz Wielki University, Poland
Katarzyna Skrzypek	University of Zielona Góra, Poland
Khrystyna Berladir	Sumy State University, Ukraine
Konrad Kobiela	Cracow University of Technology, Poland
Krzysztof Kalinowski	Silesian University of Technology, Poland
Krzysztof Nowacki	Silesian University of Technology, Poland
Krzysztof Żywicki	Poznan University of Technology, Poland
Leonilde Varela	University of Minho, Portugal
Lucia Knapčíková	Technical University of Košice, Slovakia
Łukasz Rymaniak	Poznan University of Technology, Poland
Magdalena Hryb	Poznan University of Technology, Poland
Magdalena Rybaczewska-Błażejowska	Kielce University of Technology, Poland
Magdalena Żukowska	Poznan University of Technology, Poland
Małgorzata Śliwa	University of Zielona Góra, Poland
Marcin Suszynski	Poznan University of Technology, Poland
Marek Kočiško	Technical University of Košice, Slovakia
Marek Macko	Kazimierz Wielki University, Poland
Maria Rosienkiewicz	Wroclaw University of Science and Technology, Poland
Marian Bartoszuk	Opole University of Technology, Poland
Marius Sorin Pustan	Technical University of Cluj-Napoca, Romania
Marks Gorohovs	Riga Technical University, Latvia
Marta Grabowska	Poznan University of Technology, Poland
Marta Harničárová	Slovak University of Agriculture in Nitra, Slovakia
Michał Jakubowicz	Poznan University of Technology, Poland
Michał Rogalewicz	Poznan University of Technology, Poland
Mihai Dragomir	Technical University of Cluj-Napoca, Romania
Milan Edl	University of West Bohemia, Czech Republic
Mircea Fulea	Technical University of Cluj-Napoca, Romania
Mukund Harugade	Padmabhooshan Vasantraodada Patil Institute of Technology, India
Natalya Shramenko	Lviv Polytechnic National University, Ukraine
Nataša Náprstková	University of Jan Evangelista in Ústí nad Labem, Czech Republic
Nguyen Hong Vi	Vietnamese German University, Vietnam
Nikola Vitković	University of Nis, Serbia
Olaf Ciszak	Poznan University of Technology, Poland
Olha Kalman	Technical University of Košice, Slovakia

Paulina Rewers	Poznan University of Technology, Poland
Paweł Buń	Poznan University of Technology, Poland
Paweł Szczygieł	Kielce University of Technology, Poland
Paweł Zmarzły	Kielce University of Technology, Poland
Pedro Torres	Polytechnic Institute of Castelo Branco, Portugal
Piotr Gąska	AGH University of Krakow, Poland
Piotr Kotlarz	Kazimierz Wielki University, Poland
Piotr Trojanowski	West Pomeranian University of Technology in Szczecin, Poland
Piotr Ziembicki	University of Zielona Góra, Poland
Przemysław Zawadzki	Poznan University of Technology, Poland
Radosław Wichniarek	Poznan University of Technology, Poland
Răzvan Păcurar	Technical University of Cluj-Napoca, Romania
Roman Kielec	University of Zielona Góra, Poland
Sandra Grabowska	Silesian University of Technology, Poland
Sebastian Saniuk	University of Zielona Góra, Poland
Sebastian Skoczypiec	Cracow University of Technology, Poland
Sergej Hloch	Technical University of Košice, Slovakia
Sławomir Kłos	University of Zielona Góra, Poland
Stanislaw Legutko	Poznan University of Technology, Poland
Stella Hrehova	Technical University of Košice, Slovakia
Sven Maricic	Juraj Dobrila University of Pula, Croatia
Sylvia Kusmierczak	University of Jan Evangelista in Ústí nad Labem, Czech Republic
Szymon Wojciechowski	Poznan University of Technology, Poland
Thanh Tran	RMIT University, Vietnam
Tomasz Bednarek	Kazimierz Wielki University, Poland
Vijaya Kumar Manupati	NITIE Mumbai, India
Vitalii Ivanov	Sumy State University, Ukraine
Wiesław Kuczko	Poznan University of Technology, Poland
Wiktor Szot	Kielce University of Technology, Poland
Wojciech Babirecki	University of Zielona Góra, Poland
Wojciech Macek	Gdańsk University of Technology, Poland
Yuliia Denysenko	Sumy State University, Ukraine
Zbigniew Wiśniewski	Lodz University of Technology, Poland
Zuzana Mitaľová	Technical University of Košice, Slovakia

Special Sessions and Special Session Organizing Committee

Efficiency-Driven Manufacturing: Tools and Trends in the Era of Industry 4.0

Justyna Trojanowska	Poznan University of Technology, Poland
Leonilde Varela	University of Minho, Portugal
Vijaya Kumar Manupati	NITIE Mumbai, India
Thanh Tran	RMIT University, Vietnam
Jozef Husár	Technical University of Košice, Slovakia

Intersecting Quality Management and Mechanical Design for Manufacturing Sustainability

Florin Popişter	Technical University of Cluj-Napoca, Romania
Diana Cristina Dragomir	Technical University of Cluj-Napoca, Romania
Corina Julieta Birleanu	Technical University of Cluj-Napoca, Romania
Mircea Claudiu Fulea	Technical University of Cluj-Napoca, Romania

Human factors in Manufacturing

Marcin Butlewski	Poznan University of Technology, Poland
Beata Mrugalska	Poznan University of Technology, Poland
Wiktoria Czernecka	Poznan University of Technology, Poland
Fryderyk Wachowiak	Poznan University of Technology, Poland
Krzysztof Hankiewicz	Poznan University of Technology, Poland

Circularity in Advanced Production Systems: Materials, Methods, and Digital Transformation

Marek Kočisko	Technical University of Košice, Slovakia
Lucia Knapčikova	Technical University of Košice, Slovakia
Jakub Kaščak	Technical University of Košice, Slovakia
Jozef Török	Technical University of Košice, Slovakia

Smart Manufacturing and Sustainability: Interdisciplinary Approaches in the Era of Industry 4.0/5.0

Radosław Maruda	University of Zielona Góra, Poland
Sebastian Saniuk	University of Zielona Góra, Poland
Katarzyna Skrzypek	University of Zielona Góra, Poland
Daria Moskwa - Bęczkowska	Kielce University of Technology, Poland
Karol Dąbrowski	University of Zielona Góra, Poland

From Additive Manufacturing to Healthcare Innovation through Digital Transformation in Biomedical Engineering

Filip Górski	Poznan University of Technology, Poland
Răzvan Păcurar	Technical University of Cluj-Napoca, Romania
Nikola Vitkovic	University of Niš, Serbia
Michał Rychlik	University of Technology, Poland

Intelligent Adaptive Systems in Integrated Production Management Processes

Mariusz Piechowski	Merito WSB University in Poznan, Poland
Małgorzata Jasiulewicz-Kaczmarek	Poznan University of Technology, Poland
Katarzyna Antosz	Rzeszow University of Technology, Poland
Ryszard Wyczółkowski	Silesian University of Technology, Poland
Arkadiusz Gola	Lublin University of Technology, Poland

Intelligent Solutions for Industry of the Future

Izabela Rojek	Kazimierz Wielki University in Bydgoszcz, Poland
Ewa Dostatni	Poznan University of Technology, Poland
Dariusz Mikołajewski	Kazimierz Wielki University in Bydgoszcz, Poland
Nataša Náprstková	University of Jan Evangelista in Ústí nad Labem, Czech Republic

Grey Systems Modeling for Engineering Solutions and Business Optimization

Magdalena Hryb	Poznan University of Technology, Poland
Agnieszka Kujawińska	Poznan University of Technology, Poland
Naiming Xie	Nanjing University of Aeronautics and Astronautics, China
Liangyan Tao	Nanjing University of Aeronautics and Astronautics, China
Rafał Mierzwiak	Poznan University of Technology, Poland

Intelligent Methods Supporting Manufacturing Systems Efficiency

Anna Burduk	Wroclaw University of Science and Technology, Poland
Damian Krenczyk	Silesian University of Technology, Poland
Dagmara Łapczyńska	Wroclaw University of Science and Technology, Poland
Kamil Krot	Wroclaw University of Science and Technology, Poland

Contents

Digital Twins and Virtual Engineering

Cognitive Industrial Twin–a Survey

Siyuan Sun[1], Jiehan Zhou[1](✉), Zhaojia Wang[1], Jinrui Wang[2], Anna Burduk[3], and Damian Krenczyk[4]

[1] School of Computer Science and Engineering, Shandong University of Science and Technology, Qingdao, People's Republic of China
jiehan.zhou@sdust.edu.cn
[2] College of Mechanical and Electronic Engineering, Shandong University of Science and Technology, Qingdao, People's Republic of China
[3] Department of Production Engineering and Management, Wrocław University of Science and Technology, Wrocław, Poland
[4] Department of Engineering Processes Automation and Integrated Manufacturing Systems, Silesian University of Technology, Gliwice, Poland

Abstract. The Cognitive Industrial Twin (CIT) has emerged as an advanced evolution of the Digital Twin (DT) paradigm for Industry 4.0 environments. By embedding learning, reasoning, and decision-making capabilities, CITs extend conventional DTs from passive monitoring toward adaptive and intelligent representations of industrial systems. This survey systematically examines the current state of CIT research, clarifies its definition, and distinguishes it from traditional DTs and other intelligent twin frameworks. A generic five-layer reference architecture is presented, encompassing data sensing, information fusion, knowledge cognition, autonomous decision-making, and feedback optimization. The paper further reviews key enabling technologies—including multi-modal data fusion, reinforcement learning, knowledge graphs, causal inference, and edge intelligence—and discusses their roles in supporting cognitive and autonomous twin functionalities. In addition, industrial applications across smart manufacturing, prescriptive maintenance, autonomous logistics, and sustainable production are analyzed, revealing a paradigm shift from open-loop monitoring to closed-loop, cognition-driven autonomy. Finally, emerging research directions—such as brain-inspired computing, large language model (LLM) integration, and hybrid physical–cognitive modeling—are outlined, along with the key challenges that must be addressed to enable the broader industrial deployment of CITs.

Keywords: Digital Twin · Cognitive Industrial Twin · Smart Manufacturing · Knowledge Graph · Edge Computing

1 Introduction

Digital Twins (DTs), which provide virtual representations of physical assets, have become an important component of Industry 4.0 and intelligent manufacturing [1–3]. Their core role is to maintain a bi-directional connection between the physical system and its digital model, supporting real-time monitoring, simulation, and operational

J. Trojanowska et al. (Eds.): MANUFACTURING 2026, LNME, pp. 3–17, 2026.
https://doi.org/10.1007/978-3-032-21654-0_1

optimization. As industrial processes grow in complexity, however, conventional DTs—which largely function as data-driven monitors or open-loop simulators—have shown limitations in adapting to dynamic and uncertain environments. Recent industrial needs increasingly call for DTs that can interpret data, respond autonomously, and adapt their internal models over time.

The Cognitive Industrial Twin (CIT) has been proposed to address these requirements [4]. A CIT augments traditional DTs with Artificial Intelligence (AI), machine learning (ML), and cognitive computing techniques to support functions such as interpretation, prediction, and decision-making [5]. By integrating streaming sensor data with historical information, CITs can construct higher-level representations of system behavior and use these representations to anticipate future states or recommend operational adjustments. This ability to update internal models and generate context-aware decisions distinguishes CITs from conventional DTs and has demonstrated value in applications such as autonomous process control, prescriptive maintenance, and real-time decision support. For instance, deep reinforcement learning (RL) applied within a DT environment has enabled autonomous optimization of a plastic molding process, improving quality and efficiency with limited human supervision [6].

These developments position CITs as an important step in the evolution of digital-twin technologies for smart manufacturing and industrial autonomy. This survey examines the conceptual foundations of CITs, provides a formal definition, and differentiates CITs from other forms of intelligent DTs. It further presents a five-layer reference architecture—from data sensing through feedback optimization—and reviews the enabling technologies associated with each layer, including multi-modal data fusion, knowledge graphs (KGs), causal inference, and edge intelligence. Subsequently, the paper analyzes representative industrial applications in smart manufacturing, predictive maintenance, logistics, and sustainability, highlighting the practical value of CITs in enabling closed-loop autonomy. The paper concludes by outlining emerging research directions, such as neuromorphic computing, the integration of large language models (LLMs), and hybrid physical–data-driven modeling approaches.

2 Definition of Cognitive Industrial Twin

A CIT extends a traditional DT by embedding AI/ML and cognitive computing principles, enabling it to perceive its environment, learn from data, reason through complex situations, and make autonomous decisions [7]. This definition is supported by formal models, which describe the CIT as a self-evolving system capable of sensing unexpected behaviors and reasoning about optimization strategies [8]. This self-evolution is the key differentiator: the CIT continuously updates not just its state but its underlying models and knowledge base. Other authors reinforce this view, describing CITs as "digital experts" or "co-pilots" that emulate human learning [9], or as DTs with "augmented semantic capabilities" that can understand model interrelationships to enhance decision-making [10]. Thus, the common themes across these definitions are a persistent knowledge layer and the capacity to self-improve—allowing the twin to handle novel situations through learning and reasoning.

CIT should also be distinguished from related concepts. While "Hybrid Twins" primarily combine physics-based and data-driven models for higher accuracy, a CIT builds upon this hybrid foundation by adding learning and reasoning capabilities. Similarly, the "Enhanced Cognitive Twin (ECT)" [11] also describes advanced cognition, but focuses specifically on distributing these capabilities across hierarchical levels for responsive decision-making. To clearly illustrate the advancements, Table 1 summarizes the key limitations of traditional DTs and the corresponding advantages of CITs.

Table 1. Comparison of Limitations of DT and Advantages of CIT

Comparison Aspect	Limitations of Traditional DT	Advantages of CIT
Core Functionality	Monitoring and open-loop simulation; functioning primarily as a data viewer	AI-augmented interpretation, prediction, and decision-making using cognitive computing
Adaptability	Static models with limited adaptability to dynamic environments	Self-evolving capabilities with continuous updates to internal models and knowledge bases
Reasoning Capability	Reliance on pre-defined physics or data models without semantic understanding	Capability for semantic reasoning and causal inference to handle novel scenarios
Decision Autonomy	Human-in-the-loop operation typically functioning as an open-loop system	Autonomous closed-loop control executing prescriptive actions without constant intervention
Outcome Focus	Representation of current states based on fixed parameters	Context-aware decision generation and anticipation of future states

3 Cognitive Industrial Twin Architecture: Five Core Layers

Implementing the CIT concept typically requires organizing it into multiple functional layers. Building on existing architectural models [12], this paper outlines a generic five-layer CIT architecture as illustrated in Fig. 1: (i) Data Sensing Layer, (ii) Information Fusion Layer, (iii) Knowledge Cognition Layer, (iv) Autonomous Decision Layer, and (v) Feedback Optimization Layer.

Compared with conventional three-layer Digital Twin architectures—commonly organized around physical, virtual, and connection layers [12], this five-layer design introduces a more fine-grained functional decomposition by explicitly separating cognitive reasoning from autonomous decision execution. In particular, the Knowledge Cognition Layer is responsible for high-level semantic understanding and inference,

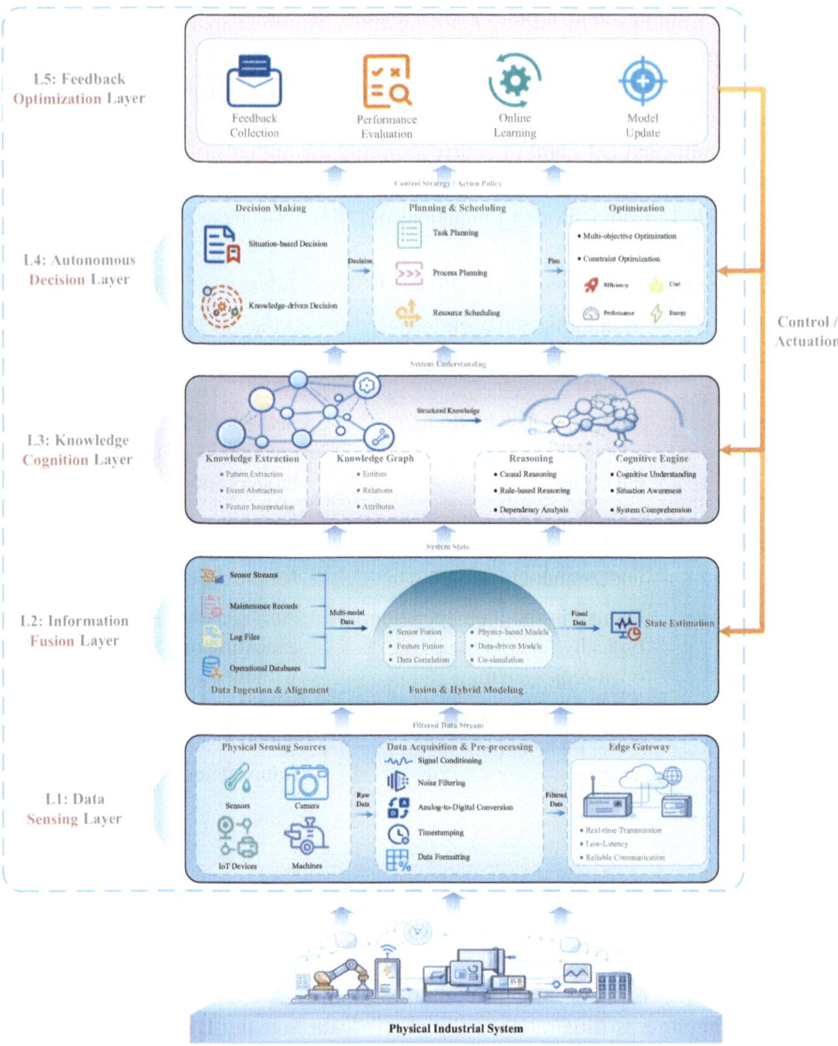

Fig. 1. The five-layer reference architecture of the CIT

while the Autonomous Decision Layer focuses on action generation and control strategies. Such a decoupled design facilitates the integration of advanced AI techniques, including knowledge graphs and learning-based reasoning models, without interfering with lower-level data acquisition and synchronization processes. As a result, the proposed architecture provides a clearer structural foundation for supporting continuous perception–cognition–decision–feedback loops, which are essential for adaptive and self-optimizing industrial systems.

3.1 Data Sensing Layer

The foundation of any DT is data from a physical world. The Data Sensing Layer encompasses all the IoT sensors, data acquisition systems, and connectivity needed to collect real-time data from a physical industrial asset or process and deliver it to its twin. In a CIT, this layer provides the raw perceptual inputs [13]. It typically includes physical sensors, IoT gateways, and communication networks that stream data into the digital model. High-bandwidth industrial Ethernet, wireless sensor networks, or even 5G can be used to ensure low-latency, reliable data flow. The goal is to create a continuously updated digital shadow of the physical system. For every significant state change or event in the physical system, data is captured and transmitted such that the twin's virtual state mirrors the real state in real-time. This layer also handles data preprocessing steps like filtering noise, performing analog-to-digital conversion, timestamping, and initial storage of time-series data streams. In advanced setups, the data sensing layer may incorporate edge devices that perform preliminary analytics at the data source. The output of this layer is a stream of multi-modal data that feeds upward into the twin. A strong data foundation is critical, as the quality and richness of sensor data largely determine the twin's fidelity and learning potential. Modern CIT implementations leverage pervasive sensor deployments and Industrial Internet of Things (IIoT) platforms to ensure comprehensive visibility.

3.2 Information Fusion Layer

The Information Fusion Layer operates above the raw data layer and is responsible for aggregating heterogeneous data sources—such as sensor streams, log files, operational databases, maintenance records, and operator inputs—into a unified representation of the system state. Typical techniques at this layer include time-series alignment, sensor fusion, state estimation, and the generation of virtual sensor outputs. It hosts the twin's core computational models and often implements a hybrid strategy in which physics-based simulations, continuously updated by sensor feedback, are combined with data-driven models that capture system behavior. For instance, a predictive model may integrate multiple sensor signals to estimate an unmeasured variable, producing a higher-level indicator for downstream decision-making. To support this integration of multi-modal data, CIT implementations commonly rely on data-ingestion frameworks such as data lakes or message brokers to synchronize and correlate streams of varying types and frequencies. These mechanisms are frequently embedded within hybrid DT platforms. The output of this layer is an up-to-date digital representation that fulfills the core role of a traditional DT. The transition to cognitive functionality begins in the next layer, where the fused information is transformed into actionable knowledge.

3.3 Knowledge Cognition Layer

The Knowledge Cognition Layer provides the core cognitive capability of the CIT. At this level, the twin organizes structured knowledge and applies cognitive algorithms to derive higher-level insights. The layer generally comprises two components: a knowledge base and a cognitive engine. The knowledge base functions as the system's semantic model, commonly implemented using a KG or ontology that captures key entities, their functions, causal dependencies, and maintenance history. It is continually enriched through knowledge extraction processes that detect patterns, anomalies, and cause–effect relations within the fused data. Historical datasets contribute predictive models, while natural language processing techniques can incorporate textual domain knowledge. The cognitive engine uses this knowledge base to perform advanced inference. Beyond identifying known patterns [14, 15], its primary role is causal reasoning—inferring relationships that explain system behavior. For example, the CIT may deduce that a drop in pump pressure triggered a downstream temperature increase. Methods such as Bayesian networks and causal inference algorithms are applied to extract these relationships, enabling explainable and trustworthy insights. The understanding produced provides the foundation for the Autonomous Decision Layer.

3.4 Autonomous Decision Layer

With a comprehensive representation of a system's state and behavior, the Autonomous Decision Layer determines the actions required to achieve operational objectives. This layer generates decisions, recommendations, or control commands, functioning as the execution component of the CIT. The nature of these actions varies by application, ranging from low-level setpoint adjustments to high-level scheduling and optimization. Several AI methodologies support this capability. RL is one commonly adopted approach, in which its twin learns control policies through interaction with its environment. For example, a manufacturing CIT may employ deep RL to tune process parameters in real time, improving product quality or throughput. The knowledge acquired at the cognition layer can guide this learning process by constraining the action space or embedding safety conditions. Rule-based or optimization-driven planning offers a complementary strategy, particularly when KG encodes causal dependencies. In such cases, the CIT can formulate actions that mitigate or prevent undesirable states. For more complex decision problems, the twin may conduct predictive simulations to evaluate alternative outcomes and support anticipatory planning. Human-in-the-loop mechanisms remain essential in safety-critical contexts. Rather than executing actions autonomously, the CIT may provide recommendations to an operator. In energy management, for instance, the twin may suggest load reduction or maintenance scheduling after detecting performance degradation [16]. In supply-chain applications, it may propose shipment rerouting or inventory adjustments in response to predicted disruptions. These decisions are typically produced in real or near-real time, supported by efficient computational models.

3.5 Feedback Optimization Layer

The final layer completes the interaction loop between a DT and its physical system. The decisions or control commands produced in the Autonomous Decision Layer are transmitted back to the physical asset, either through direct actuator interfaces or through operator intervention, thereby implementing adjustments to system behavior. This establishes the closed-loop control mechanism [17] that distinguishes full DT implementations from open-loop digital shadows. Feedback also propagates internally within the CIT. Once an action is executed, the resulting system response is captured by the data layer, enabling the twin to assess the effectiveness of its decisions. This process supports continual refinement: model parameters, system assumptions, and control strategies are updated as new evidence becomes available. Through this iterative feedback process, the CIT maintains alignment with real-world conditions and adapts its computational models accordingly. Activities at this layer may include triggering periodic retraining of ML models, adjusting simulation parameters, or reconciling discrepancies between predicted and observed system states. The feedback optimization layer therefore enables ongoing performance improvement and ensures the long-term reliability of the CIT.

4 Key Technologies Enabling Cognitive Industrial Twin

Developing a fully operational CDT relies on a coordinated set of enabling technologies. This section reviews the major technological components that support CIT functionalities, including multi-modal data fusion techniques, ML and RL for decision-making, KG–based reasoning, and edge computing approaches for real-time execution.

4.1 Multi-Modal Data Fusion and Integration

Industrial environments produce diverse and heterogeneous data, including time-series sensor measurements, images, control signals, maintenance records, operator inputs, and enterprise resource information. Integrating these sources is essential for constructing a comprehensive representation of system conditions within a CIT. Achieving this integration requires interoperability and standardized data formats, often aligned with frameworks such as the Industrial Internet Reference Architecture [18]. Data fusion can be performed at varying degrees of abstraction, typically categorized as signal-level, feature-level, or decision-level fusion. In practice, CIT implementations frequently rely on data lakes or streaming platforms to ingest raw data and use stream-processing frameworks to synchronize and merge multimodal inputs [19]. For example, combining vibration and temperature measurements can provide a more reliable indication of developing bearing faults than either signal independently [20]. Methods such as sensor-fusion neural networks and transformer-based architectures can learn how to weight and integrate heterogeneous inputs [21]. Semantic integration is also supported by KGs, which map data elements to a shared ontology and provide structure for higher-level reasoning. Despite these advances, heterogeneity remains a core challenge—arising from variations in sampling rates, data quality, and measurement units. Effective multimodal fusion ultimately produces a consistent and informative representation of the system state, forming the foundation for the cognitive functions of the twin.

4.2 Machine Learning for Decision-Making

ML plays a central role in enabling the cognitive and decision-making capabilities of a CIT [22]. Within the Knowledge Cognition Layer, supervised and unsupervised learning techniques are widely applied to develop predictive models, anomaly detectors, and pattern-recognition functions using historical operational data. Examples include deep neural networks for remaining useful life estimation and clustering methods for identifying previously unknown operating regimes. In the Autonomous Decision Layer, RL has become a prominent approach for learning control strategies. The CIT provides a simulation environment in which RL agents can explore and refine policies without disrupting physical operations, enabling faster and safer training compared with direct deployment on industrial equipment [23]. Other optimization-oriented learning methods—such as Bayesian optimization and genetic algorithms—are also used for tasks involving scheduling or parameter tuning. To maintain long-term performance, many CIT implementations incorporate online learning or employ transfer learning to reduce the data requirements associated with deploying models across different machines or production contexts [24]. These learning-based techniques enable the transition from descriptive analytics to prescriptive and autonomous decision-making. A continuing challenge, however, is ensuring that learned policies exhibit robustness and safety. CIT systems often address this by integrating ML-based policies with rule-based safety mechanisms or model-based constraints, which helps balance adaptability with reliability in complex industrial environments.

4.3 Knowledge Graphs and Semantic Reasoning

KGs and ontologies are widely used in CIT architectures to represent system knowledge in a structured and machine-interpretable form. A KG models entities as nodes and their relationships as edges, enabling the explicit encoding of equipment hierarchies, functional dependencies, and higher-level concepts such as failure modes and maintenance procedures. Within an industrial twin, this representation links operational data, engineering models, and expert knowledge to the components or processes they describe. KGs provide two principal benefits: integration and reasoning. For integration, they supply a unified schema through which heterogeneous data—sensor readings, simulation outputs, or domain rules—can be aligned within a common ontology. For reasoning, KGs support logical inference using semantic technologies and rule engines, allowing the system to derive conclusions such as identifying a likely fault source by traversing known causal relationships. They may also store historical cases to facilitate case-based reasoning. Earlier studies have reported these capabilities. Zheng et al. [25] highlight the role of semantic modeling in capturing relationships among a twin's virtual components. Lim et al. [26] develop a graph-enabled twin that uses an industrial KG to support causal inference and provide human-interpretable maintenance explanations. Constructing and maintaining a comprehensive KG remains challenging, as it often requires substantial expert input and the extraction of information from unstructured data sources. Recent studies have begun exploring the use of LLMs to assist KG construction. Mandal et al. [27], for example, employed LLMs to integrate multimodal city data into a KG for a smart-city twin. These approaches indicate a potential pathway for reducing manual effort in building the semantic layer of CITs. Once developed, the KG forms a foundational semantic layer that connects raw data with higher-level knowledge representations, supporting explainability and enabling CITs to progress beyond pattern recognition toward knowledge-guided reasoning and decision support [28].

4.4 Edge Computing and Distributed Intelligence

A key requirement for CIT deployment is the ability to execute intelligence close to the data sources on the factory floor. Edge intelligence addresses this need by enabling selective CIT functions—such as data preprocessing, anomaly detection, and fast control decisions—to run on edge devices rather than relying exclusively on cloud-based computation. This approach mitigates latency and bandwidth limitations that arise when all sensor data must be transmitted to a central server for analysis. For instance, processing vibration or temperature data on an edge gateway allows the system to detect deviations and initiate local actuator responses within milliseconds, avoiding delays introduced by remote communication.

Many CIT implementations adopt a hierarchical architecture in which latency-sensitive tasks are performed at the edge, while higher-level reasoning, model training, and long-term optimization are handled by Cloud Computing or on-premise platforms. This division of responsibilities also supports data protection, as sensitive operational data can remain on-site. Furthermore, edge deployments facilitate distributed learning strategies such as federated learning, in which global models are updated across multiple devices without requiring raw data to be centralized [29].

Edge intelligence is particularly relevant in real-time industrial settings, including autonomous mobile robots and fast-response control loops [30]. Its integration with emerging communication technologies, such as 5G, provides additional support for distributed CIT operation. In predictive maintenance applications, a lightweight twin deployed on individual equipment—such as a wind turbine tower—can continuously monitor local conditions and issue timely alerts without depending on continuous connectivity to a central system. This distributed structure improves scalability by allowing each machine or production cell to operate a localized twin, while higher-level CIT layers aggregate information for a global system optimization.

5 Industrial Applications of Cognitive Industrial Twin

Traditional DT applications are typically limited to monitoring, simulation, or predictive analytics, and often operate in an open-loop or human-in-the-loop manner. In contrast, CIT-based applications introduce an explicit perception–cognition–decision–feedback paradigm, enabling proactive and optimized human-in-the-loop autonomy [1].

Across different industrial contexts, CIT applications exhibit three common characteristics: (i) context-aware perception, achieved through multi-modal sensing and information fusion; (ii) realtime data and knowledge – driven reasoning, where structured knowledge and causal models support interpretation and explanation; and (iii) autonomous or semi-autonomous decision execution, supported by learning-based or rule-constrained decision engines.

Together, these characteristics support a fundamental shift from reactive, rule-based optimization toward proactive adaptive and self-evolving industrial operations. The following subsections analyze how this paradigm is instantiated in different industrial domains.

5.1 Smart Manufacturing and Autonomous Process Control

In discrete manufacturing, CITs enable autonomous process control by integrating real-time sensor data with decision-making algorithms. Unlike traditional monitoring systems that require manual intervention, CITs can actively adjust production parameters to optimize performance. For example, Khdoudi et al. [6] developed a CIT for a plastic molding process. This system utilized Deep Reinforcement Learning (DRL) to learn optimal process parameters—such as temperature and injection pressure—through trial-and-error in a virtual environment. The CIT successfully minimized defects and improved cycle times with limited human supervision. Similarly, in complex assembly environments, Xu et al. [31] applied a cognitive twin framework to facilitate multi-robot collaborative manufacturing. By perceiving the workspace state in real-time, the twin autonomously coordinated the actions of multiple robots to avoid collisions and optimize production tasks, demonstrating the system's adaptability to dynamic operational requirements.

From a CIT perspective, the key contribution of such systems lies not merely in parameter optimization, but in the continuous alignment between virtual cognition and physical execution. The cognitive twin maintains an internal representation of process behavior through learned models and structured process knowledge, learns from historical and real-time data, and adapts its decision policies as production conditions evolve. This enables the manufacturing system to respond autonomously to disturbances—such as material variability or equipment aging—without requiring manual retuning of control rules.

5.2 Intelligent Predictive and Prescriptive Maintenance

While traditional predictive maintenance focuses on forecasting when a failure might occur, CITs extend this capability to prescriptive maintenance—explaining why a fault happened and recommending corrective actions. This is achieved by combining data-driven predictions with causal reasoning. For instance, Lim et al. [26] proposed a graph-enabled CIT that uses an industrial KG to support causal inference. By mapping failure modes to their root causes within the graph, the twin can provide human-interpretable explanations for maintenance decisions, rather than functioning as a "black box.“ Furthermore, hybrid modeling approaches have proven effective in diagnosing complex component degradation. Regis et al. [20] demonstrated that combining physics-based wear models with data-driven observation improves the diagnostic accuracy of bush bearings compared to using either method alone. Such cognitive capabilities allow maintenance strategies to shift from reactive repairs to proactive, knowledge-guided interventions.

In industrial practice, this prescriptive capability is particularly valuable for reducing maintenance decision latency and improving trust in automated recommendations. By explicitly linking observed symptoms to causal explanations, CIT-based maintenance systems support both autonomous intervention and human decision-making, enabling a collaborative human–AI maintenance workflow rather than a purely automated black-box solution.

5.3 Smart Logistics and Material Handling

In logistics and material handling, CITs address the challenge of dynamic scheduling in uncertain environments. Automated Guided Vehicles (AGVs) often face unexpected obstacles or changing production priorities that static rules cannot handle efficiently. To address this, Lee et al. [23] proposed a CIT-driven approach for real-time optimization of AGV systems. By leveraging DRL, the twin continuously monitors the shop floor status and autonomously replans AGV paths to avoid bottlenecks and deadlocks. This self-adaptive capability ensures that material flow remains synchronized with production needs, significantly enhancing overall logistics efficiency compared to conventional heuristic-based dispatching systems.

From a cognitive twin perspective, logistics-oriented CITs continuously update their internal state representations by combining real-time observations with learned operational knowledge. This enables anticipatory decision-making under uncertainty, allowing the system to adapt not only to current disturbances but also to predicted future congestion or task conflicts.

5.4 Sustainable Production and Energy Management

Beyond efficiency and quality, CITs are increasingly applied to support sustainable industrial goals. By optimizing resource usage and minimizing waste, CITs contribute to greener production ecosystems. Padovano et al. [17] introduced a cognitive digital twin framework specifically designed for waste mitigation in manufacturing. The system monitors production anomalies that lead to material waste and prescribes adjustments to mitigate them, effectively acting as a portfolio management tool for sustainability. Additionally, in broader smart city and industrial park contexts, approaches like those explored by Mandal et al. [27] utilize multi-modal data fusion to model complex energy consumption patterns. These applications illustrate how the cognitive reasoning capabilities of CITs can be harnessed to balance operational performance with environmental responsibility.

A distinguishing feature of CIT-based sustainability applications is their ability to reason about trade-offs between productivity, energy consumption, and environmental impact. Rather than optimizing a single objective in isolation, the cognitive twin evaluates competing goals within a unified decision framework, enabling informed compromises that align operational efficiency with long-term sustainability targets.

6 Future Research Directions

Although CITs present great promise, the field is still in its infancy [31]. Several research directions are taking shape, including brain-inspired and neuromorphic approaches for CIT [32], the integration of LLMs and advanced AI with DTs [27], and hybrid modeling that combines physics-based and data-driven methods to improve accuracy and robustness [20, 33, 34]. These directions are not isolated; in many cases they intersect and together could lead to more powerful and versatile CIT systems.

6.1 Brain-Inspired and Neuromorphic Computing for Cognitive Industrial Twin

Brain-inspired computing has emerged as a potential direction for enhancing the efficiency and adaptability of CITs. Research in this area draws on principles observed in biological neural systems—such as event-driven processing and low-power operation—which are attractive for real-time industrial applications [31]. At the algorithmic level, spiking neural networks (SNNs) and associative memory models offer alternative mechanisms for processing temporal sensor streams and identifying patterns. At the hardware level, neuromorphic processors provide low-latency, energy-efficient computation suitable for edge deployment, with early studies demonstrating substantial energy savings in tasks such as real-time IoT anomaly detection.

Beyond specialized hardware, several works explore broader cognitive architectures inspired by human information-processing models. These frameworks aim to support more flexible reasoning and adaptive learning within the CIT, particularly under changing operating conditions. Approaches linked to embodied cognition—which emphasize the interaction between computational models and the physical environment—may further improve the robustness of on-site learning, although such methods remain at an early stage.

An important long-term goal of this research direction is to enable continual or lifelong learning within CITs. Industrial systems evolve over extended periods, requiring models that can incorporate new information without degrading previously learned knowledge. Neuromorphic systems and certain cognitive architectures are well suited to incremental adaptation, making them promising candidates for supporting persistent learning in CIT deployments. Advancements at the intersection of neuroscience-inspired algorithms, AI, and digital-twin technologies may ultimately contribute to more efficient and adaptive CIT implementations.

6.2 Integration of Large Language Models with Digital Twins

The integration of LLMs into DT systems is emerging as a promising research direction. One practical application lies in automated knowledge extraction, where LLMs can process unstructured sources—such as manuals, work orders, or technical reports—to assist in constructing the structured KGs used by cognitive twins. Mandal et al. [27] demonstrate that such techniques can substantially reduce the manual effort associated with developing a twin's semantic layer. Another application is natural-language interaction: LLMs can provide an interface that enables operators to query or interpret twin information using everyday language, improving the accessibility of complex model outputs.

Beyond these functions, LLMs may support higher-level analytical tasks. They can contribute to scenario analysis by generating candidate explanations for observed faults based on patterns learned from technical corpora, or assist knowledge transfer by drawing analogies across related industrial contexts. Initial work has also explored using LLMs to generate code or model components for twin subsystems, indicating potential roles in the broader engineering workflow. Integrating these capabilities with the quantitative modeling strengths of DTs may enable more comprehensive decision-support tools.

However, incorporating LLMs into safety-critical industrial environments presents substantial challenges. A primary concern is the reliability of generated responses, as LLMs may produce outputs that are syntactically plausible yet inconsistent with physical constraints. Ensuring correctness requires grounding LLM outputs in validated operational data and domain rules, for example through Retrieval-Augmented Generation (RAG) or constraint-aware reasoning. Addressing these issues is essential if LLM-enabled interfaces are to be deployed in CITs, particularly in contexts where interpretability and operational safety are critical.

6.3 Hybrid Physics-Data and Cognitive Modeling

Ensuring accuracy and robustness in CITs will increasingly require hybrid modeling approaches that integrate physics-based methods with data-driven ML [33]. Purely data-driven models can perform poorly when operating outside their training distributions, whereas purely physics-based models often struggle to represent nonlinear or context-dependent behaviors encountered in industrial settings [34].

Physics-Informed Machine Learning (PIML) embeds physical constraints, such as conservation laws, into training objectives. This ensures representations remain consistent with system dynamics while retaining data-driven flexibility, improving both accuracy and interpretability. Furthermore, hybrid strategies tightly integrate simulation and learning: simulators generate synthetic data for rare events to improve model coverage, while ML calibrates simulations using real-time data. By combining physics-based extrapolation with learning-based adaptability, hybrid modeling supports multi-scale representation and robust CIT performance under unseen operating conditions.

7 Conclusion

This paper presented a comprehensive survey of CIT, covering their conceptual foundations, architecture, enabling technologies, industrial applications, and future research outlook. Unlike conventional DTs that primarily monitor and simulate physical processes, CITs integrate AI and cognitive computing to enable perception, learning, reasoning, and autonomous decision-making, thereby fostering continuous self-evolution and optimization of industrial systems.

The five-layer CIT architecture establishes a closed-loop intelligence framework, from data sensing and information fusion to knowledge cognition, autonomous decision-making, and feedback optimization, providing a structured pathway toward truly intelligent industrial operations. Key enabling technologies, including multi-modal data fusion, RL, KGs, causal inference, and edge intelligence, empower CITs to achieve real-time cognition and trustworthy decision-making in complex and dynamic environments. Real-world applications demonstrate CIT's transformative potential in smart manufacturing, supply chain intelligence, predictive maintenance, and digital factories, offering significant gains in efficiency, resilience, autonomy, and cost-effectiveness.

Despite notable progress, several challenges remain, including data heterogeneity and sharing, explainability of cognitive reasoning, cross-domain generalization, real-time performance, and security. Looking ahead, breakthroughs in brain-inspired computing, the convergence of DTs with LLMs, cross-domain knowledge transfer, and hybrid physical–cognitive modeling are expected to advance CIT toward higher levels of intelligence and maturity. Overall, CIT represents the next milestone in the evolution of DT and is poised to become a foundational enabler for Industry 5.0, enabling human–machine symbiosis, sustainable operation, and self-evolving industrial ecosystems.

References

1. Zhou, J., Zhang, S., Gu, M.: Revisiting digital twins: origins, fundamentals, and practices. Front. Eng. Manage. **9**(4), 668–676 (2022)
2. Zhou, J., Wang, J., Hryb, M., Kujawińska, A.: Digital twins-driven sustainable industrial supply chain management: a survey. In: Burduk, A., et al. (eds.) Intelligent Systems in Production Engineering and Maintenance IV, pp. 115–124. Springer, Cham (2026)
3. Javaid, M., Haleem, A., Suman, R.: Digital Twin applications toward Industry 4.0: a review. Cogn. Robot. **3**, 71–92 (2023)
4. Eirinakis, P., et al.: Cognitive digital twins for resilience in production: a conceptual framework. Information **13**(1), 33 (2022)
5. Al Faruque, M.A., Muthirayan, D., Yu, S.-Y., Khargonekar, P.P.: Cognitive digital twin for manufacturing systems. In: Design, Automation & Test in Europe Conference & Exhibition (DATE), pp. 440–445. Grenoble, France (2021)
6. Khdoudi, A., Masrour, T., El Hassani, I., El Mazgualdi, C.: A deep-reinforcement-learning-based digital twin for manufacturing process optimization. Systems **12**(2), 38 (2024)
7. Kreuzer, T., Papapetrou, P., Zdravkovic, J.: Artificial intelligence in digital twins—A systematic literature review. Data Knowl. Eng. **151**, 102304 (2024)
8. Abburu, S., Berre, A.J., Jacoby, M., Roman, D., Stojanovic, L., Stojanovic, N.: COGNITWIN–Hybrid and cognitive digital twins for the process industry. In: IEEE International Conference on Engineering, Technology and Innovation (ICE/ITMC), pp. 1–8. Cardiff, UK (2020)
9. Fernández, F., Sánchez, Á., Vélez, J.F., Moreno, A.B.: Symbiotic autonomous systems with consciousness using digital twins. In: Ferrández Vicente, J.M., et al. (eds.) IWINAC 2019. LNCS, vol. 11487, pp. 33–42. Springer, Cham (2019)
10. Lu, J., Zheng, X., Gharaei, A., Kalaboukas, K., Kiritsis, D.: Cognitive twins for supporting decision-making of IoT systems. In: Industry 4.0 Model for Advanced Manufacturing. Springer, Cham (2020)
11. Eirinakis, P., et al.: Enhancing cognition for digital twins. In: IEEE International Conference on Engineering, Technology and Innovation (ICE/ITMC), pp. 1–7. Cardiff, UK (2020)
12. Villegas, L.F., Macchi, M., Polenghi, A.: Digital twins in manufacturing: a three-layer heat-map analysis. IFAC-PapersOnLine **58**(19), 646–651 (2024)
13. Viola, J., Chen, Y.: Digital twin enabled smart control engineering as an industrial AI: a new framework and case study. In: Industrial Artificial Intelligence (IAI), pp. 1–6. Shenyang, China (2020)
14. El Mokhtari, K., Panushev, I., McArthur, J.J.: Development of a cognitive digital twin for building management and operations. Front. Built Environ. **8**, 856873 (2022)
15. Stavropoulou, G., et al.: Digital twin meets knowledge graph for intelligent manufacturing processes. Sensors **24**(8), 2618 (2024)
16. Padovano, A., Sammarco, C., Balakera, N., Konstantinidis, F.: Towards sustainable cognitive digital twins: a portfolio management tool for waste mitigation. Comput. Ind. Eng. **198**, 110715 (2024)
17. Park, H., Easwaran, A., Andalam, S.: TiLA: twin-in-the-loop architecture for cyber-physical production systems. In: IEEE International Conference on Computer Design (ICCD), pp. 82–90. Abu Dhabi, UAE (2019)

18. Industrial Internet Consortium (IIC). The Industrial Internet Reference Architecture (IIRA), Version 1.10. Technical Report (2022)
19. Dapkute, A., Siozinys, V., Jonaitis, M., Kaminickas, M., Siozinys, M.: Digital twin data management: framework and performance metrics of cloud-based ETL system. Machines **12**(2), 130 (2024)
20. Regis, A., Arroyave-Tobon, S., Linares, J., Mermoz, E.: Physic-based vs data-based digital twins for bush bearing wear diagnostic. Wear **526–527**, 204888 (2023)
21. Cai, C., et al.: Study of deep multimodal information fusion–based digital twin method for gearbox fault diagnosis. Int. J. Adv. Manuf. Technol. **138**, 3529–3542 (2024)
22. Chen, C., Fu, H., Zheng, Y., Tao, F., Liu, Y.: The advance of digital twin for predictive maintenance: the role and function of machine learning. J. Manuf. Syst. **71**, 581–594 (2023)
23. Lee, D., Kang, Y.S., Noh, S.D.: Digital twin-driven deep reinforcement learning for real-time optimisation in dynamic AGV systems. Int. J. Prod. Res. 1–19 (2025)
24. Maschler, B., Braun, D., Jazdi, N., Weyrich, M.: Transfer learning as an enabler of the intelligent digital twin. Procedia CIRP **100**, 127–132 (2021)
25. Zheng, X., Lu, J., Kiritsis, D.: The emergence of cognitive digital twin: vision, challenges and opportunities. Int. J. Prod. Res. **60**(24), 7610–7632 (2021)
26. Lim, K.Y.H., et al.: Graph-enabled cognitive digital twins for causal inference in maintenance processes. Int. J. Prod. Res. **62**(13), 4717–4734 (2023)
27. Mandal, S., O'Connor, N.E.: LLMasMMKG: LLM-assisted synthetic multi-modal knowledge graph creation for smart city cognitive digital twins. In: AAAI Symposium Series (2024)
28. Meyers, B., et al.: Knowledge graphs in digital twins for manufacturing—Lessons learned from an industrial case at Atlas Copco Airpower. IFAC-PapersOnLine **55**(10), 13–18 (2022)
29. Zhou, J., Lu, Q., Dai, W., Herrera-Viedma, E.: Guest editorial: Federated learning for industrial IoT in Industry 4.0. IEEE Trans. Ind. Inform. **17**(12), 8438–8441 (2021)
30. Xu, W., Yang, H., Ji, Z., Ba, M.: Cognitive digital twin-enabled multi-robot collaborative manufacturing: framework and approaches. Comput. Ind. Eng. **194**, 110418 (2024)
31. Attaran, M., Celik, B.G.: Digital twin: benefits, use cases, challenges, and opportunities. Decision Anal. J. **6**, 100165 (2023)
32. Ren, J., Ahmad, R., Li, D., Ma, Y., Hui, J.: Industrial applications of digital twins: a systematic investigation based on bibliometric analysis. Adv. Eng. Inform. **65**(B), 103264 (2025)
33. Malik, A.A., Brem, A.: Man, machine and work in a digital twin setup: a case study (2020). arXiv preprint arXiv:2006.08760
34. Hadi, M.A., et al.: Implementing cognitive technologies in an assembly line based on two case studies. Procedia CIRP **97**, 520–525 (2021)

Cross Reality for Computer Control of Technical Systems in Mechanical Engineering

Myroslav Omelianenko[1(✉)], Jozef Husár[1], and Justyna Trojanowska[2]

[1] Faculty of Manufacturing Technologies With a Seat in Prešov, Department of Industrial Informatics and Applied Mathematics, Technical University of Košice, Prešov, Slovak Republic
myroslav.omelianenko@tuke.sk

[2] Faculty of Mechanical Engineering, Poznan University of Technology, Poznan, Poland

Abstract. The digital transformation of manufacturing within Industry 4.0 and 5.0 emphasizes the integration of physical and virtual systems using extended reality (XR) technologies. Although XR is increasingly adopted for visualization and training, its application in industrial computer control remains fragmented and lacks standardized evaluation. This paper presents a structured synthesis of XR integration in mechanical engineering control systems and proposes an interoperable framework based on open standards, including OpenXR, OPC UA, and IEC 61499. The study follows a PRISMA-based systematic review of literature published between 2015 and 2025 in the Scopus and Web of Science databases. The results indicate that XR-enabled computer control can improve maintenance efficiency, safety, and system understanding, with reported reductions in mean time to repair (15–30%) and improved operator response times. The proposed framework connects XR interfaces with industrial control and IoT platforms, providing guidance for practical deployment in smart manufacturing environments.

Keywords: Extended Reality (XR) · Internet of Things (IoT) · Industrial Automation

1 Introduction

Digitalization and automation are fundamentally transforming the nature of industrial production. Within the Industry 4.0 framework, special attention is devoted to the integration of physical and virtual systems, enabling more efficient control, monitoring, and optimization of manufacturing processes. One of the key technologies contributing to this transformation is XR—a convergence of AR, VR, and MR—which introduces new possibilities for human–machine interaction.

In the field of computer-based control of technical systems, XR serves as an advanced interface providing an intuitive and visual form of information exchange between humans and machines. The integration of XR with IoT platforms such as ThingWorx enables real-time visualization of data from physical devices directly within a 3D environment, accessible via smart glasses or tablets. Such systems can increase operator efficiency, reduce human error, and enhance decision-making processes.

J. Trojanowska et al. (Eds.): MANUFACTURING 2026, LNME, pp. 18–29, 2026.
https://doi.org/10.1007/978-3-032-21654-0_2

The main objective of this work is to explore the potential of XR technologies to support computer control in mechanical engineering production. The research focuses on the design and implementation of an integration between the ThingWorx platform and an XR application, as well as on the identification of benefits, technical challenges, and the potential for practical application of XR technologies in modern industrial control systems.

Definitions and frameworks

Cutting-edge technologies such as VR, AR, MR, and XR offer unique ways to combine digital elements with real ones, but they do so in different ways. VR provides a fully immersive experience where users interact in a computer-generated environment using specialised devices, completely isolating them from the real world to create a strong sense of presence [1, 2]. AR overlays digital objects onto the real world, allowing users to see and interact with a combination of real and virtual elements without losing touch with their surroundings. MR goes even further, not only overlaying but also anchoring virtual objects in the real world, allowing interaction with both digital and physical elements simultaneously. XR serves as an umbrella term covering the full spectrum of interactions between digital and physical environments [3].

Augmented reality (AR)

The term AR refers to a kind of virtual environment that adds to reality by enabling the user to view the real world with virtual items superimposed or mixed. AR preserves the user's link to reality, producing an experience where virtual and real items seem to live in the same place, in contrast to VR, which fully submerges the user in a synthetic environment. AR falls somewhere in between being entirely synthetic and being entirely real [3, 4].

Mixed reality (MR)

MR is a notion that falls under the category of reality-virtuality, wherein it creates new and enriched experiences by fusing aspects of virtual and real-world surroundings. Real and virtual items can coexist and influence one another in a space created by this technology, which enables digital objects and elements to be merged and communicated with the real environment at the same time. By offering a balance between reality and virtuality, MR enhances the user's perception and virtual integration with the environment [3, 4].

Extended reality (XR)

XR enables varying degrees of interaction between digital content and the physical environment and is increasingly applied in industrial contexts as an advanced human–machine interface for visualization, monitoring, and interaction with cyber-physical systems and digital twins (see Fig. 1).

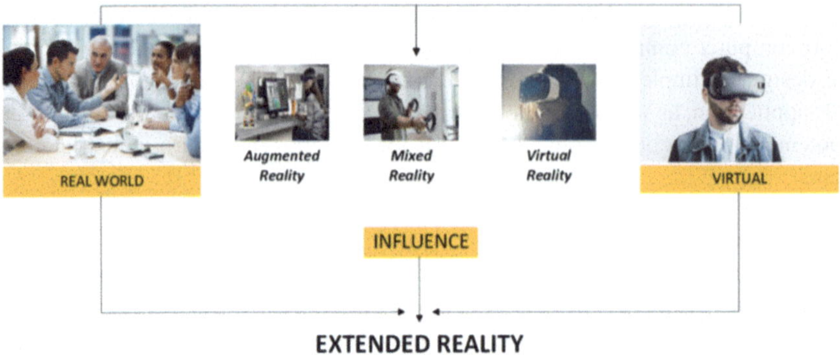

Fig. 1. Conceptual representation of the XR spectrum illustrating the transition and interaction between the real and virtual environments.

Problem and contribution to the overview
As of right now, no thorough overview of integration strategies and architectural patterns for integrating XR systems with digital twins (DT) and PLC/SCADA/MES levels exists. Existing research only considers discrete integration fragments (such as XR–MES or MES–DT), which makes developing a cohesive technique more difficult. A disconnect between scientific methods and industrial practice is further caused by the absence of defined measures for evaluating how XR technologies affect real-time control.

2 Methodology

The PRISMA-like technique served as the foundation for this systematic review, which concentrated on the openness and reproducibility of the selection, assessment, and synthesis of literature. With respect to their technological maturity and integration into industrial environments, the main goal was to identify and evaluate the most recent trends and uses of XR technologies in the framework of computer control of technical systems in mechanical engineering.

Objectives and RQ
In mechanical engineering, XR technologies are becoming more and more relevant for computer-based control of technical systems nevertheless, integration with digital twins and PLC/SCADA/MES layers is still lacking. Current research does not create standardized measures for evaluating XR impact in real-time control, nor does it offer a cohesive overview of architectures and standards like OpenXR, OPC UA, IEC 61499/61131, or ISO 23247. In addition to summarizing important performance indicators (latency, registration accuracy, OEE, FPY, and MTTR), this article attempts to synthesize integration patterns and identify open difficulties and limitations for XR deployment on the shop floor. By developing an organized information base to direct future studies and the commercial application of XR in smart manufacturing, the review contributes (Fig. 2).

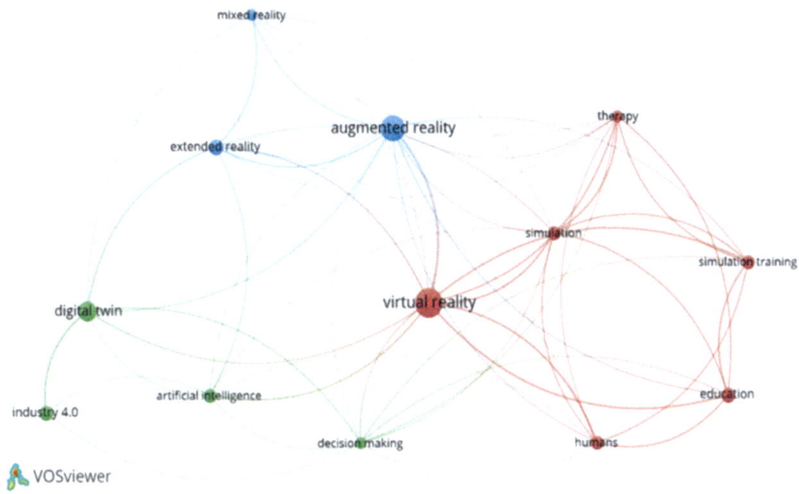

Fig. 2. Visualization of research clusters highlighting the relationships between Virtual, Augmented, and Mixed Reality with adjacent technologies.

2.1 Literature Review

A uniform, interoperable, and scalable framework for linking physical systems and their digital representations is becoming more and more necessary as digital twins and XR gain traction in industrial applications. Four globally accepted standards—ISO 23247, OPC UA, OpenXR 1.1, and IEC 61499—have been synthesized to offer a strong technique for developing such a framework and guaranteeing its applicability across various manufacturing levels [5, 6].

The conceptual underpinnings of digital twins in manufacturing are established by ISO 23247, which specifies how real objects—like machinery, procedures, or employees—be mapped to their digital counterparts. The organization and connectivity of these representations inside the manufacturing infrastructure are likewise specified by this framework. The structural underpinning of XR is ISO 23247, which makes it possible to directly display and control operational data in augmented or virtual space, improving operator overview, efficiency, and interaction [6, 7].

Between information systems and operational technology, OPC UA offers an integration layer. Because it makes data transfer from control units, sensors, SCADA, or MES systems to XR interfaces safe, standardized, and dependable, it is essential for XR systems [8]. Working with semantically rich data models, event-based notification systems, and low-latency communication are some of its advantages. Because of this, the data shown in XR is always correct, current, and useful for making important decisions [9].

OpenXR 1.1 serves as a technology bridge connecting XR hardware and software. This standard makes it possible to develop XR apps as multi-platform solutions without requiring device modifications [10]. This implies that, independent of the brand or model of headset used in an industrial setting, interactive XR apps and visualizations can be dispersed throughout the entire organization. This makes it much easier to scale, maintain, and deploy XR solutions in an industrial setting [9].

An event-based control architecture appropriate for dispersed control systems is represented by IEC 61499. Its function block approach allows for high control flexibility and modularisation, making it perfect for dynamic and complicated production processes. When used in conjunction with XR, this standard can facilitate the application of the "human-in-the-loop" idea by acting as a link between user inputs in the XR environment and actual control operations. This indicates that through XR contact, humans—an essential component of the system—directly affect how control algorithms behave [9].

As a result, a new technological realm has emerged where intelligent XR-supported production systems that can actively regulate and optimise the manufacturing process in addition to visualising it can be designed. In addition to being a technological potential, ISO 23247, OPC UA, OpenXR 1.1, and IEC 61499 together represent a strategic route to bringing engineering into the age of intelligent, adaptable, and visually assisted automation (Table 1).

Table 1. Synthesis and interconnection of components

Component	Function in XR System	Integration with the Framework
ISO 23247	Defines the digital twin and its structure	Provides the basis for data and XR representation
OPC UA	Ensures interoperability and data transfer	Feeds data into the XR environment
OpenXR 1.1	Unified access to XR devices	Enables deployment of XR visualization system
IEC 61499	Event and process control	Ensures feedback between XR and OT systems

2.2 Overview of XR Applications

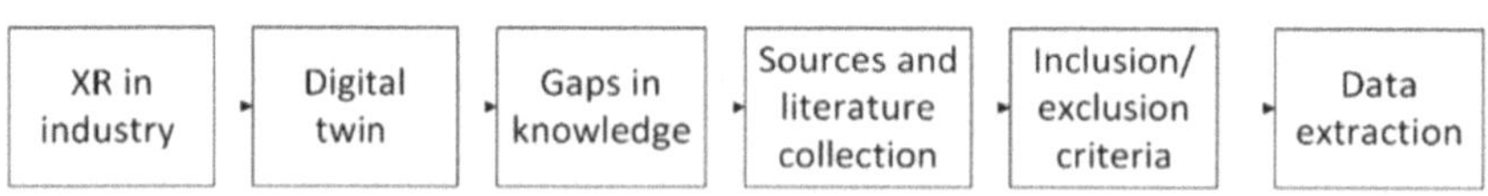

Fig. 3. Research methodology.

XR in industry—systematic reviews and trends

Part of XR technologies, AR is emerging as a key instrument for industry digitisation. Systematic reviews confirm their importance in areas such as maintenance, SCADA, installation and training. In maintenance, AR is used to display work instructions in real time and for remote assistance, where an expert guides a technician through an

AR interface [11, 12]. In training, AR reduces the time it takes for new employees to adapt and reduces the number of errors. In assembly, AR helps visualize steps directly on components, increasing accuracy and efficiency. Integration with SCADA systems increases visibility for the operator, but at the same time creates technical challenges, such as the need for accurate tracking, low latency and secure data transmission (OPC UA)Ensuring standardized interfaces (like OpenXR) and the dependability of solutions in a demanding industrial environment are crucial for successful implementation (see Fig. 3) [13, 14].

Digital twin and virtual commissioning

Digital twins are virtualised versions of real-world systems that use physical system data for detection, optimisation, manipulation, and interaction. They facilitate safe commissioning, quick reconfiguration, and optimisation in the context of flexible manufacturing cells without causing production halts. As demonstrated, ISO 23247 offers a methodical framework for creating and deploying a digital twin in a manufacturing setting, along with integration with REST API and OPC UA software [14, 15]. Configurations recorded in XML format and OPC UA servers enable the connection to XR visualisation. This makes it possible to simulate paths, detect collisions, and update visualisation situations in real time. This is especially useful for low-frequency changes, such as changes in product types (see Fig. 3) [16, 17].

When designing a production system, co-simulation—which connects mechanical, electrical, and control models—is essential [17, 18]. The FMI/FMU standard is used for this purpose, which enables parallel model execution and supports interoperability between simulation tools [19, 20].

XR visualisation complements co-simulation as an interactive interface. Real-time visualisation of simulated trajectories, collisions, and equipment statuses is achievable using 3D models (such as those from Siemens NX or RobotStudio). The XR scene and simulation results can be synchronised thanks to a data connection made possible by OPC UA and REST API [20, 21].

Gaps in knowledge

There are a number of enduring gaps in the current body of research on XR technology in industrial settings. The lack of standardised benchmarks for evaluating XR applications in actual production workshops is one of the primary ones. The majority of investigations are carried out in lab settings, which makes it challenging to apply the findings in real-world settings. The inadequate comprehension of how interactive XR interventions (also known as "read-write") affect safety and overall production efficiency (OEE) is another issue. The influence of XR on performance and operational reliability has not been systematically quantified, despite its increasing use in maintenance, training, and system tuning. Additionally, methods for assessing user experience vary.

Sources and literature collection

The literature was collected from Scopus (Elsevier) and Web of Science (Clarivate), focusing on peer-reviewed publications related to extended reality (XR) and industrial control systems. Boolean search queries targeted the intersection of XR technologies, industrial standards, and manufacturing applications: ("XR" OR "Extended Reality"

OR "AR" OR "VR" OR "MR") AND ("OPC UA" OR "OpenXR" OR "IEC 61499" OR "ISO 23247") AND ("virtual commissioning" OR "digital twin" OR "predictive maintenance" OR "teleoperation").

The study selection followed a PRISMA-style systematic review approach. The refined search identified 17 records. After removing duplicates, 14 studies remained for title and abstract screening, of which six were excluded due to insufficient industrial relevance or lack of technical validation.

To support structured synthesis, the included studies were categorized according to application domain, XR modality, integration level, and reported evaluation metrics, as summarized in Table 2.

Table 2. Classification of included XR studies in industrial control systems

Application domain	XR modality	Integration level	Main metrics reported
Monitoring	AR / MR	PLC / SCADA	Latency, registration accuracy
Virtual commissioning	VR	Digital Twin	Setup time reduction
Maintenance	AR	MES / Digital Twin	MTTR, SUS
Training	VR / MR	Simulation / Digital Twin	Task time, NASA-TLX

Inclusion/exclusion criteria

Studies focusing on industrial XR applications having a technological readiness level of at least 4, or prototype maturity, and published between 2015 and 2025 met the inclusion criteria. In addition to showcasing integration with operational technology and industrial automation standards, the chosen works had to incorporate quantitative control or user experience measures like latency, accuracy, or NASA-TLX. Studies focused exclusively on healthcare, medical rehabilitation, occupational therapy, gaming, or educational applications without direct relevance to industrial control systems or manufacturing processes were excluded.

Data extraction

From each included study, structured data were extracted according to several key dimensions. These included the integration architecture, covering the communication protocols used (e.g., OPC UA, MQTT) and connections to digital twins (DT) or simulation models, as well as the system's functional role, such as monitoring, configuration or tuning, maintenance (predictive or reactive), and teleoperation. In addition, reported performance and user experience metrics were recorded, including technical indicators, user-centered scores, and implemented security mechanisms [22].

A combination of thematic coding and tabular comparison was used to summarize the results. Articles were grouped by task type (e.g., configuration versus maintenance) and level of control (machine, cell, line, or plant). The results were presented as a weighted descriptive summary that considered both the importance and complexity of each system.

Every study was assessed for reproducibility and bias. The selection of participants or scenarios (e.g., learning effects) and the reliability of the measurements (e.g., latency measurement techniques) were the focus of the bias evaluation. The availability of setups, datasets, and experimental situations, together with the release of source code or simulation models, were taken into consideration while evaluating reproducibility. Low-quality studies were those with unclear methods or no reproducibility.

Each study was examined considering four major standardisation frameworks in order to evaluate the technological maturity and practicality of XR solutions in industrial settings. The first was ISO 23247, which outlines the layers of representation, integration, control, and security that make up the digital twin architecture for smart manufacturing [15]. The second was OPC UA, which was designed to guarantee safe and dependable data transfer between software components and heterogeneous devices in an industrial Internet of Things setting. IEC 61499, the third framework, is centred on function block-based control for complex automation systems that is decentralised, modular, and reusable. Lastly, OpenXR promotes interoperability and standardisation in extended reality applications by offering a single API for XR development across several hardware platforms and OS systems.

3 Results

According to recent publications, the application of XR technologies in industrial management focuses on four main scenarios: monitoring, configuration and virtual commissioning, predictive maintenance, and remote technical support (see Fig. 3). Data transfer between XR solutions and the OT level is usually implemented via OPC UA or MQTT protocols, while OpenXR is increasingly being used at the client level as a cross-platform framework for XR interfaces [21]. XR integration into production systems is achieved using three common architectural models. First, XR is connected to a digital twin based on co-simulation via FMI/FMU, which allows training or testing without interfering with real objects [23]. Second, direct connection to an OPC UA server is possible, providing access to parameters, events, and alarms [24]. Finally, XR can act as an HMI component in environments based on IEC 61499, using event-driven control logic that allows for flexible and modular system deployment [25].

Quantitative studies show that the implementation of XR solutions can reduce the time required to set up and start production systems by 20–40%, and the mean time to repair (MTTR) by 15–30%. However, these indicators largely depend on tracking technologies, user training levels, and environmental conditions [26, 27]. From a technical point of view, the main performance indicators for XR solutions are latency within 20–100 ms, spatial registration accuracy in the range of 1–30 mm, and user experience rating—SUS (System Usability Scale) usually ranges between 68 and 82 points. On the NASA-TLX scale (mental workload assessment), exceeding the threshold of 60 in complex scenarios indicates a need to simplify interfaces and optimise interaction. According to review papers, industrial standards are being increasingly implemented in practice. OpenXR 1.1 provides a unified approach to developing XR applications for different devices, reducing fragmentation among platforms [28]. ISO 23247 serves as a reference model for digital twins in smart manufacturing [29]. OPC UA remains the primary standard for ensuring interoperability between devices, controllers, and MES/ERP levels. IEC 61499 provides modularity and scalability in event-driven systems, which is important for XR-oriented HMIs [25].

4 Discussion

Because XR can integrate hands-free interfaces, contextual data, and spatial semantics, it shows great promise for use in virtual commissioning and diagnostics of industrial systems. This reduces operator cognitive strain and minimizes errors during system design or diagnostics in industrial settings by enabling more effective real-time access to operational data [28, 30]. Additionally, spatial visualization greatly lessens the necessity for operators to continuously shift their attention between digital panels and actual equipment, which is common in conventional HMI systems.

However, using XR in read-write modes, which allows the user to actively change parameters in a live production setting, presents additional safety and risk management difficulties. Clear regulations for role-based access control, user activity auditing, and authentication are necessary due to the high level of XR integration with the OT layer. For distant or XR-mediated equipment, the OPC UA specification's multi-layered security stack, which includes role-based controls, certificate management, and TLS-encrypted sessions, is essential. XR systems should be automatically limited to read-only mode in high-risk industrial zones. Any escalation to control functions should require dual confirmation and authorization from the OT layer [31].

Regarding transferability, the effectiveness of XR systems remains highly dependent on network infrastructure (notably 5G or TSN support), the accuracy and currency of the digital twin (DT), and the precision of sensor subsystems. For successful delivery of contextual data from the physical environment to the XR interface, it requires a consistent, real-time match between the physical object and its digital representation, as well as minimal communication latency. Without reliable DT modelling, XR may present operators with misleading information about equipment status—posing serious operational risks [23, 29].

A broader literature review confirms that many of XR's advantages in maintenance, diagnostics, and training are widely supported in the academic community. Specifically, it shows that most XR-based remote maintenance studies report improvements in task efficiency, reduced downtime, enhanced precision, and better user experience. However, the authors also highlight a lack of consistency in how performance metrics are reported—ranging from latency and jitter to cognitive workload—often without standardized measurement methodologies. This lack of uniformity makes it difficult to compare outcomes across studies [14].

5 Conclusion

According to this evaluation, XR technologies are a promising technique for facilitating computer-based industrial system control. XR applications can drastically cut commissioning time and mean time to repair (MTTR) while upholding strict safety and operational integrity when properly integrated, especially through OpenXR for interface standardisation, OPC UA for secure interoperability, ISO 23247 for digital twin modelling, and IEC 61499 for distributed control logic. These advantages are particularly noticeable in hybrid OT/IT settings where system feedback and user reaction are improved by contextual visual information.

From the standpoint of practical execution, a phased approach is recommended. To help operators become accustomed to augmented spatial interfaces, organizations should start with read-only XR visualizations based on OPC UA server data (parameters, events, and alarms). Strict role-based access, audit trails, and latency monitoring systems should all be in place to facilitate the gradual transition to regulated read-write interactions. To guarantee system stability and security, investments in network latency diagnostics and scene registration accuracy are crucial.

The creation of validated methods for XR-assisted control that adhere to industrial safety standards and standardized XR benchmarks relevant to shop floor circumstances are top priority for the next three to five years of study. Furthermore, publicly accessible datasets for XR application training and validation, as well as open and repeatable digital twin scenarios (such as FMI-based models), are urgently needed. Assessing return on investment (ROI) and total cost of ownership (TCO) for extensive XR installations, especially in manufacturing contexts, is the last research gap.

Acknowledgements. This publication was created within the project Research of Advanced Technologies for Increasing the Efficiency of Multivalent Energy Systems for Sustainable Industrial Development (ITMS21+ code: 401101C504), co-funded by the European Union under the Programme Slovakia. The work was also supported by the Slovak Research and Development Agency under Contract No. APVV-23–0591, and by the projects VEGA 1/0317/25 and KEGA 028TUKE-4/2025 funded by the Ministry of Education, Science, Research, and Sport of the Slovak Republic.

References

1. Hrehova, S., Husár, J., Duhančík, M.: Characteristics of selected software tools for the design of augmented reality models. In: 2024 25th International Carpathian Control Conference (ICCC). pp. 1–6. IEEE (2024). https://doi.org/10.1109/ICCC62069.2024.10569576
2. Zidek, K., Maxim, V., Pitel, J., Hosovsky, A.: Embedded vision equipment of industrial robot for inline detection of product errors by clustering–classification algorithms. Int. J. Adv. Robot. Syst. **13** (2016). https://doi.org/10.1177/1729881416664901
3. Ziker, C., Truman, B., Dodds, H.: Cross Reality (XR): challenges and opportunities across the spectrum. Presented at the (2021). https://doi.org/10.1007/978-3-030-58948-6_4
4. Cortes, D., Bermejo, B., Juiz, C.: The use of CNNs in VR/AR/MR/XR: a systematic literature review. Virtual Real. **28**, 154 (2024). https://doi.org/10.1007/s10055-024-01044-6
5. Nazim, A., Kondrát, M., Zidek, K., Pitel, J.: Methodology of object reconstruction by photogrammetry and structured-light scanning for industrial 3D visualisation. Sensors **25**, 7177 (2025). https://doi.org/10.3390/s25237177
6. Guo, J., et al.: Industrial metaverse towards Industry 5.0: Connotation, architecture, enablers, and challenges. J. Manuf. Syst. **76**, 25–42 (2024). https://doi.org/10.1016/j.jmsy.2024.07.007
7. Buń, P., Gapsa, J., Husár, J., Kaščak, J.: Mixed reality training in electrical equipment operating procedures. Presented at the (2022). https://doi.org/10.1007/978-3-030-99310-8_24
8. Buń, P., Husár, J., Kaščak, J.: Hand tracking in extended reality educational applications. Presented at the (2022). https://doi.org/10.1007/978-3-030-99310-8_25

9. Yang, C., et al.: Extended reality application framework for a digital-twin-based smart crane. Appl. Sci. **12**, 6030 (2022). https://doi.org/10.3390/app12126030
10. Ghinea, M., Deac, G.C., Ivanescu, R., Deac, C.N., Deac, G.A.: Secure Cloud-Streaming Digital Twin (SCSDT): a cloud-first metaverse architecture for industrial training (2025). https://doi.org/10.20944/preprints202508.0883.v1
11. Adamczak, M., Kolinski, A., Trojanowska, J., Husár, J.: Digitalization trend and its influence on the development of the operational process in production companies. Appl. Sci. **13**, 1393 (2023). https://doi.org/10.3390/app13031393
12. Antosz, K., Knapčíková, L., Husár, J.: Evaluation and application of machine learning techniques for quality improvement in metal product manufacturing. Appl. Sci. **14**, 10450 (2024). https://doi.org/10.3390/app142210450
13. Hrehova, S., Husár, J., Lazorík, P., Trojanowski, P.: The concept of an intelligent mobile application with elements of augmented reality and PLM software for value stream mapping. Ann. Oper. Res. (2024). https://doi.org/10.1007/s10479-024-06006-4
14. Breitkreuz, D., Müller, M., Stegelmeyer, D., Mishra, R.: Augmented reality remote maintenance in industry: a systematic literature review. Presented at the (2022). https://doi.org/10.1007/978-3-031-15553-6_21
15. Shao, G., Frechette, S., Srinivasan, V.: An analysis of the new ISO 23247 series of standards on digital twin framework for manufacturing. In: Volume 2: Manufacturing Equipment and Automation; Manufacturing Processes; Manufacturing Systems; Nano/Micro/Meso Manufacturing; Quality and Reliability. American Society of Mechanical Engineers (2023). https://doi.org/10.1115/MSEC2023-101127
16. Stasa, P., Benes, F., Svub, J.: Augmented Reality-based approach for interpretation and visualization of methane flow behaviour. Int. J. Min. Reclam. Environ. **32**, 449–462 (2018). https://doi.org/10.1080/17480930.2017.1388337
17. Wallner, B., Zwölfer, B., Trautner, T., Bleicher, F.: Digital twin development and operation of a flexible manufacturing cell using ISO 23247. Procedia CIRP. **120**, 1149–1154 (2023). https://doi.org/10.1016/j.procir.2023.09.140
18. Demčák, J., Židek, K., Krenický, T.: Digital twin for monitoring the experimental assembly process using RFID technology. Processes **12**, 1512 (2024). https://doi.org/10.3390/pr12071512
19. Nazim, A., Židek, K., Balog, M., Kalman, O., Svetlík, J.: Implementation of SmartTechLab digital twin to AR/VR technology for educational purposes. Presented at the (2024). https://doi.org/10.1007/978-3-031-59238-6_2
20. Elsotohy, M., Jaeschke, J., Sehr, F., Schneider-Ramelow, M.: Mission profile-based digital twin framework using functional mock-up interfaces for assessing system's degradation behaviour. Microelectron. Reliab. **150**, 115086 (2023). https://doi.org/10.1016/j.microrel.2023.115086
21. Velesaca, H.O., Holgado-Terriza, J.A., Gutierrez-Guerrero, J.M.: Industrial process automation through machine learning and OPC-UA: a systematic literature review. Electronics (Basel). **14**, 3749 (2025). https://doi.org/10.3390/electronics14183749
22. Zidek, K., Vasek, V., Pitel, J., Hosovsky, A.: Auxiliary device for accurate measurement by the smart vision system. MM Sci. J. **2018**, 2136–2139 (2018). https://doi.org/10.17973/MMSJ.2018_03_201722
23. Weber, J., Stäbler, M., Thielen, S., Paetzold, K.: Modularity as key enabler for scalability of final assembly units in the automotive sector. Procedia CIRP. **57**, 224–228 (2016). https://doi.org/10.1016/j.procir.2016.11.039
24. Li, C., He, F., Hao, N.: Verification and design of zero-sum potential games. IFAC-PapersOnLine. **53**, 16932–16937 (2020). https://doi.org/10.1016/j.ifacol.2020.12.1236

25. Colabufo, G.G., Dower, P.M., Shames, I.: Newton's method: sufficient conditions for practical and input-to-state stability. IFAC-PapersOnLine. **53**, 6334–6339 (2020). https://doi.org/10.1016/j.ifacol.2020.12.1764
26. Trojanowska, J., Żywicki, K., Pająk, E.: Influence of selected methods of production flow control on environment. Presented at the (2011). https://doi.org/10.1007/978-3-642-19536-5_54
27. Aivaliotis, P., Arkouli, Z., Georgoulias, K., Makris, S.: Methodology for enabling dynamic digital twins and virtual model evolution in industrial robotics-a predictive maintenance application. Int. J. Comput. Integr. Manuf. **36**, 947–965 (2023). https://doi.org/10.1080/0951192X.2022.2162591
28. Azamfirei, V., Psarommatis, F., Lagrosen, Y.: Application of automation for in-line quality inspection, a zero-defect manufacturing approach. J. Manuf. Syst. **67**, 1–22 (2023). https://doi.org/10.1016/j.jmsy.2022.12.010
29. Lu, Y., Liu, C., Wang, K.I.-K., Huang, H., Xu, X.: Digital Twin-driven smart manufacturing: connotation, reference model, applications and research issues. Robot. Comput. Integr. Manuf. **61**, 101837 (2020). https://doi.org/10.1016/j.rcim.2019.101837
30. Mazumder, A., et al.: Towards next generation digital twin in robotics: trends, scopes, challenges, and future. Heliyon **9**, e13359 (2023). https://doi.org/10.1016/j.heliyon.2023.e13359
31. Lee, C., Kwon, H., Lee, Y.: OPC UA-based three-layer architecture for aggregated microgrids integrating edge cloud computing and IEC 62264. J. Ind. Inf. Integr. **48**, 100965 (2025). https://doi.org/10.1016/j.jii.2025.100965

Possibilities of Using Generative AI in AI-Based Digital Twins for Industry 5.0/6.0

Izabela Rojek[1](✉), Natasa Naprstkova[2], and Dariusz Mikołajewski[1]

[1] Kazimierz Wielki University, Bydgoszcz, Poland
izabela.rojek@ukw.edu.pl
[2] Jan Evangelista Purkyně University, Ústí Nad Labem, Czech Republic

Abstract. The latest advances in Industry 5.0/6.0 paradigms are driving the more advanced use of generative artificial intelligence (genAI) in AI-based digital twins (DTs). This approach allows for the automatic collection and selection of data resources acquired by AI-based digital twins (DTs) and their use to generate a variety of documents useful in automated production processes and the supply chain. This provides greater flexibility and scalability of the solutions used, as well as the ability to further aggregate data from DTs over time (for change monitoring and predictive maintenance) and from multiple DTs simultaneously (e.g., from production line quality control points). From an operational perspective, this also ensures faster event handling and a broader selection of pre-prepared scenarios within the operator-cyber-physical system interface than traditionally possible. These genAI solutions represent another step toward supporting operators of complex industrial processes and systems, especially those operating under multiple constraints. The contribution to knowledge of this article is the development of genAI applications in technological transformation technologies in such a way as to facilitate the creation of human-centric, self-evolving technological transformation ecosystems aligned with the goals of sustainability, resilience, and cognitive autonomy of Industry 5.0 and 6.0.

Keywords: Industry · Digital Twin · Artificial Intelligence · Generative AI

1 Introduction

Generative artificial intelligence (GenAI) offers revolutionary opportunities to enhance AI-based digital twins (DTs) in Industry 5.0 and 6.0, enabling more intelligent, adaptive, and autonomous virtual representations of physical systems. Unlike traditional DTs, which rely on static or predefined models, GenAI can dynamically generate new designs, scenarios, and responses based on real-time data. This allows DTs to simulate complex environments, predict system behavior, and optimize industrial processes with greater accuracy [1]. In Industry 5.0, where human-machine collaboration is crucial, GenAI-based DTs can personalize interfaces, workflows, and production layouts based on operator preferences and performance. GenAI can generate synthetic data to securely train AI models, reducing reliance on costly or sensitive real-world datasets. It

J. Trojanowska et al. (Eds.): MANUFACTURING 2026, LNME, pp. 30–40, 2026.
https://doi.org/10.1007/978-3-032-21654-0_3

can also create realistic simulations to test cybersecurity resilience, enabling proactive defense against emerging cyber threats. In the manufacturing industry, GenAI-based DTs can design optimized production processes, suggest energy-efficient configurations, and reduce waste through continuous learning [2]. As part of predictive maintenance, these systems can predict failures by generating possible failure scenarios and recommending preventative actions [3]. In Industry 6.0, where systems are expected to achieve cognitive autonomy, GenAI will enable self-evolving DTs capable of independent decision-making and adaptation. GenAI can facilitate multi-domain integration, enabling digital twins to coordinate activities in supply chains, logistics, and smart cities within a unified ecosystem. GenAI's creative capabilities support co-design processes between humans and AI, fostering innovation in product development and sustainability strategies [4]. However, the use of GenAI poses challenges related to explainability, model bias, and control over the generated results, which must be addressed through ethical AI governance. Data integrity and authenticity are crucial, as maliciously generated data or manipulated models can mislead decision-making processes [5]. Despite these risks, GenAI-based DTs promise greater adaptability, energy optimization, and user-centric personalization than ever before. The combination of GenAI, DTs, and the vision of Industry 5.0/6.0 lays the foundation for intelligent, human-centric, and self-regulating industrial ecosystems of the future [6]. Industry 6.0 is a future paradigm in which production is carried out in symbiosis between humans and advanced technology, focused on creating value in a holistic, efficient and regenerative way, while respecting the environment and social values.

The aim of this article is to demonstrate the use of genAI in AI-based DTs for Industry 5.0/6.0 in such a way as to facilitate the creation of human-centric, self-evolving DT ecosystems aligned with the goals of sustainability, resilience, and cognitive autonomy of Industry 5.0 and 6.0. We formulated the following research questions (RQs):

- RQ1: What are the current and emerging research concepts in this area?
- RQ2: What is the roadmap for their implementation?
- RQ3: What are the implications of these changes?
- RQ4: What are the key areas for further research?

Figure 1 provides a conceptual diagram illustrating how genAI interacts with AI-based DTs and contributes to the multi-layered evolution of Industry 5.0 and 6.0 by combining physical data, digital intelligence, and human-machine collaboration within a unified system [7, 8].

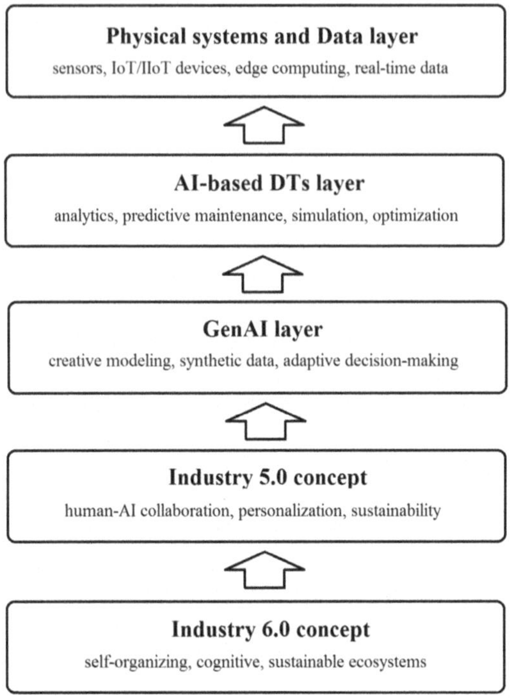

Fig. 1. Conceptual framework of genAI interaction with DTs within Industry 5.0/6.0 paradigm (own elaboration).

2 Current and Emerging Concepts

We showed our current concept of optimizing physical processes using DT on the example of 3D printing processes within the Industry 4.0 paradigm in [9]. Based on our research, we demonstrated at the time that AI-based DTs require further, more thorough research due to the nature of their semi-automatic or automatic creation and the need for oversight. However, even in the initial stages of our research, we demonstrated that adopting and mastering this technology would lead to significant improvements in performance, reliability, and profitability. However, it required investment in transforming traditional approaches to AI-based data processing technologies and training specialists from various industries who would collaborate with scientists to develop them.

The contribution to knowledge of this article is the development of genAI applications in technological transformation technologies in such a way as to facilitate the creation of human-centric, self-evolving technological transformation ecosystems aligned with the goals of sustainability, resilience, and cognitive autonomy of Industry 5.0 and 6.0.

As part of the work on this publication we developed aforementioned concept, typical applications of genAI in Industry 5.0/6.0 were examined, including those useful for AI-based DTs and collaborative humans: from the generated production line report:

- all numerical values for DTs were automatically extracted;

- the equipment status was determined based on these values, both absolute and relative (e.g., % of service life to the next inspection);
- report shortcuts were created for operators and the technologist;
- the date and time of the next required report were entered (Fig. 2).

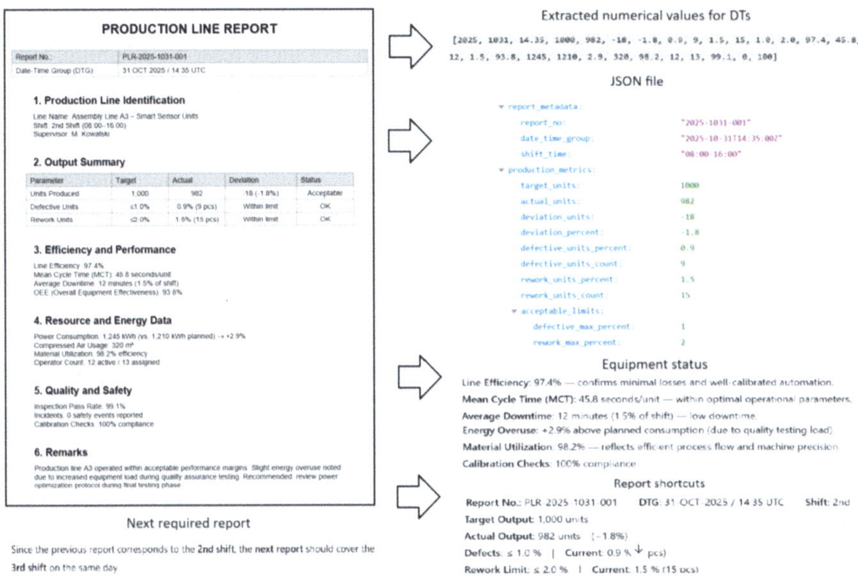

Fig. 2. Documents and data generated for this study (examples).

The above examples only offer a selection of capabilities that can be scaled both temporally and spatially. GenAI helps DTS monitor, classify, manage, and predict complex, nonlinear relationships on the production line, including modeling complex problems (both sequential and concurrent) that traditional analysis and control methods cannot solve. GenAI ensures the consistency of individual documents (including by securing them with blockchain or its successor within post-quantum technologies), while also facilitating the detection of anomalies in reports. This also allows for earlier and more precise definition of action scenarios before changes (both intentional resulting from control strategies and pathological changes resulting from, for example, component wear) become visible in the physical system (by which time it may be too late for effective preventative intervention).

GenAI helps automate and accelerate decision-making by providing opinions and recommendations based on DT analyses and multi-variant simulations of future developments at various levels of detail (technologist, operator, technical control). This data, in various forms, processed or generated by DTs can be more easily and quickly integrated with other DTs or superior AI systems. This allows for a clearer picture of overall operation, including automatic and semi-automatic monitoring and control of multiple connected systems by a limited number of operators, predicting alerts and warnings, and proposing emergency response plans.

The scalability of the connected GenAI-based DTs system enables their near-automatic implementation in increasingly complex systems, with minimal manual intervention and adaptation by specialists or expert knowledge required to implement data updates. Continuous training of these systems not only improves the DTS's performance but also gradually and progressively accumulates knowledge within the system.

The study showed that GenAI-based DT represents a new generation of accurate, advanced tools for supporting human operators in managing complex systems, optimizing and maintaining them in the short and long term. For these reasons, planning, building, implementing, operating, and modernizing/decommissioning GenAI-based DT is already a future skill among Industry 5.0/6.0 specialists, as well as technologists and operators of industrial systems. This stems from the fact that an operator supported by this group of solutions will be more effective than their counterpart without such support. It is therefore a valuable and worthwhile investment. However, it still requires refinement if it is to be used to solve ever-new, perhaps not yet fully identified, problems, such as energy efficiency, the carbon footprint of production (the degree of use of renewable energy sources or recycled products), or ESG.

It is important to quickly scale to the number of parameters and employees, as well as to generate a larger number of realistic scenarios as a second-opinion system, especially in situations with tight production schedules, delivery schedules, and potential line maintenance requirements. This approach allows for the optimization of the above-mentioned tasks based on individual human preferences and predictive maintenance, but requires further refinement within the framework of the proposed roadmap below. This may allow for improving production efficiency or testing/finding new ways of manufacturing products without compromising the actual production process, etc.

Broader adoption of our proposed genAI-based DTs concept in Industry 5.0/6.0, instead of traditional approaches, could become standard practice, seeking creative solutions to the most challenging production problems, which will help accelerate this development and redirect efforts to more challenging issues.

The roadmap for the broader implementation of generative artificial intelligence (GenAI) in AI-based DTs for Industry 5.0/6.0 begins with fundamental research and standardization, focusing on interoperability, model validation, and data management frameworks:

- Phase 1 should emphasize the development of scalable DT architectures capable of integrating GenAI into simulation, design optimization, and adaptive learning;
- Phase 2 encompasses the enhancement of data infrastructure through edge and cloud computing, ensuring real-time connectivity between physical resources and their virtual counterparts;
- Phase 3 focuses on creating high-quality, diverse datasets for GenAI training, supported by privacy-preserving methods such as federated learning (and, in the future, post-quantum cryptography);
- Phase 4 requires the implementation of understandable and trustworthy GenAI models to address transparency and accountability issues;
- Phase 5 includes pilot implementations in key sectors (manufacturing, energy, healthcare, etc.) to validate performance, safety, and scalability;

- Phase 6 should address cybersecurity and ethical risk management by incorporating AI auditing, bias detection, and responsible use principles into the implementation framework;
- Phase 7 prioritizes workforce development through targeted training programs that equip operators and engineers with AI, data, and simulation skills;
- Phase 8 strengthens collaboration between industry, academia, and policymakers to establish regulatory standards, certification processes, and international cooperation.

Achieving widespread adoption of GenAI in industrial transformation technologies will require a balanced approach that combines technological innovation, ethical management, and human-centered design to realize the intelligent, sustainable, and adaptive vision of Industry 5.0 and 6.0.

3 Discussion

GenAI-enhanced DTs within Industry 5.0/6.0 are adaptive, semantic models of physical processes/objects that generate scenarios, designs, and optimizations using generative networks and large models—while maintaining human-machine collaboration and sustainable development goals (ESG). Technically, this means a hybrid of physical-based and data-driven models, a distributed edge/cloud architecture, real-time telemetry streams, and mechanisms for generating production variants and energy forecasts via GenAI. From a regulatory perspective, DTs must be designed with privacy and data protection (GDPR), operational resilience and incident reporting (NIS2, DORA for the financial sector), and information security management in line with ISO/IEC 27001. Additionally, ESG reporting and metrics must be tracked and auditable.

To meet these requirements, implementations should employ privacy-by-design (e.g., DPIA, pseudonymization, data minimization), model explainability mechanisms (model cards, decision logging), and strong model and data management policies (MLOps with auditable pipelines). Security technologies include edge computing (processing and pre-aggregating data locally to reduce data transfer and the risk of breaches), federated learning (training models without centrally collecting raw data), and blockchain/ledger for immutability and auditability of key events, SLAs, and ESG metrics.

Additional layers of protection include end-to-end encryption, HSM/KMS for key management, secure execution environments (TEEs), and differential privacy mechanisms or homomorphic operations where higher privacy is required. From a compliance perspective, it is worthwhile to map each DT component to specific requirements: for example, NIS2-compliant logging and retention, DORA-compliant resilience and BC/DR testing, and ISO27001-compliant access policies and contion controls. Formal organizational roles are recommended: data owner, model owner, DPO, SecOps/SOC team, and procedures for responsible disclosure of model errors and retraining processes. Regular external audits and penetration tests increase credibility. GenAI-DT implementation must also address ethical and social risks (bias, autonomous decisions affecting employees), introduce human-in-the-loop oversight mechanisms, and policies that compensate for environmental impacts for ESG purposes. In summary: a secure and compliant GenAI-based digital twin is a combination of technical and organizational measures—edge + federated learning + auditability (blockchain) + strong encryption and governance—that together ensure compliance with GDPR, NIS2, DORA, ISO27001, and the achievement of ESG goals in Industry 5.0/6.0.

Developers use Amazon Bedrock to interpret customer requests and suggest intelligent solutions. British energy provider Octopus Energy uses a bot that does the work of 250 people and achieves higher satisfaction rates than human agents. Motif FoodWorks uses genAI to design and test plant-based foods, taking into account factors such as regional taste preferences, dietary requirements, and even seasonal ingredient availability. The newspaper Die Presse uses genAI to generate interview questions, article ideas, and social media headlines. KSAT-TV uses AI to transcribe videos into text. Beamery's TalentGPT AI assistant uses genAI to help employees explore potential career paths and help companies hire the best-suited talent. Walmart's chatbots use intelligent AI to respond to customer inquiries in a more natural and intuitive way, improving the customer experience. GenAI also helps banks like Wells Fargo suggest optimal investment strategies and create personalized portfolios based on individual risk tolerance.

3.1 Implications

Integrating GenAI with AI-based DTs introduces advanced technical capabilities, such as autonomous system design, real-time optimization, and the generation of synthetic data for continuous learning. However, these systems also present technical challenges related to model interpretability, computational demands, and ensuring the reliability of generated simulations. Maintaining data integrity in GenAI-based DTs is complex, as small model deviations can cause significant operational errors in real-world processes [10, 11]. Economically, GenAI increases productivity by enabling predictive maintenance, reducing downtime, and accelerating innovation cycles. However, the high costs of implementing computing infrastructure, data management, and training a skilled workforce can limit adoption, particularly in small and medium-sized enterprises. Over time, GenAI-based automation can deliver cost savings and increased return on investment through operational optimization and energy savings. At the societal level, these technologies promote collaboration between humans and AI, aligning with the Industry 5.0 vision of empowering workers and personalizing production environments. However, increased automation could displace some employee roles, requiring retraining initiatives to maintain an inclusive workforce. Ethical implications focus on the transparency and accountability of GenAI decisions, as self-generating models can perform in ways that are difficult for humans to understand. Ethical risks also include model bias, data misuse, and the generation of misleading or harmful synthetic results, which can distort industrial decision-making. Privacy concerns are heightened when GenAI systems process sensitive industrial and personal data for optimization and predictive analytics. From a legal perspective, the use of GenAI challenges existing intellectual property frameworks, as it is unclear who owns AI-generated designs or simulation results. Determining liability becomes problematic if a GenAI-generated model leads to system failures, security incidents, or regulatory noncompliance. Regulators will need to establish new standards for certification, auditing, and accountability of GenAI-based digital twins. While GenAI has enormous potential to propel Industry 5.0 and 6.0 towards smart, sustainable, and human-centric ecosystems, its responsible implementation requires a careful balance of innovation, ethics, and governance (Table 1).

3.2 Limitations

One of the key limitations of using GenAI in AI-based digital twins is the lack of model transparency, as many generative models act as black boxes, making interpretation and decision justification difficult. This necessitates the broader use of post-hoc XAI in this area. The quality and reliability of generated results are highly dependent on training data, and poor or biased datasets can lead to inaccurate simulations or unsafe system recommendations. GenAI models often require significant computational resources, which poses challenges for real-time implementation in complex industrial environments, including virtual reality (VR) and augmented reality (AugR) [12]. Ensuring data security and integrity becomes difficult when synthetic data or automatically generated models are fed into critical industrial processes. There is a risk of error propagation, where a single incorrect generative prediction can cascade across interconnected DTs, leading to large-scale operational disruptions. Ethical and regulatory gaps persist, as current regulations struggle to define responsibility and ownership for AI-generated projects or decisions. Maintaining and retraining generative models requires continuous data input and expert supervision, which increases operational complexity and costs. Integrating GenAI with existing digital twin architectures poses compatibility and interoperability challenges, especially across industrial systems from different vendors. Generating synthetic or simulated data raises concerns about authenticity and traceability, which can undermine trust in industrial decision-making. While GenAI offers enormous potential for Industry 5.0 and 6.0, its limitations underscore the need for robust validation, governance, and human oversight to ensure safe and responsible implementation. Key mitigation strategies are presented in Table 1.

Table 1. Selected limitations and mitigation strategies for genAI in AI-based DTs within Industry 5.0/6.0 paradigms.

Limitation	Description	Proposed mitigation
Lack of transparency	GenAI models often act as "black boxes," making it hard to explain outputs or decisions	Implement XAI methods and model interpretability tools to increase understanding and trust
Data quality dependence	Poor, biased, or incomplete data can lead to inaccurate or unsafe generative outcomes	Use data validation pipelines, bias detection, and diverse datasets to improve input quality
High computational demand	Large-scale generative models require powerful infrastructure and energy resources	Employ edge-cloud hybrid architectures and optimize models for efficiency (e.g., pruning, quantization)
Data security and integrity risks	Synthetic data generation and sharing increase exposure to cyber threats	Apply federated learning, encryption, and blockchain for secure data management

(continued)

Table 1. (*continued*)

Limitation	Description	Proposed mitigation
Error propagation	Incorrect outputs can cascade across interconnected digital twin systems	Introduce redundancy checks, real-time validation, and human-in-the-loop oversight
Regulatory and ethical uncertainty	Lack of clear laws on ownership, liability, and ethical use of GenAI outputs	Develop AI governance frameworks, and align with emerging AI ethics and data protection standards
Continuous retraining needs	Models require ongoing updates to remain accurate, increasing cost and complexity	Use automated retraining pipelines and adaptive learning mechanisms to reduce human workload
System interoperability issues	Integrating GenAI across different platforms and vendors can cause compatibility conflicts	Promote open standards, modular architectures, and cross-platform APIs
Authenticity and traceability of synthetic data	Difficult to verify the source and reliability of AI-generated data	Use digital watermarking, provenance tracking, and data lineage frameworks
Overreliance on automation	Excessive trust in autonomous GenAI decisions can reduce human vigilance	Maintain human supervision, decision audits, and clear accountability mechanisms

3.3 Directions of Further Research

Future research should focus on developing understandable and reliable generative AI models to ensure transparency and accountability in AI-based DTs decision-making. There is a need to explore hybrid modeling approaches that combine generative AI with physics-based and symbolic-based inference models to provide more robust industrial simulations [13, 14]. Further research should address energy-efficient and lightweight GenAI architectures suitable for real-time applications on edge devices in industrial environments [15]. Research must also address data management and security frameworks that ensure the secure generation, validation, and sharing of synthetic data in distributed systems. Integrating human-in-the-loop mechanisms to balance automation with human oversight should be explored, particularly in safety-critical industrial contexts. Progress is needed in multi-agent DTs ecosystems, where GenAI enables coordination, negotiation, and adaptation in interconnected industrial networks. Further research into ethical and regulatory frameworks is essential to define ownership, liability, and responsible use of GenAI-generated results. Research should include cross-sectoral applications such as smart energy, healthcare, and logistics to understand how generative digital twins can be generalized across domains. Quantum computing and neuromorphic processors can be explored to enhance the computational efficiency and scalability of GenAI-based twins. Future research should aim to create human-centric, self-evolving digital twin ecosystems that align with the sustainability, resilience, and cognitive autonomy goals of Industry 5.0 and 6.0 [16].

4 Conclusions

GenAI offers transformative capabilities for AI-based digital twins in Industry 5.0 and 6.0, enabling continuous learning, adaptation, and innovation in industrial systems. It allows DTs to autonomously generate new designs, optimize processes, and simulate complex scenarios beyond human-defined parameters. With its creative and predictive capabilities, genAI enhances human-machine collaboration, supporting Industry 5.0's goals of personalization, safety, and sustainability. As Industry 6.0 evolves toward self-organizing and cognitive industrial ecosystems, genAI becomes a key enabler of the autonomy and intelligence of digital twins. Ultimately, its integration marks a shift from reactive modeling to proactive, co-evolving systems that combine artificial creativity with human experience, driving the next industrial revolution.

References

1. Greasley, A., Edwards, J.S.: Enhancing discrete-event simulation with big data analytics: a review. J. Oper. Res. Soc. **72**, 247–267 (2021)
2. de Paula Ferreira, W., Armellini, F., De Santa-Eulalia, L.A.: Simulation in industry 4.0: a state-of-the-art review. Comput. Ind. Eng. **149**, 106868 (2020)
3. Agalianos, K., Ponis, S.T., Aretoulaki, E., Plakas, G., Efthymiou, O.: Discrete event simulation and digital twins: review and challenges for logistics. Procedia Manuf. **51**, 1636–1641 (2020)
4. Aretoulaki, E., Ponis, S.T., Plakas, G., Tzanetou, D.: Discrete event simulation and Digital Twins in warehouse logistics: a bibliometric and content analysis-based systematic literature review. Int. J. Comput. Integr. Manuf. **37**, 1376–1403 (2024)
5. Tsinarakis, G., Sarantinoudis, N., Arampatzis, G.: A discrete process modelling and simulation methodology for industrial systems within the concept of digital twins. Appl. Sci. **12**, 870 (2022)
6. Rojek, I., Mikołajewski, D., Piszcz, A., Małolepsza, O., Kozielski, M.: Role of generative AI in AI-Based digital Twins in Industry 5.0 and evolution to Industry 6.0. Appl. Sci. **15**, 10102 (2025)
7. Akpan, I.J., Etti, G.E.: Computer simulation everywhere: mapping fifteen years evolutionary expansion of discrete-event simulation and integration with digital twin and generative artificial intelligence. Symmetry **17**, 1272 (2025)
8. Akhavan, A., Jalali, M.S.: Generative AI and simulation modeling: how should you (not) use large language models like ChatGPT. Syst. Dyn. Rev. **40**, e1773 (2024)
9. Rojek, I., Marciniak, T., Mikołajewski, D.: Digital Twins in 3D printing processes using artificial intelligence. Electronics **13**, 3550 (2024)
10. Wilson, R., Mercier, P.H., Navarra, A.: Integrated artificial neural network and discrete event simulation framework for regional development of refractory gold systems. Mining **2**, 123–154 (2022)

11. Atalan, A., Şahin, H., Atalan, Y.A.: Integration of machine learning algorithms and discrete-event simulation for the cost of healthcare resources. Healthcare **10**, 1920 (2022)
12. Akpan, I.J., Offodile, O.F.: The role of virtual reality simulation in manufacturing in industry 4.0. Systems **12**, 26 (2024)
13. Frydenlund, E., Martıınez, J., Padilla, J.J., Palacio, K., Shuttleworth, D.: Modeler in a box: how can large language models aid in the simulation modeling process? SIMULATION **100**, 727–749 (2024)
14. Akpan, J.I., Razavi, R., Akpan, A.A.: Evolutionary trends in decision sciences education research from simulation and games to big data analytics and generative artificial intelligence. Big Data **13**(5), 416–437 (2025)
15. Rojek, I., Mikołajewski, D., Dostatni, E., Cybulski, J., Kozielski, M.: Personalization of AI-Based Digital Twins to optimize adaptation in industrial design and manufacturing—Review. Appl. Sci. **15**, 8525 (2025)
16. Akpan, I.J., Kobara, Y.M., Owolabi, J., Akpan, A.A., Offodile, O.F.: Conversational and generative artificial intelligence and human–chatbot interaction in education and research. Int. Trans. Oper. Res. **32**, 1251–1281 (2025)

Modeling the Root Causes of Production Line Stoppages by Leveraging Downtime History and Buffer Dynamics: Simulation-Based Evaluation

Tomasz Bartkowiak[1](✉), Jędrzej Gembiak[1], Jan Dąbrowski[2], and Dariusz Dąbrowski[3]

[1] Poznan University of Technology, Poznan, Poland
tomasz.bartkowiak@put.poznan.pl
[2] PACKGURU Sp. z o. o., Piaseczno, Poland
[3] Ilabo Sp. z o. o., Piaseczno, Poland

Abstract. Accurately identifying the true drivers of production stoppages is critical for improving availability, maintenance prioritization, and throughput in discrete multi-stage manufacturing systems. In serial lines with finite buffers, disturbances propagate with time delays, often causing downstream blocking far from their origin. Conventional diagnostic approaches, such as OEE-based classification and bottleneck detection heuristics, do not account for this lag and therefore misattribute symptoms to non-causal stations. This study introduces a proportional multi-source root-cause attribution method that distributes blocked time among all feasible upstream internal disturbances occurring within buffer-dependent propagation windows. Contributions are weighted by disturbance duration and historical downtime, reflecting cumulative production loss rather than event proximity. The method is evaluated using a FlexSim discrete-event simulation of cosmetic packaging. Compared to a simplified single-source backtracking approach, the proposed method significantly improves causal resolution. In the studied case, blocked time attributed to the filler via the cartoner decreased from 9.56 min to 8.27 min, with reductions exceeding 75% in selected scenarios. The method successfully captures the influence of remote and short-duration micro-stoppages ignored by simpler logic. These results demonstrate that accounting for distributed causality yields more realistic diagnostic output suitable for integration with MES/SCADA and digital-twin systems.

Keywords: Root-cause · machine failures · Gantt chart · OEE

1 Introduction

Modern automated production systems comprise tightly coupled sequences of processing stations linked by finite-capacity buffers. While local reliability and efficiency of machines affect performance, the global behaviour of such systems is governed by complex spatio-temporal interactions among machines, conveyors, and buffers. Small disruptions at one station may propagate along material-flow paths and cause downstream

J. Trojanowska et al. (Eds.): MANUFACTURING 2026, LNME, pp. 41–54, 2026.
https://doi.org/10.1007/978-3-032-21654-0_4

blocking or upstream starvation after non-negligible delays. These effects complicate the diagnosis of production stoppages and consequently hinder responsive maintenance, continuous improvement, and production planning.

The behaviour of unreliable serial lines with intermediate buffers has been extensively investigated since seminal analytical studies on throughput sensitivity and blocking/starvation interactions [1–5]. Finite buffers were shown to decouple variability but also to enable delayed causal propagation, introducing ambiguity into the interpretation of observed machine states. In practice, downstream symptoms often do not coincide temporally with the originating fault, which makes naive event-based attribution unreliable in large systems.

Research on bottleneck detection has evolved from static to dynamic approaches. Classical methods such as the Shifting Bottleneck (SB) / Active Period method [6], Inter-Departure-Time Variance (IDTV) [7], and Bottleneck Walk [8] have demonstrated value in identifying active throughput constraints. More recent comparative studies [9, 10] highlight their differing sensitivities to product mix, machine reliability, and layout complexity. However, these techniques focus on locating the current bottleneck rather than identifying the root cause of a stoppage. Since bottlenecks may shift dynamically and may not coincide with the causal origin of a disturbance, classical indicators are insufficient in scenarios involving multiple upstream disturbances or substantial buffer-related lag.

Industrial monitoring practices typically rely on machine-state classification (e.g., OEE categories), which distinguish downtime from productive time but rarely distinguish internal disturbances (machine-level faults) from external disturbances (propagated blocking/starvation). Consequently, events may be attributed to stations that are only symptomatic, rather than causal. This misalignment can misdirect corrective actions, reduce maintenance effectiveness, and prolong systemic inefficiencies.

Recent developments in data-driven and model-driven diagnostics have increased the use of discrete-event simulation (DES) and Digital Twins (DT) for production-system analysis [11–14]. DES offers structured means to reproduce propagation behaviour and evaluate diagnostic logic under controlled scenarios, while DT integrations enable real-time reasoning. Nevertheless, systematic allocation of causal responsibility remains an unresolved challenge, particularly in long serial lines with heterogeneous failure dynamics.

Existing bottleneck and diagnostic techniques typically assume single-source causality and lack mechanisms to integrate (i) buffer-dependent propagation behaviour with (ii) the historical sequence of upstream downtime. This limitation is critical in cases involving (a) multiple short-duration disturbances, (b) long spatial separation between cause and effect, or (c) large buffers inducing time-lagged propagation. Under such conditions, simplistic nearest-preceding-event assignment leads to systematic misclassification.

This work extends a previously published root-cause identification method by introducing a proportional attribution model that accounts for the historical downtime of upstream machines. Instead of assigning each blocked interval to a single candidate source, the proposed method distributes causal weight across all feasible upstream contributors within a time window determined by buffer propagation characteristics. We evaluate the method using a high-fidelity FlexSim DES of an eight-station cosmetic packaging line, for which failure and repair times follow statistically fitted distributions. A baseline simplified assignment method is used for comparison.

The main contributions are:

1. A proportional multi-source root-cause attribution framework incorporating downtime history and buffer-dependent propagation.
2. A detailed FlexSim-based DES simulation model reflecting industrial failure-time distributions.
3. A comparative evaluation demonstrating improved sensitivity to short-duration and remote disturbances, especially under large buffer capacities.

To the authors' knowledge, proportional event attribution supported by buffer-dependent propagation modelling has not been previously applied to RCA in discrete serial manufacturing systems. The results indicate that the proposed method provides improved diagnostic resolution, enabling more targeted maintenance strategies and offering a viable pathway toward integration within MES/SCADA and DT environments.

It is important to note that the objective of this work is not to directly improve system-level performance indicators such as throughput, cycle time, or throughput time through changes in machine logic or buffer configuration. Instead, the proposed method addresses the diagnostic problem of root-cause attribution by improving how downstream blocking is interpreted with respect to upstream disturbances. All operational parameters in the simulated system remain unchanged; therefore, performance indicators are invariant across attribution methods. The value of the proposed approach lies in enabling more accurate interpretation of downtime causes, which in practice can support better-informed maintenance and improvement decisions.

2 Methods

2.1 System Overview

The target system is a serial production line composed of N processing stations connected by finite-capacity conveyors operating under FIFO assumptions. Material flows unidirectionally from the first process to the final station. Each station can undergo intentional or unintentional downtime, including mechanical failure, operator intervention, or short disruptions arising from upstream or downstream dependency. Intermediate buffers absorb variability; however, they also induce propagation lag, meaning that the impact of an upstream failure may only manifest downstream after a measurable time.

Let the set of machines be $M = \{M_1, M_2, \dots, M_N\}$, and let B_i denote the buffer between stations M_i and M_{i+1}. Each machine cycles through discrete operational states over time. The temporal sequences of machine and buffer states provide the foundational input for downstream inference of root-cause relationships.

The method introduced in this work relies on the machine-level event timeline and on buffer-specific propagation margins, which define feasible causal windows for tracing back blocking and starvation events.

2.2 Machine and Buffer States

Each station is modeled with a finite alphabet of operational states:

- **RUN**—productive operation at nominal speed;
- **INT**—internal downtime caused by a station-specific disturbance (e.g., failure, setup, operator stop);
- **BLK**—blocked, unable to output due to downstream congestion;
- **STR**—starved, unable to process due to lack of upstream material;
- **IDLE**—available but not producing for non-constraining reasons (not critical in RCA).

Internal downtime (INT) is considered a potential causal source, whereas BLK and STR represent propagated states. The RCA objective is to assign (wholly or proportionally) periods of BLK to upstream INT events.

Buffers B_i are assumed to have finite capacity C_i. Their occupancy levels influence how long propagation takes to reach a downstream station; therefore, they are integral to defining feasible causal relationships. This contrasts with classic bottleneck identification, which typically targets the machine with the largest share of active blocking or starvation time without explicitly coupling such information to upstream history.

2.3 Baseline Procedure

When a station M_j enters BLK at time t_0, the algorithm searches upstream for the most recent INT event within a global backward horizon L.

1. Detect onset of $\left(M_j, \text{BLK}, t_0\right)$.
2. Traverse upstream machines $M_{j-1}, M_{j-2}, \ldots, M_1$.
3. For each M_k, find the most recent INT event at time $t_e \leq t_0$ occurring within $[t_0 - L, t_0]$.
4. Assign full M_j blocked duration to the *first* event encountered.

This approach is computationally efficient but forces single-source attribution. If multiple INT events lie within the feasible time region, only the first identified receives all attribution, producing systematic bias when buffers are large or INT events overlap. This method was directly adopted from the research by Dąbrowski et al. [15].

2.4 Proposed Method: Proportional Multi-Source Attribution

To address the limitations of exclusive assignment, we propose a proportional allocation method, whereby a blocked event at station M_j is assigned fractional contributions from all upstream internal events occurring within a physically plausible time window.

2.4.1 Temporal Causality Window

For each conveyor B_i, we define an empirical propagation margin Δ_i, representing the maximal expected delay between the occurrence of an upstream INT event at M_i and the appearance of a BLK state downstream. The propagation margin accounts for buffer capacity, conveyor speed, product accumulation dynamics, and station cycle times.

For a blocked interval beginning at time t_0, the causality window for machine M_k upstream of M_j is:

$$\left[t_0 - \Delta_{k\to j},\ t_0\right], \tag{1}$$

where:

$$\Delta_{k\to j} = \sum\nolimits_{u=k}^{j-1} \Delta_u. \tag{2}$$

This ensures that attribution is restricted to events that could physically propagate to M_j by time t_0. The propagation windows are not free tuning parameters but are physically constrained by buffer capacities, conveyor speeds, and station cycle times. They represent the maximum feasible delay with which an upstream disturbance can manifest as downstream blocking. This ensures that attribution is limited to events that are causally admissible given the material-flow dynamics of the line, rather than relying on purely temporal proximity.

2.4.2 Candidate Weak Source Set

All internal-down events occurring within the causality window of each upstream machine form a candidate set:

$$C_j = \bigcup\nolimits_{k=1}^{j-1} \left\{e \in \mathrm{INT}_k | t_e \in \left[t_0 - \Delta_{k\to j},\ t_0\right]\right\}. \tag{3}$$

If $C_j = \emptyset$, the blocked period is labeled as unattributable (rare, but possible, e.g. due to the missing data).

2.4.3 Weighting and Allocation

Each event $e \in C_j$ receives a weight w_e proportional to its duration:

$$w_e = d_e \tag{4}$$

and normalized:

$$\hat{w}_e = \frac{w_e}{\sum_{e' \in C_j} w_{e'}}. \tag{5}$$

Attribution for BLK interval of duration D_j is:

$$A_{e\to M_j} = \hat{w}_e \cdot D_j. \tag{6}$$

Aggregated over all blocked intervals, the method yields fractional assignment of responsibility to each machine.

Duration-based weighting is adopted because downstream blocking is driven by cumulative lost production time rather than by the mere count of events. A sequence of short micro-stoppages can collectively deplete buffers and cause blocking even if no individual event is dominant. While alternative weighting schemes (e.g., recency-based or hybrid formulations) are possible, the present study focuses on demonstrating the conceptual advantage of multi-source attribution over exclusive backtracking. The evaluation therefore emphasizes causal coverage rather than optimization of weighting parameters.

2.4.4 Simulation Model of the Manufacturing Line

The method was evaluated on a detailed discrete-event simulation (DES) model of an industrial cosmetic packaging line supplied by a multinational manufacturer. The line consists of eight sequential processing stations connected by conveyor-based transport with finite intermediate buffers. The stations represent a typical high-volume filling and secondary packaging flow, including: (1) bottle unscrambling, (2) filling, (3) capping, (4) inspection, (5) labeling, (6) cartoning, (7) case packing, and (8) palletizing. Each machine was modeled with stochastic processing times and failure characteristics based on empirical log-data collected on the shop floor. Historical failure and repair durations were fitted using ExpertFit [13], with distributions chosen based on goodness-of-fit criteria. Dominant failure-time families followed Weibull and Johnson-SU distributions [16], consistent with typical industrial repairable processes. Workstations operated cyclically on a single-unit flow basis, with deterministic or lightly variable cycle times depending on station type. The same model was a subject of the previously published studies [17–19].

Material between stations was transported using conveyor segments with capacity limits, forming finite buffer regions. These buffers enable partial decoupling of operations, allowing downstream machines to continue operation during short upstream disturbances. Buffer capacities ranged from small (tens of units) to large (hundreds of units), producing varying propagation-delay characteristics across the line.

Stations were represented in one of four discrete operating states: RUN: processing normally, INT: internal disturbance (e.g., fault, jam), BLK: blocked due to downstream unavailability, STR: starved due to upstream unavailability.

Transitions between states followed discrete occurrence rules: internal failures triggered INT states until repaired, and lack of input/output capacity triggered STR/BLK states respectively. Machine logic was modeled using independent reliability parameters; no state feedback or predictive controls were assumed.

Observations from early analysis (e.g., Gantt reconstruction) highlighted that even short INT episodes could cause BLK events at distant downstream stations after significant delay, often hundreds of seconds later. This time separation depended primarily on buffer inventory at the moment of disturbance. For example, the maximum propagation window between the filler and labeler exceeded 200 s, demonstrating substantial temporal decoupling and motivating the proportional attribution method used in this study.

The DES model was implemented in FlexSim, which allowed visualization of time-state evolution, material transport, and failure propagation. Event logs were exported with timestamped station-state records, enabling reconstruction of blocking intervals and upstream disturbance histories used in the attribution analysis. The overall system layout is shown in Fig. 1, which highlights the linear material flow, buffer positions and machines.

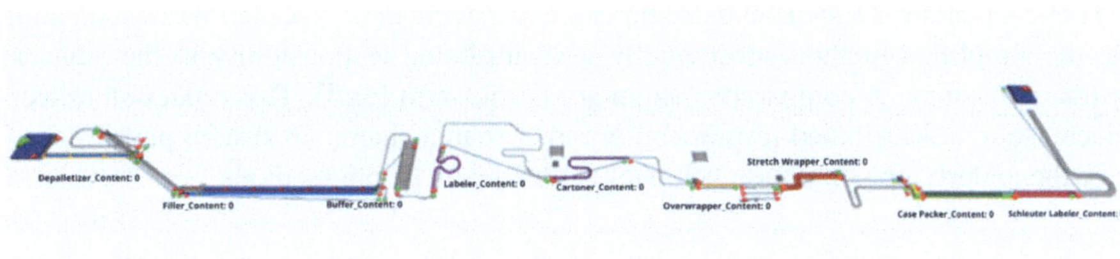

Fig. 1. Overview of the manufacturing line model as presented in FlexSim

3 Results

The performance of the proportional multi-source attribution method was evaluated using a discrete-event simulation model of an eight-station cosmetic packaging line. The simulation incorporated empirical failure/repair distributions and cycle-time variability measured in industrial conditions.

A ground-truth causal stream was recorded from the simulation to enable direct comparison between observed downstream blocked intervals and their true upstream disturbance origins. Both the simplified single-source backtracking approach and the proposed proportional allocation method were applied to the event log for evaluation.

Because the study is conducted in a discrete-event simulation environment, the causal structure of blocking propagation is explicitly defined by the model logic, buffer dynamics, and material-flow rules. Each blocked state arises deterministically from upstream internal disturbances filtered through buffer capacities and transport delays. This embedded causal structure serves as the ground truth against which attribution methods are evaluated. The comparison therefore does not assess predictive accuracy in a statistical sense, but rather the degree to which each method reconstructs the true upstream contributors to observed downstream blocking.

In examining system dynamics, state-timeline visualization revealed how blocking propagates gradually toward the line head, often long after upstream events have ended. Figure 2 illustrates a representative Gantt-style state profile across several stations. The time lag and multi-event layering clearly show that multiple independent internal stoppages can contribute to eventual downstream blocking.

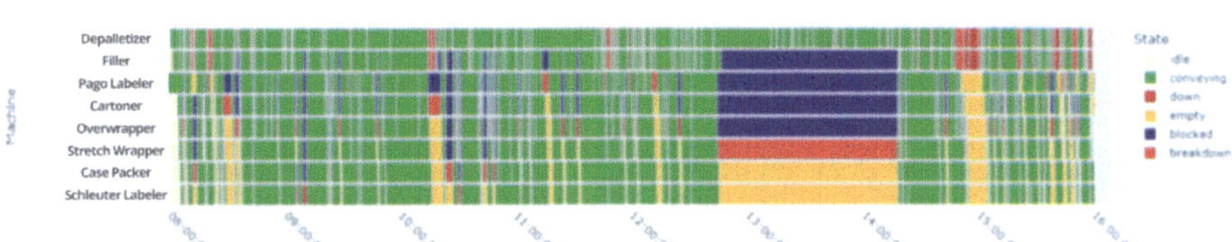

Fig. 2. Gantt chart of the machine states occurring over a period of 8 h

The proportional attribution method demonstrated improved agreement with ground truth compared with the simplified baseline. For blocked intervals at the filler originating at the cartoner, simplified attribution assigned an average of 9.56 min of blocked time, whereas proportional attribution assigned 8.27 min, yielding an average reduction of 17.9% (median -11.5%). In extreme cases, the reduction exceeded 87%, indicating that the simplified method substantially over-attributed responsibility to the adjacent upstream machine. A comparative summary is shown in Fig. 3. This reduction reflects a decrease in misattributed responsibility rather than a change in system performance, since the underlying simulation behavior is identical for both methods.

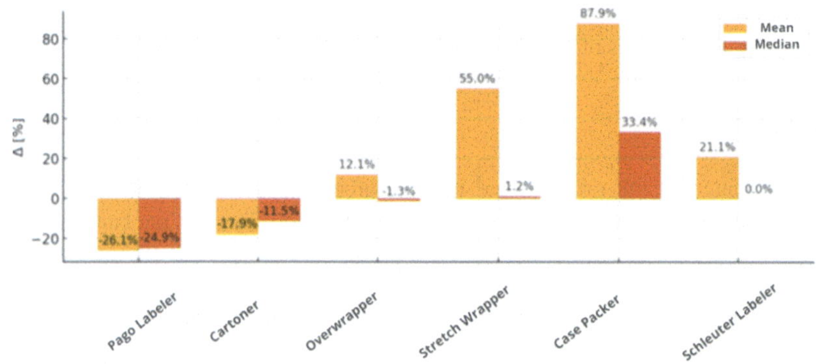

Fig. 3. Average and median nominal differences between simplified and multi-source attribution methods which assigned machine being a root-cause of the entire line downtime

The introduction of duration-weighting enabled effective detection of clusters of short micro-stoppages. Despite individual durations of less than ~ 15 s, groups of such disturbances collectively influenced downstream material flow. The simplified method ignored these short-duration events, whereas the proportional model successfully detected their influence and awarded corresponding fractional responsibility. A representative decomposition of a blocked interval into proportional contributions is shown in Fig. 4.

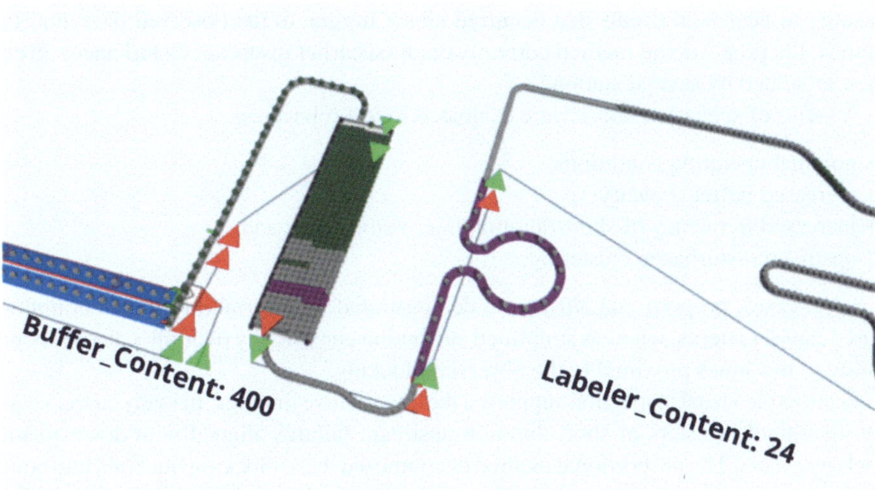

Fig. 4. Visualization of fractional contributions as seen color-coded items on buffer; the presence of each same-color item in the buffer is due to the same historic INT state of the downstream machines; separated colors are used to differentiate between different INT events

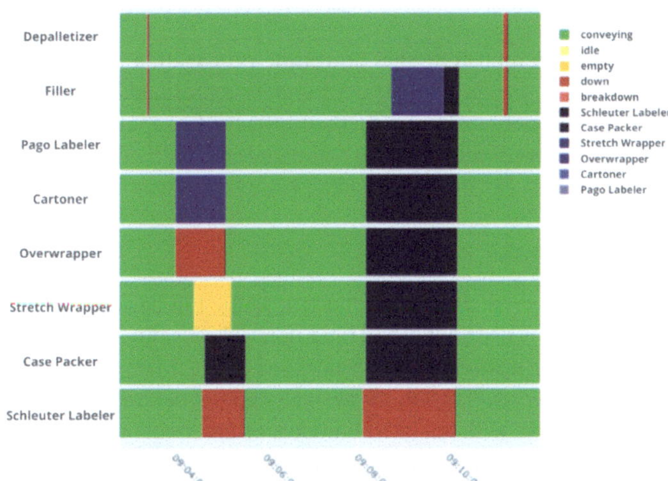

Fig. 5. Multi-source attribution as visualized on the Gantt chart. Failure of Filler is caused both by Pago Labeller and Schleuter Labeller

Propagation-window estimation confirmed that buffer structure strongly influences temporal causality. Measurements indicated that maximum propagation margins exceed 200 s for some station pairs, most notably between the filler and labeler, making it possible for far-upstream stoppages to cause downstream blocking after substantial time delay (see Fig. 5). Under these conditions, the simplified method frequently misattributed causality to near-field events that occurred closer in time to the observed blocking. In contrast, the proportional method correctly credited earlier upstream disturbances, even when separated by several stations.

A series of scenario classes were evaluated to test robustness:

(1) nominal operating conditions;
(2) increased buffer capacity;
(3) increased frequency of short-duration micro-stoppages; and.
(4) upstream disturbance clustering.

In all cases, proportional attribution demonstrated superior ability to detect multi-source causal patterns, whereas simplified attribution consistently over-allocated responsibility to machines proximal to the observed blocking.

Gantt-style visual inspection supported the quantitative findings. In many cases, visually identifiable clusters of short-duration upstream failures aligned with downstream blocking events. The proportional method decomposed these blocking intervals into contributions consistent with the temporal and spatial ordering seen in the event log. The simplified approach instead attributed such intervals to a single event, typically chosen based on a first-match rule, ignoring other plausible contributors.

Although the proportional method incurs slightly greater computational cost than the simplified baseline, runtime remained well within practical limits. Full evaluation of the 8-station, 14-day simulation trace required under two seconds on a standard workstation, suggesting that the method is suitable for integration within MES/SCADA or digital-twin infrastructure.

Overall, the results demonstrate that downtime propagation in buffered serial systems is rarely singular in origin and more accurately represented as a distributed phenomenon. The proportional attribution method effectively captures this structure and offers significantly improved causal fidelity compared to simplified diagnostic heuristics.

4 Discussion

The results of this study show that downtime propagation in buffered serial production systems is characterized by distributed and time-delayed causality. Because finite buffers introduce latency between upstream disturbances and downstream effects, a failure at one station can cause blocking several stations away, long after the original event has ended. This behavior is consistent with analytical models of unreliable transfer lines [1–3, 5], which established that finite buffers allow temporary decoupling while still transmitting disruption once accumulation limits are reached. However, while these models explained system-level dynamics, they did not provide practical guidance for root-cause attribution (RCA) based on industrial event logs.

The proportional attribution method presented here improves RCA performance by distributing blocked-time among all upstream failures falling within a buffer-dependent propagation window. This approach significantly outperforms simplified single-source backtracking, particularly when short micro-stops occur in clusters or when buffers are large. For example, the blocked time assigned to the filler due to cartoner-related disturbances decreased from 9.56 min to 8.27 min, a mean reduction of ~ 17.9%

(median 11.5%) and up to − 78% in extreme cases, showing that simplified attribution overestimates local effects and neglects cumulative small disturbances.

These findings support earlier theoretical work stating that downstream performance is influenced by the accumulated variability of upstream stations [1, 2]. Yang and Gershwin [3] showed that blocking and starvation patterns move along the line in delayed waves, while Yamashita and Altiok [5] demonstrated that buffer size strongly affects throughput sensitivity. In the studied system, propagation delays exceeded 200 s, confirming that event proximity is not a reliable RCA criterion.

Traditional bottleneck detection methods—such as Active Period [6], interdeparture time variance [7], and simulation-based identification [20]provide valuable insight into throughput constraints, and later comparative assessments [9, 10, 21] highlighted their adaptability in dynamic conditions. However, bottleneck location is not equivalent to causal origin. A machine may appear to be the bottleneck despite having minimal internal failure time; blocking may be driven by upstream effects rather than local performance. The present results confirm that bottleneck detection alone does not correctly assign causal responsibility.

A key strength of proportional attribution is its ability to detect collective influence of micro-stoppages, which often go unnoticed individually yet significantly affect flow. Because micro-stops accumulate and affect material availability downstream, duration-weighted attribution is better suited than deterministic backtracking. This agrees with observations that transient disturbances play an outsized role in throughput loss [1, 5].

Gantt-based qualitative validation showed that the proportional method provides highly interpretable causal decomposition. This aligns with current digital-twin research emphasizing the integration of temporal machine-state information with system-level modeling [15, 22–26]. The method therefore fits into emerging data-centric RCA frameworks combining event logs, machine telemetry, and buffer states [15].

Practically, proportional attribution improves maintenance prioritization. OEE-based summaries often overemphasize direct downtime and underrepresent system-level impact. Fractional causal allocation reveals machines whose recurring small disturbances materially degrade system performance, enabling better asset prioritization and justification for maintenance or investment.

Although classical system-level indicators such as throughput, cycle time, and throughput time are central to production system analysis, they are unaffected by the proposed attribution logic because no operational parameters are altered. For this reason, such KPIs are not used as comparative outcome measures in this study. Instead, the contribution lies in improving how downtime is interpreted: by reallocating blocked time from symptomatic machines to true upstream sources, the method provides a more reliable diagnostic basis for maintenance prioritization and improvement planning. In practical settings, such improved attribution is a prerequisite for effective actions that may subsequently enhance throughput or reduce flow time.

Nonetheless, several limitations exist. Propagation windows were static; dynamic windows based on real-time WIP could improve filtering. This method also assumes FIFO flow; re-entrant or branching networks require generalization. Finally, simulation-based validation must be supplemented with industrial field trials to address data quality issues, a challenge previously highlighted in DES input-data studies [18, 27].

The weighting scheme and propagation windows reflect physically motivated design choices rather than exhaustive optimization. Alternative formulations may be explored in future work; however, the present results already demonstrate the fundamental advantage of proportional multi-source attribution over single-source backtracking in buffered serial systems.

Overall, this work reinforces three insights: (1) RCA in buffered lines must assume distributed causality; (2) propagation windows effectively constrain feasible causes; and (3) proportional weighting provides more realistic attribution than single-source heuristics. These characteristics make the method a strong candidate for integration into digital-twin and MES/SCADA ecosystems and a promising foundation for future developments in production-system RCA.

5 Conclusion

This work introduced a proportional multi-source attribution method for diagnosing the origins of downtime in multi-stage serial production lines with finite buffers. Unlike simplified backtracking approaches that assume a single causal event, the proposed method distributes responsibility across all feasible upstream disturbances based on duration, recency, and historical behavior. Evaluation using a high-fidelity DES model of an industrial cosmetic packaging line demonstrated that the method effectively captures the cumulative influence of micro-stoppages and long-range propagation effects. Quantitatively, attribution accuracy improved substantially, reducing over-assignment to nearby stations by an average of ~18% and by up to ~80% in selected scenarios.

These findings confirm that downtime in buffered systems exhibits distributed causality and cannot be reliably interpreted using proximity-based heuristics. The proposed method offers clearer insight into systemic behavior, supports more informed maintenance prioritization, and is readily compatible with MES/SCADA and digital-twin infrastructures. Future work will explore dynamic buffer-state–aware attribution and probabilistic causal inference to further enhance diagnostic resolution.

Acknowledgements. This research was financially supported by the National Center for Research and Development, project titled: "Self Learning Factory–Recommendation Engine built based on artificial intelligence with video based production events recognition and utilizing global production machines libraries" (POIR.01.01.01-00-0542/20). Tomasz Bartkowiak's work was also partially funded by the Polish Ministry of Science and Higher Education as a part of annual subsidy–project 0614/SBAD/1603 and 0614/SBAD/1586.

References

1. Dallery, Y., Gershwin, S.B.: Manufacturing flow line systems: a review of models and analytical results. Queueing Syst. **12**, 3–94 (1992). https://doi.org/10.1007/BF01158636
2. Gershwin, S.B.: Modeling and analysis of three-stage transfer lines with unreliable machines and finite buffers. Oper. Res. **31**, 354–380 (1983). https://doi.org/10.1287/opre.31.2.354
3. Yang, S., Gershwin, S.B.: Modeling and analysis of multi-stage transfer lines with unreliable machines and finite buffers. IIE Trans. **32**, 1027–1038 (2000). https://doi.org/10.1023/A:1018944411591

4. Bartkowiak, T., Gessner, A.: Modeling performance of a production line and optimizing its efficiency by means of genetic algorithm. In: ASME 2014 12th Biennial Conference on Engineering Systems Design and Analysis, ESDA 2014 (2014). https://doi.org/10.1115/ESDA2014-20141
5. Yamashita, H., Altiok, T.: Buffer capacity allocation for a desired throughput. Ann. Oper. Res. **93**, 141–156 (1998). https://doi.org/10.1023/A:1007564620313
6. Roser, C., Nakano, M., Tanaka, M.: Shifting bottleneck detection. In: Proceedings of the Winter Simulation Conference. pp. 1079–1086. IEEE. https://doi.org/10.1109/WSC.2002.1166360
7. Betterton, C.E., Silver, E.A.: Detecting bottlenecks in serial production lines: a focus on interdeparture time variance. Int. J. Prod. Res. **50**, 4158–4174 (2012). https://doi.org/10.1080/00207543.2011.596847
8. Roser, C., Lorentzen, K., Deuse, J.: Reliable shop floor bottleneck detection for flow lines through process and inventory observations: the bottleneck walk. Logist. Res. **8**, 7 (2015). https://doi.org/10.1007/s12159-015-0127-2
9. West, N.: Comparative study of methods for real-time bottleneck detection. In: Intelligent Human Systems Integration (2022). https://doi.org/10.1007/978-3-031-08530-7_1
10. Skoogh, A.: Others: throughput bottleneck detection in manufacturing: systematic review. Prod. Manuf. Res. **11**, 1–27 (2023). https://doi.org/10.1080/21693277.2023.2283031
11. Koyuncuoğlu, M.U.: Buffer allocation design for unreliable production lines. Sustain. Manuf. Ser. Econ. 1–12 (2021). https://doi.org/10.1016/j.susmat.2020.100020
12. Kassoul, K.: Buffer allocation in unreliable lines via GA + FPA. Int. J. Prod. Res. 1–19 (2022). https://doi.org/10.1080/00207543.2021.1909169
13. Law, A.M., McComas, M.G.: How the ExpertFit distribution-fitting software can make your simulation models more valid. In: Winter Simulation Conference (2000). https://doi.org/10.1109/WSC.2000.899726
14. Bartkowiak, T., Ciszak, O., Jablonski, P., Myszkowski, A., Wisniewski, M.: A simulative study approach for improving the efficiency of production process of floorboard middle layer (2018). https://doi.org/10.1007/978-3-319-68619-6_2
15. Dąbrowski, J., Bartkowiak, T., Wierzchowski, P., Dąbrowski, D.: Manufacturing line-level root cause analysis and bottleneck detection using the digital shadow concept and cloud computing. Presented at the (2024). https://doi.org/10.1007/978-3-031-56444-4_8
16. Weibull, W.: A statistical distribution function of wide applicability. J. Appl. Mech. **18**, 293–297 (1951). https://doi.org/10.1115/1.4010337
17. Jasiulewicz-Kaczmarek, M., Bartkowiak, T.: Improving the performance of a filling line based on simulation. IOP Conf. Ser. Mater. Sci. Eng. **145**, 042024 (2016). https://doi.org/10.1088/1757-899X/145/4/042024
18. Pawlewski, P., Jasiulewicz-Kaczmarek, M., Bartkowiak, T., Hoffa-Dabrowska, P.: Validation of simulation model of the filling line failures. J. Konbin. **38** (2016). https://doi.org/10.1515/jok-2016-0023
19. Bartkowiak, T., Pawlewski, P.: Reducing negative impact of machine failures on performance of filling and packaging production line-a simulative study. In: 2016 Winter Simulation Conference (WSC). pp. 2912–2923. IEEE (2016). https://doi.org/10.1109/WSC.2016.7822326
20. Sengupta, S., Das, D., Dhekane, M.: A new method for bottleneck detection. In: Winter Simulation Conference (2008). https://doi.org/10.5555/1516744.1517049
21. Ericson, A.: Bottleneck detection in manufacturing a throughput case study at SCANIA CV (2017)
22. Lu, Y., Liu, C., Wang, K., Huang, H., Xu, X.: Digital Twin-driven smart manufacturing. Robot. Comput. Integr. Manuf. **61**, 101837 (2020). https://doi.org/10.1016/j.rcim.2019.101837

23. Cimino, C., Negri, E., Fumagalli, L.: Review of digital twin applications in manufacturing. Comput. Ind. **113**, 103130 (2019). https://doi.org/10.1016/j.compind.2019.103130
24. Aivaliotis, P.: Digital Twin for predictive maintenance. Int. J. Comput. Integr. Manuf. **32**, 1–20 (2019). https://doi.org/10.1080/0951192X.2019.1686173
25. Wang, K., Liu, C., Lu, Y.: Ensemble Bayesian Network for root-cause analysis of product defects. J. Manuf. Syst. **75**, 102–115 (2024). https://doi.org/10.1016/j.jmsy.2024.06.001
26. Kumbhar, M.: DT-based framework for detection, diagnosis and improvement of bottlenecks. J. Manuf. Syst. **68**, 230–246 (2023). https://doi.org/10.1016/j.jmsy.2022.10.011
27. Skoogh, A., Johansson, B., Stahre, J.: Automated input data management for DES. SIMULATION **88**, 556–570 (2012). https://doi.org/10.1177/0037549712443404

Conceptual Use of a Digital Twin in a Smart Factory Model

Karolina Bendowska(✉) and Krzysztof Żywicki

Faculty of Mechanical Engineering, Poznan Univeristy of Technology, Poznan, Poland
karolina.bendowska@put.poznan.pl

Abstract. Digital twins (DT) are one of the key pillars of industrial transformation toward the Industry 4.0 concept. By integrating process modeling, data analytics, and real-time communication, DTs enable accurate virtual representations of physical systems. Recently, their application has expanded beyond industrial practice to research and education, particularly within smart factory and learning factory demonstrators. These laboratories bridge theory and practice, supporting the development of engineering competencies and experimentation with cyber-physical systems, simulation, and control. In academic environments, digital twins play a dual role as both educational tools and research platforms. They enable controlled, repeatable experiments and systematic evaluation of process parameters, supporting research in automation, robotics, and data-driven decision-making. When connected to real-time data from sensors and PLC controllers, DT-based models achieve a high level of realism, facilitating studies on system interoperability, lifecycle management, and real-time operational optimization. This article provides a concise overview of the role of digital twins in modern demonstrator laboratories, highlighting their importance for research, innovation, and engineering education. In addition, it presents a conceptual application of a digital twin within a smart factory model, serving as an integrated experimental and educational platform for testing advanced manufacturing and control solutions.

Keywords: Digital twin · Production process · Smart Factory · Industry 4.0

1 Introduction

In the last decade, the concept of the digital twin (DT) has become one of the main paradigms in the development of modern manufacturing systems. The term describes a dynamic, data-driven digital representation of a physical object or process that enables its observation, analysis, and control in real time [1]. In the industrial environment, DT serves both as a diagnostic and prognostic tool, supporting process optimization, root-cause analysis of failures, and planning of predictive maintenance [2].

The development of digital twins is made possible by advances in information technologies such as the Industrial Internet of Things (IIoT), edge computing, and the integration of MES and SCADA systems. The combination of sensor data with simulation models enables continuous synchronization of the physical and virtual states, making DT a key tool for real-time decision-making [3].

J. Trojanowska et al. (Eds.): MANUFACTURING 2026, LNME, pp. 55–65, 2026.
https://doi.org/10.1007/978-3-032-21654-0_5

In industrial practice, digital twins have found applications in many areas, including product design (virtual prototyping), production process planning, supply chain management, equipment diagnostics and predictive maintenance, as well as energy and quality optimization [4, 5]. Thanks to the ability to perform what-if scenarios on the digital model, companies gain a safe tool for assessing the effects of process changes, reducing downtime, and shortening time-to-market. Implementation examples show that well-designed twins can significantly increase operational efficiency and reduce asset maintenance costs [4].

In educational and research contexts, the role of smart factory demonstrators is growing, as they allow digital twin solutions to be tested in environments with controlled complexity. Such laboratories, often referred to as learning factories, enable the integration of physical educational production lines with 3D models and PLC control systems. This allows users to design, implement, and test DT models under real conditions while minimizing cost and operational risk [6].

From a didactic perspective, DT-based laboratories enable learning through practice, which enhances the understanding of industrial processes and integrates knowledge from production engineering, mechatronics, automation, computer science, and data analytics [7]. Studies show that incorporating DT into teaching processes increases learning effectiveness and develops key digital competencies among students, such as data interpretation, CPS system design, and virtual process optimization [8].

rom a scientific perspective, DT demonstrators enable experimental verification of research concepts, including adaptive control algorithms, data-driven prediction methods, and strategies for synchronizing simulation models with real processes [9]. For example, Nezzi et al. [10] developed an educational digital twin of a storage stacker crane, enabling remote control and real-time visualization of its operation by integrating Unity technology with a real PLC controller.

At the same time, the literature points to significant challenges associated with DT implementation in demonstrator environments. These include the lack of unified data exchange standards, difficulties in validating simulation models, limited scalability of solutions, and the need for continuous model updates based on real data [1, 2]. Despite these limitations, smart factory demonstrators remain a key element in digital twin research, enabling technological development in a controlled, transparent, and repeatable manner.

In summary, digital twins in educational and research environments currently constitute not only a didactic tool but also an experimental platform for testing concepts of future production systems. As methods of modeling, data integration, and DT lifecycle management continue to evolve, smart factory laboratories will play an increasingly important role in shaping the competencies of Industry 5.0 engineers and in supporting knowledge transfer between academia and industry.

2 Application of Digital Twin

Digital twins have become one of the central concepts of industrial transformation in the era of Industry 4.0. By integrating data from sensors, automation systems, and business sources, a digital twin enables both retrospective event analysis and predictive forecasting, as well as experimentation in a safe virtual environment [11].

The technologies behind digital twins combine the Internet of Things (IoT), Big Data, machine learning methods, multiphysics simulations, and advanced integration and visualization systems. Sensors and edge devices collect operational data, which is transmitted to the storage and processing layer. There, analytical algorithms and simulation models transform raw measurements into useful knowledge. A key feature is the feedback loop. It is not merely a static model, but a system that adapts as new data arrives, enabling continuous improvement of predictions and automatic adjustment of control parameters [12].

The literature distinguishes several implementations of digital twin demonstrators in the context of smart factories, which show both their architecture and practical applications in real production systems. Analyzing these cases provides valuable guidance for designers and researchers planning the deployment of DT solutions in industrial facilities.

In the article by Ferrario et al., a small multifunctional smart factory is presented as an educational and research platform. It combines manufacturing technologies (3D printing, milling, laser processing) with an advanced monitoring system: vision systems, precision sensors, and real-time simulations. The factory's digital twin allows virtual tracking of machine and process status, while students and researchers can experiment with automation, optimization, and diagnostic concepts. The local educational infrastructure supports training "from the ground up," from classical automation (PLC, MES, SCADA) to advanced Industry 4.0 topics: simulations, IoT, and DT. As a result, the facility functions both as a "training factory" and as a research laboratory, enabling the testing of new technologies in realistic but controlled conditions [13].

Garcia et al. presented the "Digital Twin Learning Ecosystem" (DTLE) framework, which integrates operator (human) knowledge with machine data in traditional manufacturing plants. The proposed solution targets small and medium-sized enterprises, where infrastructure may be limited. The authors propose a lightweight, non-invasive method for integrating DT with existing (legacy) assets. The framework includes IIoT data collection, simulation models, and learning-layer mechanisms (e.g., decision and interaction layers). Tests were conducted using real data from a milling machine and an induction furnace. The results demonstrated, among other benefits, a 9% improvement in energy efficiency for the furnace and enhanced maintenance processes. The authors emphasize the importance of such an architecture for improving workforce competencies and increasing the competitiveness of traditional factories [14].

In their article, Baalsrud Hauge et al. present a digital twin testbed (demonstration laboratory) focused on production logistics. Its main component is a real-time locating system (RTLS), which tracks the positions of resources such as pallets, carts, or components within the facility. The location data is continuously transmitted to the digital layer, where simulations and analyses support optimization of material flow, reduction of downtime, and improved warehouse management. The authors present logistic usage scenarios and demonstrate specific benefits, including increased flexibility and better performance of internal transport control [15].

A novel digital twin architecture focused on integrating legacy machines with modern equipment within a smart factory using IIoT and a large language model (LLM) is presented by Gautam et al. The approach aims to enable companies with traditional, older machines that were not originally designed with digitalization in mind to create their digital representations without costly infrastructure redevelopment. Data from IIoT sensors is collected and transmitted to the digital twin platform, where it is used both for monitoring the condition of the physical machines and for generating predictions of their behavior. Additionally, the authors integrate an LLM that supports semantic analysis and the generation of operational recommendations (e.g., maintenance-related) based on the collected data. The system enables a closed information loop: the digital model not only reflects the physical state but can also drive decisions related to process optimization or maintenance. The demonstrative implementation shows that such a solution can significantly increase the flexibility and predictability of operations in factories containing a mix of modern and legacy machines, while also reducing modernization and maintenance costs [16].

Despite the clear benefits, the implementation of digital twins faces significant challenges. The most important include interoperability between heterogeneous industrial systems and protocols, ensuring the security and privacy of transmitted and stored data, scalability of solutions in highly complex environments, and the validation and verification of simulation models [12, 17]. In addition, organizational issues arise. Integrating digital twins with existing decision-making processes requires changes in technical culture and business workflows, as well as investments in competencies related to data and systems engineering.

3 Assumptions for the Application of a Digital Twin

3.1 Production System Model

At the Smart Factory Laboratory of Poznań University of Technology, an integrated production system operates that is designed to support research and teaching activities related to modern industrial solutions. Its infrastructure makes it possible to replicate key operations occurring in real manufacturing plants, enabling the testing of smart factory concepts under controlled conditions.

The demonstrator (Fig. 1), in addition to an automated assembly line equipped with a controller, also includes a storage area for parts and finished products, as well as racks serving individual workstations.

In each production scenario, the manufacturing of specific product types is planned. These products are assembled on an automated conveyor line and transported on dedicated pallets. The assembly line includes six operator stations, where successive product components are added. The pallets move automatically along the line, while their release can be performed either automatically based on the control program settings or manually by an operator.

The products assembled on the line are built from blocks of different sizes and colors. These components are stored in bins located on workstation racks. Assembly takes place on special trays that are placed on the pallets moving along the line. Each pallet has a unique identification number, which is read by sensors at every workstation, enabling the execution of the program stored in the controller's memory.

Communication between devices is carried out using industrial AS-Interface and Profinet networks. The operation of the entire system is supported by IT modules of

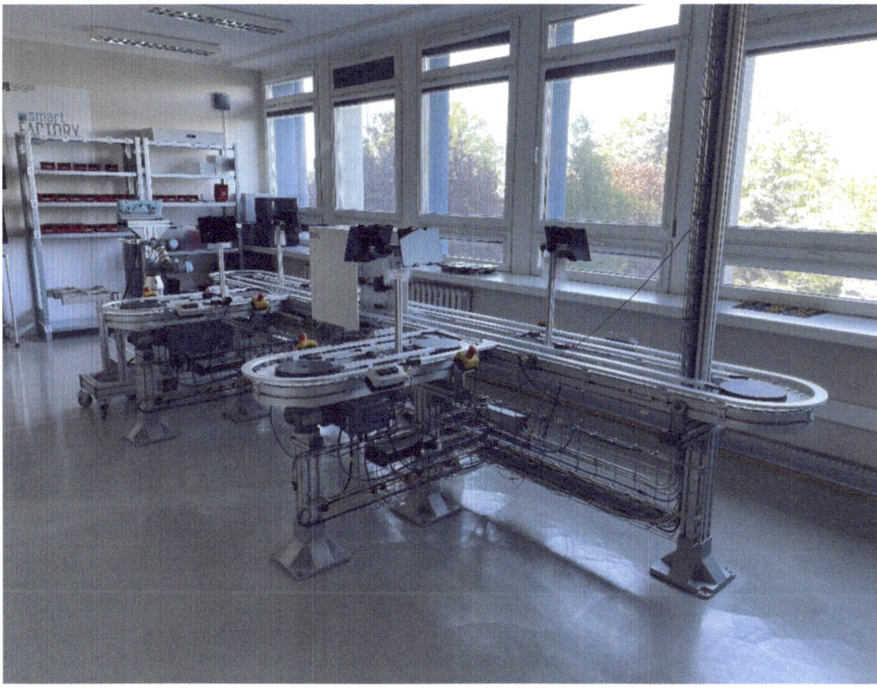

Fig. 1. Automated assembly line in the Smart Factory laboratory.

the 4Factory platform. This software is responsible for planning and coordinating the production process, monitoring material flow, and managing the control of the assembly line, enabling comprehensive supervision of the demonstrator's operation.

3.2 Digital Twin of the Production Line-Model

A digital twin enables detailed representation of both the physical layout and the logical process dependencies. Its structure includes a model of the production line, transport devices, workstations, and other elements of the manufacturing infrastructure. In addition to representing physical components, the digital twin also reproduces control algorithms, material flow rules, and information related to production scheduling. Such a comprehensive structure allows experiments to be carried out without the risk of disrupting the operation of the real production line.

One of the most commonly used environments for building digital twins of manufacturing systems is Tecnomatix Plant Simulation, an advanced simulation tool developed by Siemens [18]. The software allows for modeling material flows, representing assembly processes, planning machine layouts, as well as analyzing system flexibility and performance. Plant Simulation uses an object-oriented approach, where each object such as a machine, conveyor, or storage unit has its own parameters, properties, and logic that can be modified and extended by the user.

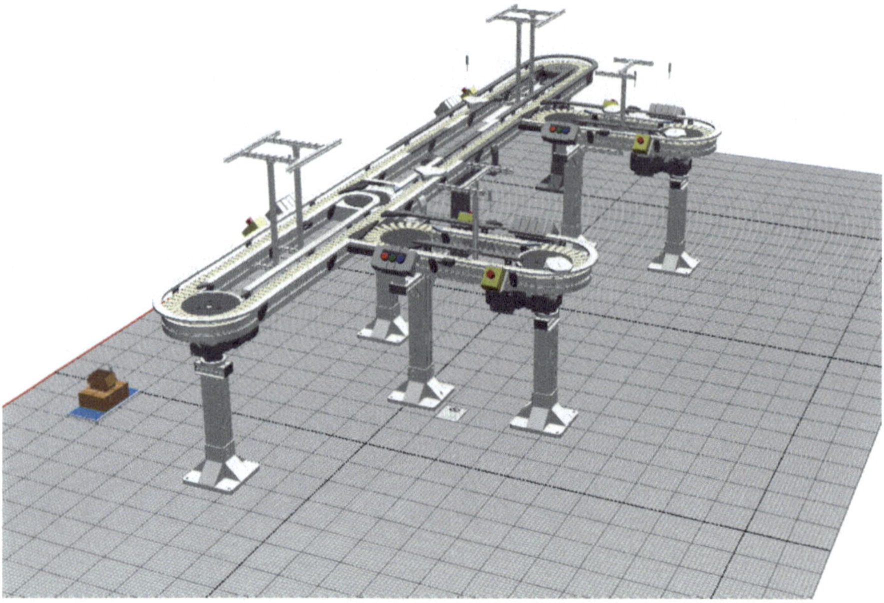

Fig. 2. Preliminary design of the assembly line in Tecnomatix Plant Simulation.

Building a digital twin in Plant Simulation begins with mapping the layout of the production line. At this stage, all infrastructure elements, their spatial arrangement, and the appropriate transport connections are defined. The next step is assigning operational parameters to the objects, such as cycle times, through-put rates, storage capacities, or flow characteristics. The model also incorporates process logic, including queuing rules, operation sequences, and resource handling strategies. The result is a simulation model of the production line (Fig. 2), which must undergo verification to ensure that the real system has been accurately represented.

An essential element of a digital twin is the ability to integrate with real production data. This is a critical aspect of the integration process, as it requires an appropriate communication infrastructure. In practice, industrial protocols such as OPC UA are used, providing standardized and secure data exchange between various IT systems and devices. Plant Simulation can utilize such interfaces for reading and writing data, which facilitates integration with existing IT infrastructure and automation systems. The software enables the import of technological data, schedules, failure information, and process statistics, allowing the current state of the production line to be accurately reflected. As a result, the model can be used not only for design analyses but also as a tool supporting operational management, for example, for real-time optimization of workstation loads or resource demand analysis.

Simulation results in Plant Simulation can be presented in the form of reports, charts, histograms, and performance indicators such as OEE, throughput, or workstation utilization. The tool also offers functions for automatic optimization of parameters using optimization algorithms, including a genetic optimizer. This enables obtaining process configurations that maximize performance or minimize production lead time.

A digital twin created in Tecnomatix Plant Simulation thus becomes not only an analytical environment but also a platform supporting the development of intelligent production management systems. It can be used to test innovative scheduling strategies, integrate artificial intelligence methods, assess line flexibility, or implement mass production concepts under conditions of product personalization.

3.3 Digital Twin of Warehouses–Model

The warehouse digital twin is a virtual representation of the logistics processes taking place in the material supply and distribution area that supports the operation of the production line. The simulation model serves as a tool for analyzing material flows, organizing deliveries, controlling inventory levels, and optimizing replenishment strategies. As a result, the warehouse digital twin becomes an integral component of the demonstrator's digital ecosystem, supporting both production planning and day-to-day operational management.

The warehouses cooperating with the production line handle several key processes, such as receiving deliveries, storing components, picking materials, and distributing them to assembly stations. The digital twin enables the representation of warehouse structures, including both their spatial layout and operational logic. The model may include racks of various types, transport paths, conveying equipment, and human resources. These objects are equipped with operational parameters such as location capacity, handling times, transport speeds, or resource availability.

Within the simulation model of the digital twin, different warehouse management strategies can be implemented, including replenishment policies (e.g., min–max, Just-in-Time, Kanban), picking rules, and location assignment strategies for stored items. This enables the comparison of different warehouse organization scenarios and assessment of their impact on material flow efficiency.

A key aspect is the integration of the warehouse digital twin with the production line model. Linking both environments allows for comprehensive representation of the dependencies between material availability and the course of the production process. In the simulation model, material flows between warehouses and the production line can be executed by logistics operators or an AGV. Production demand data, such as quantity and composition of components, order schedules, or material picking times, are transferred to the warehouse model, which generates the corresponding logistics tasks.

This integration can operate in real time or based on historical data, enabling both predictive analyses and reconstruction of past system behavior. In practice, communication between the models, as well as between the digital twin and real warehouse systems, is carried out using communication protocols such as OPC UA. This ensures synchronization of operating parameters, inventory levels, and production demand across the entire environment. The architecture diagram illustrating the data flow and integration between the physical system, the digital twin of the production line, and the warehouse digital twins within the conceptual application is presented in Fig. 3.

By integrating warehouses with the production line, the digital twin encompasses the entire material flow within the demonstrator, allowing for analyses that cover the full internal logistics chain. The model enables identification of bottlenecks in the supply process, evaluation of transport system efficiency, analysis of the impact of logistics delays

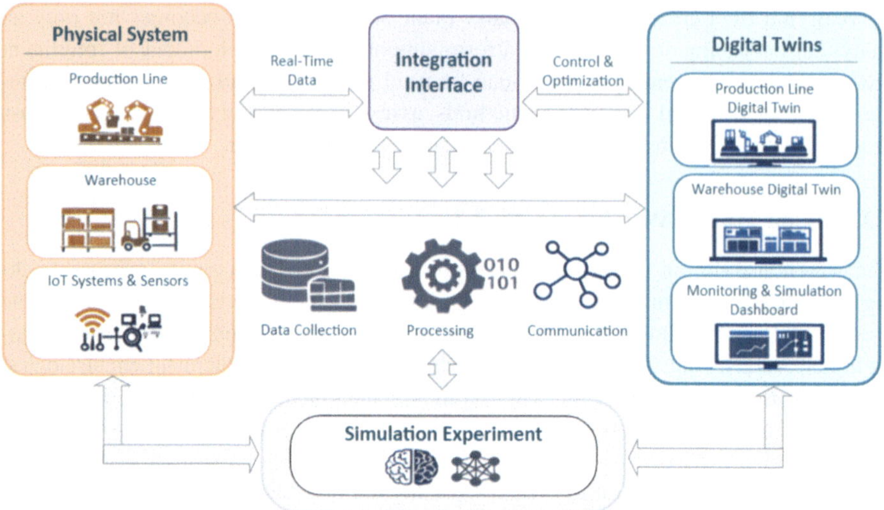

Fig. 3. Architecture diagram.

on order completion times, and testing of various warehouse reorganization scenarios. Additionally, the digital twin makes it possible to assess the influence of production changes, such as shortened production series or modified schedules.

The warehouse digital twin cooperating with the production line becomes a valuable tool supporting strategic and operational logistics decision-making. Integrating both systems within a simulation model allows for achieving a higher level of predictability, stability, and flexibility across the entire production-logistics flow.

3.4 Application of the Digital Twin in Production Flow Control

A digital twin is an advanced virtual model that replicates the real processes taking place in a manufacturing system, enabling analysis, monitoring, and optimization of plant operations. It is one of the key tools of the Industry 4.0 concept, combining simulation technologies, data analytics, and real-time integration with the physical system. The functioning of a digital twin is based on bidirectional information exchange between the model and the production line, which makes it possible to obtain an up-to-date view of system performance and to predict its future behavior.

From a functional perspective, the digital twin acts as a central analytical environment integrating data from control systems, planning systems, and the logistics infrastructure. It enables the representation of the physical flow of materials and information, tracking of current line utilization and inventory levels, simulation of system reactions to changes in operating parameters, and identification of bottlenecks and inefficient interactions between resources. At the operational level, the digital twin functions as a "virtual control center," allowing real-time monitoring, analysis, and optimization of workflow.

One of the key applications of the digital twin is supporting production scheduling. In traditional systems, schedules are often static and do not account for rapidly changing operating conditions. The digital twin enables the generation of dynamic schedules updated based on the current state of the production line, analysis of alternative production scenarios, verification of schedule feasibility before implementation, and synchronization of warehouse and supply activities with production needs. In the simulation environment, it is possible to test different job sequencing strategies, resource allocation methods, and operation prioritization rules. This allows for adapting the schedule to current conditions, ensuring better machine utilization and reduced order lead times.

The digital twin also enables the creation of operational forecasts based on historical, real-time, and predictive data. This functionality includes forecasting machine and workstation load, predicting queues, waiting times, and throughput, estimating order completion times, and assessing the risk of disruptions in production flow. The predictive model makes it possible to detect emerging problems early, such as growing delays, persistent resource shortages, or excessive variability of process parameters. This enables corrective actions to be taken before failures or drops in productivity occur.

Machine availability is a critical element of stable production flow, which is why a digital twin must accurately reflect it. The model can incorporate both planned downtime and unexpected failures, allowing for a better understanding of their impact on order completion times and queue formation. Simulation shows which workstations are most vulnerable to disruptions and how schedule changes, buffers, or load balancing can mitigate the effects of stoppages. Additionally, integrating the digital twin with diagnostic data from the real production line enables prediction of potential failures and planning maintenance in a way that minimizes the risk of unplanned downtime.

Using a digital twin to analyze the impact of material constraints on production flow is a key element in making correct operational decisions. The simulation model allows the representation of both material availability and potential delays in deliveries, making it possible to predict moments when production may be stopped or slowed. The digital twin also enables analysis of how material shortages affect workstation load and schedule execution, identifying areas particularly vulnerable to disruptions. The resulting forecasts support early implementation of corrective actions, such as adjusting the production plan, reallocating tasks among resources, or modifying order priorities. This helps maintain process continuity while minimizing losses arising from material-related downtime.

The use of a digital twin in controlling production flow enables the combination of real-world data with simulation-based analysis, enhancing the efficiency and stability of processes. The model serves as a predictive and flexible tool not only for analyzing ongoing disruptions but also for long-term planning and operational decision-making under changing production conditions.

4 Conclusions and Future Work

Digital twin demonstrators in smart factory laboratories are today an important component of both research on modern manufacturing systems and the education of future engineers. They enable realistic replication of processes occurring in a real factory, allowing experimentation with production line configuration, material flow, scheduling, and control without the risk of disrupting actual operations. Such environments make it possible to test new optimization methods, assess the effects of failures, simulate various operational scenarios, and understand the complexity of contemporary manufacturing systems. From an educational perspective, digital twins support the development of practical skills by combining theory with hands-on experience, giving users the opportunity to work with technologies used in Industry 4.0.

In the coming years, digital twin demonstrators may evolve toward more advanced integration with laboratory infrastructure and expanded functionality related to data analysis. One natural direction for development is the full acquisition of data from machines, sensors, and control systems, followed by its use to continuously update the simulation model. Another area is the integration of artificial intelligence tools, which can significantly enhance the capabilities of the demonstrator. Machine learning models can be used for failure prediction, production load forecasting, anomaly detection in machine operations, or recommending optimal process parameters. The combination of historical, real-time, and simulated data opens the way for creating autonomous decision-support systems capable of predicting the outcomes of actions before they are executed.

The development of such laboratories could lead to fully interactive environments in which the digital twin, data analytics, and artificial intelligence work together as a cohesive system. These demonstrators would become not only a tool for supporting education but also a research platform enabling the creation, testing, and refinement of technologies for the future of intelligent manufacturing.

Acknowledgments. The presented results derive from a scientific statutory research conducted by Chair of Production Engineering, Faculty of Mechanical Engineering, Poznan University of Technology, Poland, supported by the Polish Ministry of Science and Higher Education.

References

1. Friederich, J., Francis, D.P., Lazarova-Molnar, S., Mohamed, N.: A framework for data-driven digital twins of smart manufacturing systems. Comput. Ind. **136**, 103586 (2022). https://doi.org/10.1016/j.compind.2021.103586
2. Pajpach, M., Pribiš, R., Drahoš, P., Kučera, E., Haffner, O.: Methodology and pilot implementation of modular educational-development platform for Interoperable Digital Twin. Sci. Rep. **15**, 28020 (2025). https://doi.org/10.1038/s41598-025-13579-y
3. Uhlemann, T.H.-J., Schock, C., Lehmann, C., Freiberger, S., Steinhilper, R.: The Digital Twin: demonstrating the potential of real time data acquisition in production systems. Procedia Manuf. **9**, 113–120 (2017). https://doi.org/10.1016/j.promfg.2017.04.043
4. Qi, Q., Tao, F.: Digital Twin and big data towards smart manufacturing and Industry 4.0: 360 degree comparison. IEEE Access **6**, 3585–3593 (2018). https://doi.org/10.1109/ACCESS.2018.2793265
5. Soori, M., Arezoo, B., Dastres, R.: Digital twin for smart manufacturing, a review. Sustain. Manuf. Ser. Econ. **2**, 100017 (2023). https://doi.org/10.1016/j.smse.2023.100017
6. Kremslehner, N., Sobottka, T., Nacsa, J., Beregi, R., Schlund, S.: Digital Twin training concept based on miniature demonstration factories. SSRN Electron. J. (2023). https://doi.org/10.2139/ssrn.4458212
7. Palmer, C., et al.: Digital Twinning remote laboratories for online practical learning. Prod. Manuf. Res. **10**, 519–545 (2022). https://doi.org/10.1080/21693277.2022.2097140

8. Ortiz, J.S., Quishpe, E.K., Sailema, G.X., Guamán, N.S.: Digital Twin-Based active learning for industrial process control and supervision in Industry 4.0. Sensors **25**, 2076 (2025). https://doi.org/10.3390/s25072076
9. Martinez, S., et al.: A *Digital Twin Demonstrator* to enable flexible manufacturing with robotics: a process supervision case study. Prod. Manuf. Res. **9**, 140–156 (2021). https://doi.org/10.1080/21693277.2021.1964405
10. Nezzi, C., De Marchi, M., Aruväli, T., Vidoni, R., Rauch, E.: Implementation of 3D simulation to foster digital twin based applications in manufacturing: an educational case study. SSRN Electron. J. (2023). https://doi.org/10.2139/ssrn.4469934
11. Grieves, M.: Digital twin: manufacturing excellence through virtual factory replication. White Pap. 1–7 (2014)
12. Fuller, A., Fan, Z., Day, C., Barlow, C.: Digital Twin: enabling technologies, challenges and open research. IEEE Access **8**, 108952–108971 (2020). https://doi.org/10.1109/ACCESS.2020.2998358
13. Ferrario, A., Confalonieri, M., Barni, A., Izzo, G., Landolfi, G., Pedrazzoli, P.: A multipurpose small-scale smart factory for educational and research activities. Procedia Manuf. **38**, 663–670 (2019). https://doi.org/10.1016/j.promfg.2020.01.085
14. García, Á., Bregon, A., Martínez-Prieto, M.A.: Digital Twin learning ecosystem: a cyber–physical framework to integrate human-machine knowledge in traditional manufacturing. Internet Things. **25**, 101094 (2024). https://doi.org/10.1016/j.iot.2024.101094
15. Baalsrud Hauge, J., et al.: Digital twin testbed and practical applications in production logistics with real-time location data. Int. J. Ind. Eng. Manage. **12**, 129–140 (2021). https://doi.org/10.24867/IJIEM-2021-2-282
16. Gautam, A., et al.: IIoT-enabled digital twin for legacy and smart factory machines with LLM integration. J. Manuf. Syst. **80**, 511–523 (2025). https://doi.org/10.1016/j.jmsy.2025.03.022
17. Jones, D., Snider, C., Nassehi, A., Yon, J., Hicks, B.: Characterising the Digital Twin: a systematic literature review. CIRP J. Manuf. Sci. Technol. **29**, 36–52 (2020). https://doi.org/10.1016/j.cirpj.2020.02.002
18. Bendowska, K., Zawadzki, P.: Development and verification of a simulation model of an automated assembly line. Appl. Sci. **13**, 10142 (2023). https://doi.org/10.3390/app131810142

Principles of Teamwork in Digital Twin Factory Projects

Pawel Pawlewski(✉) and Patrycja Hoffa-Dąbrowska

Faculty of Management Engineering, Poznan University of Technology, Poznan, Poland
pawel.pawlewski@put.poznan.pl

Abstract. Creating a digital twin of a factory is a major undertaking. Developing a comprehensive simulation model requires mapping physical buildings, technical resources, and all operational and management processes. When integrated, these elements form a cohesive digital twin. While traditional simulation models often focused on solving isolated problems, the rapid evolution of technology and digitalization now requires companies to analyze the entire system holistically. Consequently, organizations are increasingly adopting digital twins to evaluate overall system performance and predict how changes in one area might impact the entire enterprise. At the Poznan University of Technology's Faculty of Engineering Management, students recently developed a digital twin of an automotive assembly area. This article outlines a teamwork methodology for large-scale digital twin projects, detailing the necessary steps based on the experiences and insights gained from that project.

Keywords: Digital twin · Teamwork · Assembling · Industry 4.0

1 Introduction

Creating a digital twin of a factory is a significant design undertaking. A primary challenge in developing large-scale simulation models is the difficulty of collaborative teamwork. Typically, these models are developed linearly, or step-by-step, as simulation software manufacturers rarely provide built-in frameworks for parallel team development [1, 2]. Instead, existing methodologies tend to favor a situational, problem-solving approach. While these programs offer object libraries and tools for building flow logic, the specific architecture of the model is left entirely to the user. Consequently, even if a project is divided into smaller sub-models for separate development, integrating them into a single, cohesive simulation remains a complex and difficult task.

Anylogic proposes improvements for large-scale simulation projects [3], but this applies to facilitations related to panning, zooming, navigation, i.e. working with the mouse in the workspace and visualization of results. They also focus on the limitations of Java (64k limit) and the suggestion of dividing large agents into smaller ones, creating user libraries.

FlexSim, on the other hand, proposes storing the model in multiple storage files [4] and using version management tools to handle changes to these files [5].

J. Trojanowska et al. (Eds.): MANUFACTURING 2026, LNME, pp. 66–77, 2026.
https://doi.org/10.1007/978-3-032-21654-0_6

Teamwork is a cornerstone of both professional and personal success. The ability to collaborate, delegate tasks, and execute activities in parallel is an essential component of modern project management. These skills are introduced in early childhood and refined throughout the educational process. Within the Logistics program at the Poznan University of Technology's Faculty of Engineering Management, teamwork and project-based learning are emphasized, reflecting the necessity for logisticians to coordinate across various departments. During a 2024 course on logistics process design, third-year students collaborated to develop a digital twin of an automotive assembly area. To manage this complex undertaking, specific cooperation principles were established, which are detailed in this article.

The article consist of five parts. In Sect. 2 the literature review about team projects, including digital twins aspects and students project is presented. The bibliometric analysis is presented. In Sect. 3 methods of conducting projects on such a large scale are described. Section 4 presents results from realized work. The conclusion and possibilities of future development are presented in the last section.

2 Literature Research

The issue of teamwork is widely described in the literature. First article at Web of Science database (accessed from the Clarivate Analytics server) are dated 1922. However, interest in teamwork has not decreased in recent years, and has remained at a level of above 2000 articles per year. Figure 1 shows the number of articles in the database for this topic in recent 10 years. Of course, the current year is not closed–the analysis was made at November 2025.

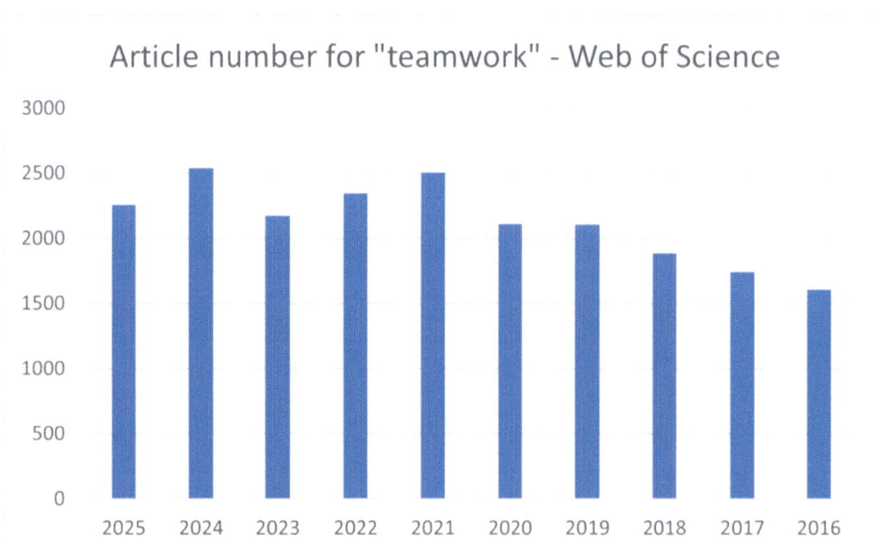

Fig. 1. Number of articles for "teamwork". (Source: Web of Science Analyze Results)

Referring to the Cambridge dictionary, teamwork means: "the activity of working together in a group with other people, especially when this is successful" [6]. Based on Oxford dictionary teamwork is: „the activity of working well together as a team" [7].

Teamwork seems to be a concept in human resource management. Teamwork allows combining competences and obtaining added value resulting from the contribution of different values (skills) from different people, which is an obvious benefit. Teamwork means also challenges related to organizing the cooperation of several (or dozens or more) people in order to achieve a defined goal. The teamwork concept it is not only benefits but also challenges. One of the main seems to be simultaneously take into account the individual preferences and goals of a single person in team and how to motivate person to co-work. Feeling of underestimation of an individual or an unequal burden of duties in team also is an important things and challenges. The importance of teamwork and awareness of the benefits resulting from it are described in many articles [8, 9]. Chmelarova & Pasiar present results from their survey about students attitudes towards teamwork, what is described at [10]. Tannenbaum et. al. Describe the nine common teamwork barriers and provide advice [11]. Konak and Kulturel-Konak describe in their article the impact of online teamwork self-efficacy on attitudes toward teamwork [12]. The teamwork in education area are popular topic in literature. Examples of articles about teamwork in education [13–19].

In order to illustrate the popularity of the discussed topics, it was decided to search the article in Web of Science database in terms of the number of articles for selected issues, as: teamwork, student teamwork and the student teamwork in simulation project or digital twin project. The obtained results are presented in Table 1. The article research was carried out in November 2025. Authors deliberately searched the database in terms of the occurrence of specific concepts in the entire area, they did not intentionally narrow the searching. This was intended to show general interest in a specific topic, while narrowing down issues searched aspects.

Table 1. Number od articles and year of first article for chosen topics. (Source: own study, based on bibliometric analysis from Web of Science database (accessed from the Clarivate Analytics server))

No.	Searching word	Number of articles	Year of first article
1	teamwork	35278	1922
2	student* teamwork	10467	1956
3	teamwork + principle*	1668	1964
4	teamwork + rule*	477	1983
5	teamwork + simulation	3248	1973
6	teamwork + simulation project	603	1991

(*continued*)

Table 1. (*continued*)

No.	Searching word	Number of articles	Year of first article
7	teamwork + simulation project + student*	**292**	**1991**
8	teamwork + simulation project + student* + logistic*	7	2012
No.	Searching word	Number of articles	Year of first article
9	teamwork + simulation project + student* + production	11	2010
10	teamwork + simulation project + student* + assembly	2	2005
11	digital twin	30090	1941
12	digital twin project	9803	1994
13	teamwork + digital twin	21	2005
14	teamwork + digital twin project	**10**	**2005**
15	teamwork + digital twin + student*	**4**	**2021**
16	teamwork + digital twin project + student*	**3**	**2021**
17	methodology + simulation project	35252	1978
18	methodology + digital twin	3125	1968
19	methodology + digital twin project	1117	2005
20	methodology + teamwork + digital twin	**2**	**2024**
21	methodology + teamwork + digital twin project	**1**	**2024**

Analyzing searching results it can be noticed that can be found many articles about student teamwork. When the search area will be narrowed to simulation project the number of articles is less then 300. Focusing on digital twin and aspect of teamwork, the number of articles is 10. It is also worth taking a closer look at the issue of digital twin project. In the searched database, the number of articles on digital twin project is almost 10,000. Analyzing the number of articles per year over the last 10 years, a significant increase in publications can be seen over the last 3–4 years (Fig. 2).

In order to more explore the topic of teamwork in simulation or digital twin projects, it was decided to check the publishing categories for the search "teamwork + simulation project + student*" (Fig. 3) and "teamwork + digital twin project" (Fig. 4). Due to the small number of articles in the case of the digital twin, it was decided to show the list for the search case without the word student. For the first searching, on the top is: (1) Education Scientific Disciplines, (2) Education Educational Research and (3) Nursing. For the second searching, on the top is: (1) Computer Science Interdisciplinary Applications, (2) Engineering Industrial.

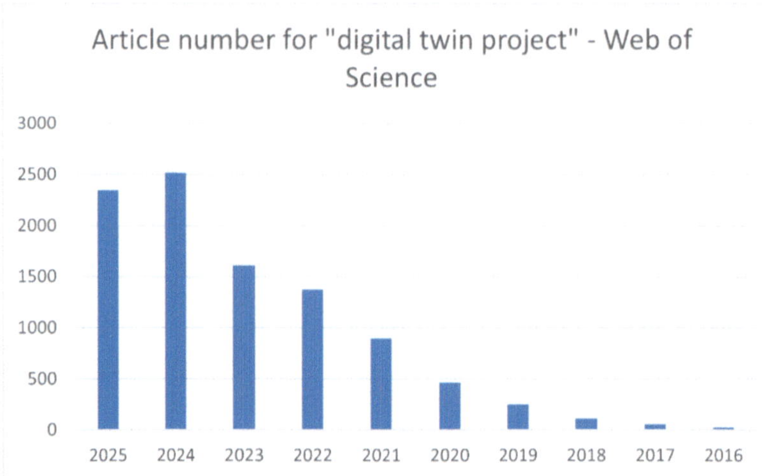

Fig. 2. Number of articles for "digital twin project". (Source: Web of Science Analyze Results)

Fig. 3. TreeMap Chart for Web of Science categories for "teamwork + simulation project + student*" searching. (Source: Web of Science Analyze Results)

A rough search of the selected database shows that the aspect of teamwork in the area of digital twin is an area that is not very well described. The oldest article for the teamwork digital twin project comes from 2005. Additionally, combining this area with the issue of student teamwork, we obtain only a few articles in the database. This additionally shows the uniqueness of the issues discussed in this article.

Fig. 4. TreeMap Chart for Web of Science categories for"teamwork + digital twin project" searching. (Source: Web of Science Analyze Results)

Beisheim et. al. [20] present a 3d developed environment, which includes various user concepts and allows multi-user access. They described different using of digital twin by various employees (department) but they do not focus of teamwork aspect and rules of this teamwork. Hu and Lykourentzou [21] in their article described an agent-based simulator designed for modeling the behavior of people in teamwork, especially hackathon. They focus on team formation. They compared three modes of collaboration. In that paper the human parameters, as: knowledge domain, expertise, nationality, and other, are modelled. The other articles described teamwork with digital twin, or with robots, or AI, or using digital twin/ simulation models for educate person, or teamwork between human-vehicle or human-robot collaboration, so they do not described people teamwork, teamwork management during realization the digital twin project.

Due to the lack of an article addressing the rules of teamwork during the realization of a digital twin project, it was decided to search for articles in the publicly available Google Sholar database. In literature articles about using Digital twin in education can be found [22, 23]. It was found article about teamwork student during realization project, which include aspect of building digital twin. Hagedorn et. al. [24] described student project. The project consist of 8–9 students and students task (each group separately) was designing and developing a mechatronic product with a Digital Twin. Authors detailed described task definition and evaluation, teaching staff and structure, methodological support for students (in which steps for project can be founded) and results of research.

3 Methods

Factory structuring and process structuring enables teamwork in factory digital twin projects. A large part of the work related to the creation of the FDT (Factory Digital Twin) is related to filling in the databases. At the same time, a notation describing the factory was proposed. Naming is limited to workstations names, PFEP (Plan For Every Part) database names, OpenTask and CycleFlow function names, and Logistics

Navigator names. Adopting a common principle for the team to create these names and respect them throughout the project is a basic principle of teamwork. The methodology of teamwork is a modified methodology presented in [25]. The methodology is presented on the example of a car assembly factory–data obtained during a student project carried out by the Faculty of Engineering Management of the Poznań University of Technology and VW Poznań were used. Figure 5 shows the layout of the assembly line. A model of the entire assembly area was built. Technical, manipulators, containers, employees are taken into account–without logistics processes, which means that parts are available at all times. It was interesting to answer the question whether by launching a specific sequence of vehicles, the assembly would be completed in time. Will the workers in this sequence have time to do their job? In addition, it was assumed that it would be possible to play with the sequence, that it would be possible to analyze employee workload charts.

The vehicles are assembled on a moving assembly line divided into 170 stations. There are several employees working at each position. Robots are also used. In total, more than 400 operators–employees and robots–work on one shift. Several variants of vehicles are installed–depending on the variant, the routes of employees change. The names of the parts are adopted according to the part IDs in the BOM (Bill of Materials).

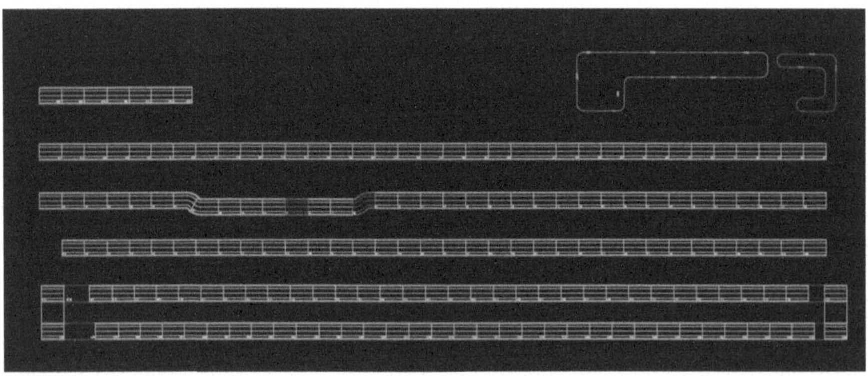

Fig. 5. A top view of the assembly hall with marked stations. (Source: own elaboration based on the project documentation)

In this case, teamwork is based on the fact that several teams build 170 job models at the same time, which in the last phase are combined into one FDT model. The first steps of the methodology are carried out by a team called G_00 (G–Group, 00–next number). Its task is to prepare a startup file that will be used by the other teams. These teams build models for the positions to which they have been assigned.

Steps of team G_00:

Step 1–Prepare a "light" layout–i.e. a layout that contains only the outline of the workstations–as shown in Fig. 1.

Step 2–Creating an empty PFEP (Plan For Every Part) database named FDT_pfep and defining the lines FA1 (name of line is original from factory), FA2, FA3, FA4, FA5, FA6, FA7 and the F000 line (the line with the stand where the moving assembly line will be

modeled). Each line contains the positions assigned to it–the structure of the name T_x where x is the bar number.
Step 3–Build a model of the entire F000 assembly line–using conveyors–and get it up and running.
Step 4–Prepare the X sequence of vehicle variants.
Step 5–Share the startup file with the other teams.

Other groups build individual stations in the starting model independently of each other. One of the factors that makes this possible is the mechanism for detecting objects on the conveyor based on a marker on the floor that the worker approaches. This marker is not an element of the conveyor, it can be an element of a station other than the one to which the conveyor belongs. When you create such a tag, you must associate it with a given conveyor by specifying the identifier of that conveyor along with the name of its station. This means that the stations to be built can be created independently of each other, and the link to the moving line is realized only by the name. This means that the markers on the floor are an element of the station being built, not the station to which the conveyor belongs. Figure 6 shows this situation where the conveyor Cnv_01 belongs to the Conv station and the marker on the floor G_1 and G_2 belongs to the T_131_station.

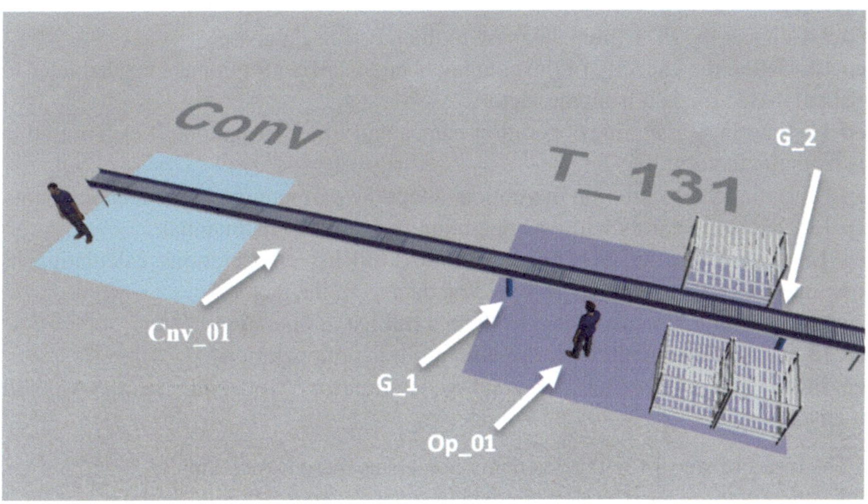

Fig. 6. An illustration of the connection between an assembly station and a moving line made of convey-ors. (Source: own elaboration)

Due to the large number of model files, it is assumed that the following file naming rule will be adopted:

Version_Tx-y_Step_x_Data_vx.fsm.

Where:

Version–is the version number of the simulation program.

Tx-y–is the name with the bar numbers that are in the file.

Step_x–The step number of the methodology.

Date–The date the model was executed.

vx-version of the file.

Example: 23.0.3.9_T1-T6_ Step 5_240210_v01.fsm which means 23.0.3.9-version number of the program.

T1-T6–The first 6 bars are included in the model file.

Step 5–indicates that the model file contains a model that matches step 5

240210–means that the model was last modified on February 10, 2024

v01_PP–indicates the serial number of the model version.

The adopted principle ensures that order is maintained in the management of the entire project.

The next steps of each group (carried out separately but in parallel, at the same time) are:

Step 6–Finding the right waorkstation T_x.
Step 7–Populate the PFEP database–FDT-pfep–with the parts that are assigned to the position–based on the data obtained at the factory (BOM).
Step 8–Checking and completing the DB (DataBase) Containers database–if there is no shape, the shape of the standard container with the appropriate dimensions is assumed, and then you can possibly make the shape in Sketchup–based on the data obtained from the factory.
Step 9–Connect the FDT_pfep database to the DB Bins database.
Step 10–Define the capacity of the containers and KanBan (optional - needed later for logistics)–based on data from the factory.
Step 11–Entering (assigning) operation names and times to stations (ticks)–based on data from the factory.
Step 12–Define the location in the station–adopt the part naming in the location name.
Step 13–Define Operators with one operation from the job list/manual.
Step 13_1–Modeling the first operator with a full list of operations–calculating the operation times without transport–based on data from the factory.
Step 13_2–Adding a second Operator with a full list of operations.
Step 13_n - Adding the nth Operator with a full list of operations.
Step 14–Add/set up a model with graph generation for Yamuzumi, Gantt, VA, Loads and Distances.

The result of step 14 will be as many models as there were teams.

The next steps are performed linearly by a single team.

Step 15–Adding (copying) containers from the models to the first model with a T_1 workstation. The result is a model with a full database of DB containers.
Step 16–Connect the FDT_pfep databases from the models to the database FDT_pfep the first model with a T_1 workstation.
Step 17–Connect all stations by filling in the table with stations and their elements from the models to the first model with a T_1 bar and generate all the bars according to the content of the DB Station Table.
Step 18–Copying positions (ticks) from the model tree or by exporting/importing from models to the model first.
Step 18_T1_T6–Model file containing bars T1–T6

Step 18_T1−T12 - Model file containing bars T1−T12

.....

Step 18_T1−T120–model file containing all bars.
Step 19–Perform experiments for different sequences.

The final result is a single model that combines all the individual models of the design teams. This model represents the entire assembly area of the company.

4 Results

The teamwork resulted in each group creating simulation models encompassing several cycles. These individual models were subsequently integrated into a single, comprehensive digital twin. The final outcome is illustrated in Fig. 7. Although Fig. 7 shows the line filled with a single car model of the same color—a scenario that does not occur in reality—the project's primary objective was not to model every vehicle variant. Instead, the goal was to simulate a large-scale assembly area and evaluate the feasibility of large-group collaboration on a project of this magnitude.

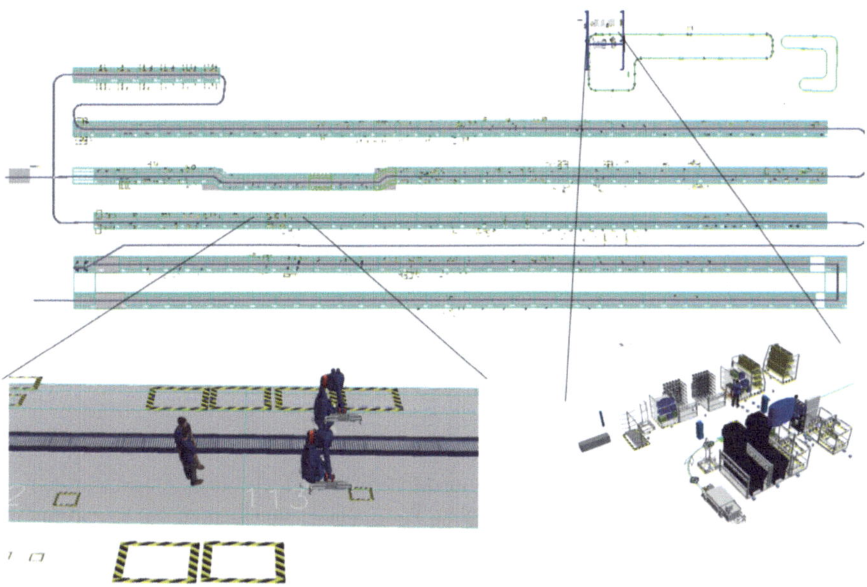

Fig. 7. Digital twin of the assembly area. (Source: own elaboration)

5 Conclusions

The objective of this project, conducted by third-year Logistics students at the Poznan University of Technology's Faculty of Engineering Management, was to build a digital twin for an automotive assembly area comprising over 170 stations. This large-scale initiative involved 91 students divided into 22 groups. On the corporate side, a main coordinator and four area-specific coordinators were appointed. The students completed the project within 3.5 months, following an initial period of data provision and methodological briefing.

Beyond the proposed project methodology, successful collaboration required robust communication and knowledge management tools. The following systems were utilized:

- Task Management: Asana and the ECourse platform.
- Knowledge Base and Documentation: ECourse and Cloud Drive.
- Synchronous Communication: Microsoft Teams.

To mitigate project risks, an analysis was performed based on the complexity of the assembly stations. The most complex stations were identified as high-risk; consequently, their implementation was assigned to the highest-performing student groups and closely monitored by project leaders. This strategic allocation minimized errors and ensured the project's overall success.

The project confirmed the initial assumptions and demonstrated that large-scale digital twins can be successfully implemented using this collaborative framework. While the core methodology was validated, certain areas for refinement were identified, particularly regarding inter-team communication, data verification, and documentation maintenance.

The resulting assembly line model offers significant potential for further development. In terms of utility, the model could be expanded to include additional vehicle variants by introducing specific operation times for different models. This would allow for experimentation with production sequences, output estimation, and the analysis of employee workloads at each station. A logical next step would be to incorporate intralogistics—specifically the delivery and collection of parts containers—which was excluded from the current scope due to time constraints and the high level of complexity for a student-led project.

From a scientific point of view, the proposed approach overcomes the limitations and linear creation of simulation models in favor of an approach that allows for hierarchical decomposition of the model and its simultaneous creation in teams, which in the final stage merges into a single model.

References

1. Borshchev A.: The Big Book of Simulation Modeling, Anylogic North America (2013)
2. Beaverstock, M., Greenwood, A., Nordgren, W.: Applied simulation. modeling and analysis using flexsim, Flexsim Software Products, Inc., Canyon Park Technology Center, Orem, USA (2017)
3. The Anylogic Company. https://www.anylogic.com/blog/anylogic-8-8-5-enhancements-for-large-scale-simulation-projects/. Access date 01 2026
4. Autodesk. https://forums.autodesk.com/t5/flexsim-forum/how-to-work-on-a-big-model-as-team-at-the-same-time/m-p/13529738. Access date 01 2026
5. Autodesk. https://docs.flexsim.com/en/22.1/Reference/DeveloperAdvancedUser/XML/XML.html. Access date 01 2026
6. Cambridge Dictionary. https://dictionary.cambridge.org/dictionary/english/teamwork, Access date: 01 2026
7. Oxford Learner's Dictionaries. https://www.oxfordlearnersdictionaries.com/definition/american_english/teamwork. Access date: 01 2026

8. Gafa, M., Fenech, A., Scerri, C., Price, D.: Teamwork in healthcare organisations. Pharm. Educ. **5**(2) (2005)
9. Korzeniecka-Bondar, A., Tołwińska, B., Borawska-Kalbarszyk, K.: Benefits of time management skills, teamwork skills and personal information management perceived by students from the Polish-Belarusian Borderland. In: Ahmedi, B., Thaqi, X., Aljimi, E. (eds.) Innovations, Technologies and Research in Education, pp. 91–107 (2018)
10. Chmelárová, Z., Pasiar, L.: Attitudes of economics students towards teamwork at university. Int. J. Eng. Pedagogy **13**(6) (2023)
11. Tannenbaum, S., Castillo, G.F., Salas, E.: How to overcome the nine most common teamwork barriers. Organ. Dyn. **52**(4), 101006 (2023)
12. Konak, A., Kulturel-Konak, S.: Impact of online teamwork self-efficacy on attitudes toward teamwork. Int. J. Inf. Technol. Proj. Manage. (IJITPM) **10**(3), 1–17 (2019)
13. Bencsik, A., Noszkay, E., Marosi, I.: Teamwork in education. Prob. Educ. 21st Century **10**, 9 (2009)
14. Navarro, J., et al.: Teamwork: Assessment of teamwork competence in higher education. In: Proceedings of the 3rd International Conference on Higher Education Advances (pp. 1024–1032). Editorial Universitat Politècnica de València (2017)
15. Malone, D.M., Gallagher, P.A.: Special education teachers' attitudes and perceptions of teamwork. Remedial Spec. Edu. **31**(5), 330–342 (2010)
16. Fathi, M., Ghobakhloo, M., Syberfeldt, A.: An interpretive structural modeling of teamwork training in higher education. Educ. Sci. **9**(1), 16 (2019)
17. Titov, E.: Teamwork as one of the major study methods in entrepreneurship education. In: ICERI2016 Proceedings (pp. 1978–1985). IATED (2016)
18. Ehrlenspiel, K., Giapoulis, A., Günther, J.: Teamwork and design methodology—Observations about teamwork in design education. Res. Eng. Design **9**, 61–69 (1997)
19. Volkov, A., Volkov, M.: Teamwork benefits in tertiary education: Student perceptions that lead to best practice assessment design. Educ. Train. **57**(3), 262–278 (2015)
20. Beisheim, N., et al.: Simultaneous interdisciplinary teamwork on digital twins in a 3D collaborative environment. In: Transdisciplinarity and the Future of Engineering (pp. 380–389). IOS Press (2022)
21. Hu, J., Lykourentzou, I.: TONIC: A teamwork simulator and digital twin for organizational innovation challenges. In: Proceedings of the 16th International Conference on PErvasive Technologies Related to Assistive Environments (pp. 584–588) (2023)
22. Hazrat, M.A., Hassan, N.M.S., Chowdhury, A.A., Rasul, M.G., Taylor, B.A.: Developing a skilled workforce for future industry demand: the potential of digital twin-based teaching and learning practices in engineering education. Sustainability **15**(23), 16433 (2023)
23. Nikolaev, S., Gusev, M., Padalitsa, D., Mozhenkov, E., Mishin, S., Uzhinsky, I.: Implementation of "digital twin" concept for modern project-based engineering education. In: Product Lifecycle Management to Support Industry 4.0: 15th IFIP WG 5.1 International Conference, PLM 2018, Turin, Italy, July 2–4, 2018, Proceedings 15 (pp. 193–203). Springer International Publishing (2018)
24. Hagedorn, L., Riedelsheimer, T., Stark, R.: Project-Based learning in engineering education-developing digital twins in a case study. Proc. Design Soc. **3**, 2975–2984 (2023)
25. Pawlewski P.: Built-In Lean Management Tools in Simulation Modeling. In: Proceedings of the 2019 Winter Simulation Conference. Mustafee, N., Bae, K.-H.G., Lazarova-Molnar, S., Rabe, M., Szabo, C., Haas, P., Son, Y.-J. (eds.) (2019)

Artificial Intelligence in Design and Management

AIRI–A Methodology for Evaluating AI Integrability in Classical Industrial Equipment

Alexandru Moldovan[1], David Chereche ș[1], Grigore Pop[1], Mircea Murar[1], Mircea Fulea[1], Cristian Moldovan[2], and Bogdan Mocan[1(✉)]

[1] Technical University of Cluj-Napoca, Cluj-Napoca, Romania
bogdan.mocan@muri.utcluj.ro
[2] Politehnica University Timişoara, Timişoara, Romania

Abstract. The AI Integrability Readiness Index (AIRI), introduced in this paper, constitutes a novel systematic methodology for quantifying Artificial Intelligence integration potential within legacy industrial equipment and production systems, aligned with the human-centric imperatives of Industry 5.0. Distinctively, while established indices-such as AI Singapore's AIRI (assessing national AI maturity) or the general AI Readiness Index (evaluating organizational strategy and talent)-operate at organizational levels, the proposed AIRI addresses a critical assessment gap by targeting individual industrial assets. This equipment-level framework evaluates AI integrability through seven specialized technical domains: Data Acquisition, Control and Interoperability, Predictive Maintenance Readiness, Hardware Adaptability, Cybersecurity, Energy Efficiency, and Human-AI Interaction. Through weighted aggregation, the methodology classifies equipment into four distinct readiness categories: Legacy, AI-Ready, AI-Integrable, or AI-Native. Comprehensive validation across heterogeneous industrial environments-encompassing automotive, chemical, metal products, and food processing sectors-confirms the framework's robustness and cross-sectoral applicability as a standardized diagnostic instrument. The AIRI provides actionable, domain-specific insights enabling engineers and managers to strategically plan modernization initiatives, optimize technological upgrades, and maximize AI investment returns within existing operational infrastructures. This contribution advances the theoretical understanding of equipment-level AI readiness while offering practical decision-support capabilities for industrial digital transformation.

Keywords: Artificial Intelligence (AI) · Industry 5.0 · AI Readiness Index · AI-Augmented Manufacturing · Digital Transformation

1 Introduction

The rapid advancement of artificial intelligence (AI) is reshaping industrial sectors, creating significant opportunities for enhanced efficiency, adaptability, and productivity, particularly within the framework of the Fourth Industrial Revolution (Industry 4.0) [1]. Presently, the focus is shifting toward Industry 5.0, which extends the 4.0 vision by

J. Trojanowska et al. (Eds.): MANUFACTURING 2026, LNME, pp. 81–95, 2026.
https://doi.org/10.1007/978-3-032-21654-0_7

foregrounding sustainability, resilience, and crucially human-centred Human-AI Interaction [3]. A pivotal challenge in this transition is the integration of AI technologies into legacy industrial equipment, which constitutes most global asset bases. Although smart retrofitting-the digital modernization of existing machinery-offers a cost-effective alternative to full equipment replacement, especially for small- and medium-sized enterprises (SMEs), the processes of assessment remain uneven.

The AI Integrability Readiness Index (AIRI) offers a unique asset-level assessment for retrofitting specific industrial equipment, addressing a critical gap left by broader organizational maturity models.

Existing AI readiness and digital maturity frameworks operate predominantly at organizational levels. AI Singapore's AIRI [1] evaluates enterprise adoption readiness through five pillars (Organizational, Business Value, Ethics & Governance, Data, and Infrastructure Readiness), classifying entities from "AI Unaware" to "AI Competent" (Table 1). Similarly, the Acatech Industrie 4.0 Maturity Index [2] assesses company-wide transformation across four structural areas through six developmental stages. While effective for strategic corporate planning, these frameworks lack the technical granularity required to evaluate individual industrial assets - a critical gap for enterprises pursuing smart retrofitting of legacy "brownfield" equipment.

Table 1. Comparative Positioning of AIRI against Established Digital Maturity Frameworks

Feature	AIRI (AI Integrability Readiness Index)	Acatech Industrie 4.0 Maturity Index	AI Subgapore AIRI
Scope of Assessment	Specific Industrial Equipment/Asset (e.g., machine tool)	Entire Organization/Business Process [3]	Enterprise/Business Unit [1, 4]
Primary Goal	Quantify AI Integrability for Retrofitting	Manage Digital Transformation towards I4.0 [3]	Assess Organizational Readiness for AI adoption [1]
Number of Domains	Domains 7 Technical and Operational Domains	4 Structural Areas (Resources, Information Systems, Culture, Structure) [3]	5 Pillars/12 Dimensions (e.g., Data, Ethics, Business Value) [1]
Inclusion of Industry 5.0	Explicit: Includes Human-AI Interaction and Energy Efficiency	Implicit (Focus on Industry 4.0)	Focuses on Ethics & Governance readiness [1]
Key Metric	AIRI Index (0–100) based on weighted sum of technical criteria	Maturity Stage (ML1–ML5/ML6) based on current/target state	Readiness Level (AI Unaware to AI Competent) [1]

The proposed AIRI addresses this gap by shifting the unit of analysis to the individual machine or production system. Seven evaluation domains capture the technical prerequisites for AI integration at the asset level: data acquisition capabilities (A), control architecture and interoperability (B), predictive maintenance readiness (C), hardware adaptability (D), cybersecurity and reliability (E), energy efficiency (F), and human–AI interaction (G). This structure maps onto the three layers of Cyber-Physical Production Systems, encompassing physical connectivity, cyber-layer processing, and human-centric considerations aligned with Industry 5.0.

The quantitative output-a weighted $0 \div 100$ index-enables benchmarking across heterogeneous equipment types and data-driven prioritization of modernization investments. This asset-level specificity positions AIRI as a complementary tool bridging organizational readiness assessments and practical engineering decisions for AI deployment in legacy industrial environments.

Table 1 systematically articulates the conceptual and methodological distinctiveness of the AI Integrability Readiness Index (AIRI) framework by performing a comparative analysis against prevailing digital maturity models, including enterprise-level Industry 4.0 frameworks (e.g., Acatech) and organizational AI readiness indices (e.g., AI Singapore AIRI).

2 Theoretical Framework

The AI Integrability Readiness Index (AIRI), proposed through this paper, is an asset-level framework designed to complement macro-level organizational readiness models. It addresses the granular evaluation required for smart retrofitting in legacy industrial environments. The framework is built upon three pillars:

- **Maturity Models and Capability Theory:** AIRI adapts Capability Maturity Model (CMM) principles, treating "AI Integrability" as a measurable capability that evolves from ad hoc stages to optimized performance. It quantifies this capability across four tiers—Legacy, AI-Ready, AI-Integrable, and AI-Native using a $0 \div 5$ scoring scale. This approach aligns with the Acatech Industrie 4.0 Index [2] and operational frameworks like S2FC MM [3, 5].
- **Socio-Technical Systems:** The methodology recognizes that AI integration is not purely technical. By including domains such as Human-AI Interaction and Energy Efficiency, AIRI accounts for the interplay between technology, workers, and organizational goals, aligning with Industry 5.0 mandates [2, 6].
- **Industrial Digital Transformation:** The framework bridges the gap between traditional Operational Technology (OT) and Information Technology (IT). It focuses on "brownfield" environments where legacy equipment must be upgraded to support modern paradigms like predictive maintenance and real-time analytics [2, 6].

3 Methodological Objectives

The AIRI methodology is built around clear objectives that turn AI integration from a theoretical idea into a practical tool for engineers, plant managers and decision-makers in digital transformation [2].

Standardized assessment framework–AIRI unifies fragmented Industry 4.0 maturity models into a single, repeatable process for evaluating AI readiness across diverse assets, filling a recognized gap in the literature [1, 2, 5, 7, 8].

Quantification of key factors–By breaking down AI deployability into data acquisition, control interoperability, hardware adaptability and cybersecurity, AIRI assigns quantitative scores to these technical and operational dimensions, turning qualitative judgments into empirical data [9–12].

Decision-support for investment and retrofitting–The methodology offers data-driven cost-benefit insight, helping SMEs decide whether to upgrade, retrofit or replace legacy equipment and to build a technology roadmap [5, 8–10, 13].

Benchmarkable AIRI Index–A single 0 ÷ 100 score enables benchmarking of machines, lines or factories, guiding organizations to prioritize investments with the highest ROI and to advance performance in the Industry 4.0/5.0 transition [2, 11].

4 Structure of the AIRI Framework

The AIRI methodology employs a multi-dimensional, hierarchical framework that evaluates AI-integration readiness across seven domains (A–G), providing a comprehensive view of both technological and operational facets that is exposed in Table 2.

Table 2. Domains of criteria

Domain	Focus	Number of Criteria	Weight
A. Data Acquisition Capabilities	Access and quality of process data	4	0.20
B. Control and Interoperability	Openness of control interfaces, protocol support	4	0.20
C. Predictive Maintenance Readiness	Sensorization and diagnostic capabilities	4	0.15
D. Hardware Adaptability	Physical and electrical expandability	4	0.15
E. Cybersecurity and Reliability	Protection mechanisms and firmware robustness	4	0.10
F. Energy Efficiency and Sustainability	Monitoring and AI-based optimization potential	3	0.10
G. Human–AI Interaction	Interface usability and operator collaboration	3	0.10

It adopts a five-point ordinal rating (0 = non-existent, 5 = fully AI-integrable), a convention widely used in maturity-model research to balance discriminative power with practical usability for managers and consultants [13].

The seven domains Data Acquisition, Control & Interoperability, Predictive-Maintenance Readiness, Hardware Adaptability, Cybersecurity & Reliability,

Energy Efficiency & Sustainability, Human-AI Interaction & Usability were selected as a minimal sufficient set that maps directly onto the three layers of a Cyber-Physical Production System (physical connectivity, cyber-layer processing, human-centred sustainability) and aligns with the essential dimensions identified in recent systematic reviews of AI-readiness frameworks [14–16].

By grounding the scale and domain selection in these well-documented conventions, AIRI inherits the methodological credibility of established maturity-assessment tools while remaining lightweight enough for rapid, asset-level deployment.

The table below summarizes the seven domains, their primary focus, the number of criteria within each, and an example weighting scheme for a typical discrete manufacturing scenario.

The normalized score for each domain *i* is obtained by dividing the raw ordinal score S_i (range 0–5) by the maximum of the scale, yielding a value in [0,1]. The overall AIRI index is then:

$$\text{AIRI Index} = \tfrac{100}{5}\sum_{i=1}^{n}(S_i \times W_i) \tag{1}$$

where:

- S_i = score assigned to each criterion (0–5 scale),
- W_i = weight of the corresponding criterion or domain.

Assumptions

- Additivity – the contribution of each domain to the final score is assumed to be approximately independent of the others, an assumption that is standard in maturity-model literature and has been empirically validated in Industry 4.0 readiness assessments [17].
- Linearity of importance – the linear relationship between weight and impact on the index reflects the practice adopted in most readiness and maturity frameworks, where higher-level capabilities are expected to bring proportionally higher value [17].

4.1 Weighting Scheme and Aggregation Model – Methodological Justification

The AIRI index aggregates the domain scores into a single synthetic metric by means of a weighted-sum model. This linear additive aggregation is the most widely adopted approach in Multi-Criteria Decision-Analysis (MCDA) because it provides transparency, interpretability and ease of benchmarking while preserving the relative importance of each dimension [18, 19].

Why a weighted sum?

- The literature on ordinal-information-based weighting methods stresses that a sum-type aggregation is appropriate when the decision-maker supplies only rank-order information and, optionally, additional strength-of-difference data [18].
- The "cardinal sum-reciprocal" method, recommended for cases with a semantic-scale input, is itself a linear combination of normalized scores [18].

- Simpler additive schemes such as the Simple Additive Weighting (SAW) have been shown to be robust across engineering, environmental and energy domains [20].

Derivation of the weights

- In practice, the weights w_i are elicited from a panel of domain experts (engineers, OT/IT specialists, production managers) using structured judgement techniques. The Analytic Hierarchy Process (AHP) is a common tool for converting expert pair-wise comparisons into a normalized weight vector, and it has been successfully applied to construct pair-wise matrices for multi-criteria weighting [21].
- When expert input is limited, the Constant Weight Ratio (CWR) method can be employed: the decision-maker provides a single additional piece of information (e.g., a weight ratio or the total weight of the most important set of criteria) and the method derives consistent weights for all criteria [22].
- Objective-weighting alternatives such as CRITIC, entropy or the newly proposed Centroidous method can be used to validate or complement expert-derived weights, ensuring that the final weight vector reflects both data-driven dispersion and expert knowledge [19, 23].

The expert panel builds a 7 × 7 pair-wise comparison matrix A using Saaty's 1–9 scale. A representative (rounded) matrix is shown in Eq. 2; the values reflect the judgement that Data Acquisition (A) is moderately more important than the other domains.

$$A = a_{ij} = \begin{pmatrix} 1 & 3 & 2 & 4 & 2 & 5 & 5 \\ 1/3 & 1 & 1/2 & 2 & 1 & 3 & 3 \\ 1/2 & 2 & 1 & 3 & 2 & 4 & 4 \\ 1/4 & 1/2 & 1/3 & 1 & 1/2 & 2 & 2 \\ 1/2 & 1 & 1/2 & 2 & 1 & 3 & 3 \\ 1/5 & 1/3 & 1/4 & 1/2 & 1/3 & 1 & 1 \\ 1/5 & 1/3 & 1/4 & 1/2 & 1/3 & 1 & 1 \end{pmatrix} \tag{2}$$

$$W_i = \frac{1}{7}\sum_{j=1}^{7} \frac{a_{ij}}{\sum_k^7 k_j} i = A, \ldots, G \tag{3}$$

Normalizing the columns of A and averaging each row yields the weight vector:

$$\mathrm{w} = (0.31, 0.14, 0.20, 0.08, 0.17, 0.05, 0.05) \tag{4}$$

The Consistency Ratio (CR) is computed as:

$$CR = \frac{CI}{RI} \tag{5}$$

with:

$$CI = \frac{\lambda_{max} - 7}{6} \tag{6}$$

and:

$$RI = 1.32 \tag{7}$$

the resulting

$$CR = 0.07 < 0.10 \tag{8}$$

confirm acceptable consistency. These weights are then used in the weighted-sum aggregation (see Eq. (1)).

5 Evaluation Domains and Criteria Description

The seven domains of the AIRI framework are designed to provide a holistic and multi-faceted evaluation of a machine's readiness for AI integration. Each domain targets a distinct aspect of the systems architecture, from fundamental data accessibility to its capacity for human collaboration.

A. **Data Acquisition Capabilities**

Assesses whether the equipment can generate and deliver high-quality, standardized data streams, the essential substrate for training and deploying AI models in an Industry 4.0 cyber-physical environment.

B. **Control and Interoperability**

Examines the openness of the control architecture and its ability to exchange bidirectional information with external AI services, a prerequisite for moving beyond passive monitoring to active optimization.

C. **Predictive Maintenance Readiness**

Maintenance Readiness–Evaluates the availability of diagnostic data and condition monitoring capabilities that enable AI algorithms to forecast failures before they occur.

D. **Hardware Adaptability**

Evaluate the physical and electrical flexibility required to incorporate additional hardware such as edge computers, gateways or sensors.

E. **Cybersecurity and Reliability**

Measures the system's resilience to digital threats and its overall operational stability, ensuring trustworthy AI through transparency, robustness and accountability.

F. **Energy Efficiency and Sustainability**

Considers the potential for AI-driven energy optimization, aligning the asset with Industry 5.0's sustainability objectives [24].

G. Human–AI Interaction and Usability

Gauges how effectively human operators can collaborate with AI-augmented machinery, reflecting the human-centered shift of Industry 5.0 towards cooperative rather than replacement-focused automation [25].

For transparency, the full operationalization of all seven domains (A–G), including quantitative criteria and weighting, is detailed in Appendix 1. This enables granular, objective assessment for smart retrofitting decisions.

6 Scoring Model and Classification

The final AIRI Index, calculated from the weighted sum of the criteria scores, is translated into a clear, qualitative classification that is presented in Table 3 with the description of each criteria score. This model provides an immediate and intuitive understanding of the equipment's current capabilities, and the strategic path required for its integration into an AI-driven ecosystem. This classification system directly supports decision-making at operational, tactical, and strategic levels by providing a standardized benchmark for evaluating assets [26]. The four distinct levels map the evolution from traditional machinery to fully autonomous, intelligent systems, reflecting the progression from reactive maintenance paradigms to predictive and prescriptive ones [27–29].

Table 3. Scoring Model and Classification

AIRI Score	Classification	Description
0 ÷ 25	Legacy/non-integrable	Very low digital maturity: old, closed-loop machines lacking sensors, data ports or communication–retrofit is usually uneconomic; replace instead
26 ÷ 50	AI-Ready (Partial)	Some digital parts (basic PLC/HMI) but only one-way data access; suitable for pilot data-gathering, but real-time bidirectional AI needs major upgrades
51 ÷ 75	AI-Integrable	Modern control, OPC-UA/MQTT support, adequate sensors; provides high-quality real-time data for predictive-maintenance, anomaly detection, optimisation; bidirectional AI possible [30]
76 ÷ 100	AI-Native/Smart Equipment	State-of-the-art, fully open architecture with built-in edge computing; ready for digital twins, prescriptive maintenance, self-learning, and resilient AI operation [31]

The quantitative thresholds defining the four readiness levels (Legacy, AI-Ready, AI-Integrable, and AI-Native) were deliberately determined by dividing the total potential score (0 ÷ 100) into four distinct, linearly ascending quartiles. This systematic division, resulting in the ranges 0 ÷ 25, 26 ÷ 50, 51 ÷ 75, and 76 ÷ 100, serves several methodological and practical purposes.

Methodologically, segmenting the assessment scale into explicit, quantifiable ranges is consistent with established best practices in Capability Maturity Model Integration (CMMI) theory [32] and multi-criteria decision-making (MCDM) frameworks [33], which posit that complex capabilities can be effectively mapped onto discrete maturity levels for systematic improvement. The use of distinct thresholds provides the essential mathematical boundary necessary to translate the composite, weighted score into actionable, qualitative strategic guidance.

Practically, this segmented approach ensures that the framework functions as an objective, intuitive decision-support tool for. By setting clearly defined ranges, the model achieves the following:

- **Standardization and Benchmarking:** a linear, four-band progression makes the numeric result comparable across assets, lines or sites, supporting clear prioritization of digital-transformation investments.
- **Transparency:** Fixed quartile thresholds give an instantly understandable interpretation of the weighted-sum score, aligning with MCDA weighting principles [33].
- **Actionable Roadmap:** Each band triggers a specific set of investment actions; for example, moving past the 50-point cut-off signals the shift from basic data collection (AI-Ready) to full-scale, bidirectional AI deployment (AI-Integrable).

7 Implementation Procedure

Figure 1 depicts a concise three-phase and six-steps workflow for evaluating AI integrability at the asset level integrated within an industrial manufacturing system.

Phase 1–Assessment (Technical Audit): data are collected and asset schematics are produced, followed by evidence-based scoring of 21 criteria (domains A-G) on a $0 \div 5$ scale, generating raw domain scores Si.
Phase 2–Aggregation (Multi-Criteria Decision Analysis): an Analytic Hierarchy Process (AHP) is applied; pairwise comparisons of the domains yield a consistency ratio $CR < 0.01$ and normalized weights Wi, which are combined in the weighted-sum Eq. (1).
Phase 3–Tier Mapping & Gap Analysis: the index is mapped to four maturity tiers (Legacy $0 \div 25$, AI-Ready $26 \div 50$, AI-Integrable $51 \div 75$, AI-Native $76 \div 100$) and a final report provides gap-analysis recommendations aligned with Industry 4.0, and 5.0 principles.

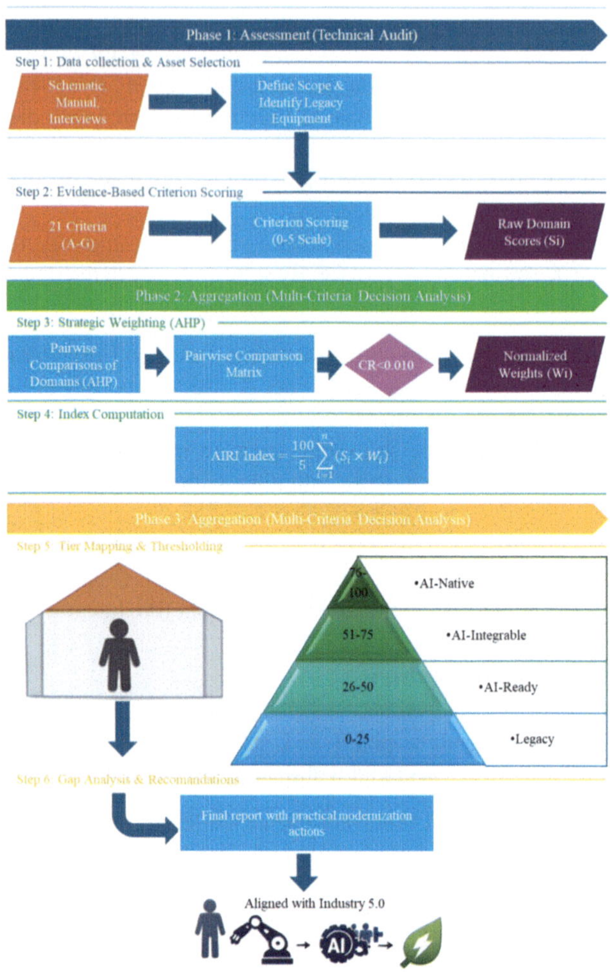

Fig. 1. AIRI workflow for AI-readiness assessment

8 Validation and Use Cases

The first study is about the multi-box gripper (metallurgical sector)–it achieved an AIRI Index of 72.5 / 100, classifying the asset as AI-Integrable (51 ÷ 75 quartile). The main technological constraint identified is that the PLC supplies only raw data and lacks native OPC-UA or MQTT interfaces, preventing the bidirectional communication required for AI-driven analytics. To remedy this, a targeted smart-retrofit is recommended: the installation of an industrial edge gateway to convert PLC signals into standard OPC-UA/MQTT protocols. The methodology further notes that future steps should include a sensitivity analysis of domain weights and the calculation of inter-rater reliability (ICC or Cohen's Kappa) to confirm the robustness of the AI-Integrable classification.

The second study is about the chain conveyor (automotive industry)–it obtained an AIRI score of 66.0 / 100, also placing it in the AI-Integrable (51 ÷ 75 quartile) range. The asset shows an exceptionally high Business Value Readiness (9/10) for Predictive Maintenance, highlighting a strong strategic need for digital augmentation. However, the composite score is limited by low ratings in Data Readiness (5/10) and Infrastructure Readiness (4/10), reflecting technical obstacles such as reliance on discrete-state data. A mandatory smart retrofit is prescribed: installation of critical temperature and vibration sensors together with an industrial gateway to transform discrete data into the high-frequency time series required by AI models.

The third study is about the Leaf-Spring Transfer Unit (automotive manufacturing) – it attained an AIRI Index of 74.0 / 100, firmly classifying it as AI-Integrable. The asset benefits from strong Infrastructure Readiness (7/10) thanks to a modern Siemens S7-1200 PLC with native OPC-UA/MQTT support, and it scores a high Business Value Readiness (9/10) due to its operational criticality. The primary constraint is low Data Readiness (5/10), as monitoring relies on discrete electrical and mechanical limiters rather than continuous, high-frequency sensor data. The recommended remediation is a mandatory smart retrofit that adds vibration sensors for the geared motor and pressure/current sensors for the hydraulic pump, thereby providing the real-time data needed for advanced prognostics.

The fourth study is about the Gantry Palletizer (chemical industry)–it achieved an AIRI Index of 82.0 / 100, placing it in the highest AI-Native / Smart Equipment tier (76 ÷ 100 quartile). This high rating stems mainly from excellent Infrastructure Readiness (9/10), with the FANUC control system natively supporting OPC-UA/MQTT, eliminating the need for major hardware retrofits. The only developmental need is to improve Data Readiness (7/10). Accordingly, a minimal smart retrofit is advised: installing supplementary high-frequency vibration sensors on specific mechanical wear components to enable immediate implementation of advanced Machine-Learning models for Remaining Useful Life prediction.

The fifth study is about the Palletizing Cell (food-processing sector)–it received an AIRI Index of 86.0 / 100, firmly categorizing it as AI-Native (76 ÷ 100 quartile). The asset exhibits maximal Business Value Readiness (10/10), excellent Infrastructure Readiness (9/10) with native OPC-UA/MQTT support, and high Data Readiness (8/10) due to dense, high-frequency sensor streams (robot torque, load-cell weight). Consequently, no major hardware retrofitting is required; the optimal strategy is to proceed directly to the deployment of advanced Predictive-Maintenance and optimization algorithms, leveraging the existing rich data stream.

9 Conclusions

The paper demonstrates that the AI Integrability Readiness Index (AIRI), proposed through this paper, provides a rigorous and systematic framework for evaluating the potential of legacy industrial equipment to support AI-driven functionalities within the context of Industry 4.0, 5.0. By integrating seven essential evaluation domains into a unified, weighted index, AIRI effectively differentiates between distinct readiness levels and enables objective benchmarking across heterogeneous industrial environments.

Validation through several case studies confirms the methodology's robustness and practical value, offering engineers and decision-makers a reliable diagnostic instrument for planning targeted modernization pathways. Overall, AIRI contributes a standardized, scalable, and industry-relevant approach that facilitates informed investment decisions and accelerates the transition toward resilient, human-centric, and AI-enabled production systems.

Furthermore, the AIRI methodology fills key digital-transformation gaps by delivering more than a technical audit. It provides:

1. **A quantitative, repeatable assessment framework**–defined domains, measurable criteria and a consistent $0 \div 5$ scoring system replace subjective, hard-to-replicate maturity checks, enabling objective, comparable evaluations across assets and over time.
2. **A common language for OT-IT collaboration**–the structured report and clear classification translate technical specifications into a shared vocabulary, bridging the "language barrier" between operational and information-technology teams and fostering consensus on AI projects.
3. **A strategic tool for Industry 5.0 upgrades** – by highlighting low-scoring domains, the method helps managers prioritize investments (e.g., sensors, edge-AI modules), build robust business cases and align upgrades with human-centred, sustainable objectives.

10 Limitations and Future Work

While the preliminary validation of the AIRI methodology provides promising results, this study has several limitations that must be acknowledged. The empirical validation is currently based on a small set of assets from a narrow industrial spectrum, which limits the generalizability of the findings. Consequently, the study is currently unable to provide statistically robust conclusions that apply across all industrial sectors, equipment generations, or varying control architectures.

Furthermore, the AIRI framework has not yet undergone retrospective outcome validation. Therefore, the extent to which specific AIRI scores or recommended retrofitting actions-such as installing protocol gateways or adding vibration sensors—translate into measurable operational or economic benefits remains unconfirmed. Demonstrating a clear correlation between the index and metrics such as reduced downtime, lower retrofitting costs, or higher ROI is essential to substantiate the framework's managerial and policy relevance.

To address these shortcomings, the next phase of the research program will focus on the following validation activities:

1. **Inter-rater reliability assessment** – Multiple independent evaluators will score the same set of assets. Reliability will be quantified using ICC for the continuous $0 \div 100$ AIRI index and Cohen's κ for the ordinal domain scores [34].
2. **Sensitivity analysis of the weighting scheme** – Domain weights will be varied systematically ($\pm 10\%$ and $\pm$ 20%) to examine the impact on asset classification (Legacy $\rightarrow$ AI-Ready, etc.). The approach follows the "weight-ratio" robustness checks advocated for ordinal-information-based weighting methods.

3. **Construct and criterion validity** – A larger sample of assets (N > 40) will be collected across different industrial sectors. Exploratory and Confirmatory Factor Analyses will verify the underlying dimensionality of the seven domains [35]. Concurrent validity will be tested by correlating AIRI scores with objective retrofitting cost data, implementation time, and ROI, as recommended for industrial AI readiness metrics [36].
4. **Cross-validation with existing AI-readiness frameworks** – AIRI results will be compared against established organizational-level AI-readiness indices (e.g., AI Singapore's AIRI, AI Readiness Index) to demonstrate convergent validity and to position AIRI as an asset-level complement [37].
5. **Documentation of the validation protocol** – All validation steps will be recorded in a publicly accessible protocol repository, following the best-practice checklist for methodological rigor in quantitative research [34].

By systematically applying these procedures, the AIRI instrument will satisfy the methodological standards outlined in contemporary quantitative-methods [37] and will provide robust, reproducible evidence of its reliability and predictive power.

Acknowledgments. This work was supported in part by a grant of the Ministry of Research, Innovation and Digitization, CCCDI-UEFISCDI, project number PN-IV-P7-7.1-PED-2024-0959, within PNCDI IV.

References

1. AI READINESS INDEX. https://airi.psdc.org.my/. Last accessed 24 Nov 2025
2. Zeller, V., Hocken, C., Stich, V.: Acatech Industrie 4.0 maturity Index–A multidimensional maturity model. In: Moon, I., Lee, G.M., Park, J., Kiritsis, D., Von Cieminski, G. (eds.) Advances in Production Management Systems. Smart Manufacturing for Industry 4.0, vol. 536, pp. 105–113. Springer International Publishing, Cham (2018). https://doi.org/10.1007/978-3-319-99707-0_14
3. Cha, C.-W., Lee, J., Jun, S., Kang, K., Chang, T.-W.: A Maturity model for diagnosing the capabilities of smart factory solution suppliers and its pilot application. Systems **11**, 569 (2023). https://doi.org/10.3390/systems11120569
4. OECD: Introducing the OECD AI capability indicators. OECD Publishing (2025). https://doi.org/10.1787/be745f04-en
5. Bueno, A., et al.: Advancing towards industry 4.0: a maturity model for smart shop floor control. Int. J. Prod. Econ. **282**, 109538 (2025). https://doi.org/10.1016/j.ijpe.2025.109538
6. Amaral, A., Peças, P.: A framework for assessing manufacturing SMEs Industry 4.0 maturity. Appl. Sci. **11**, 6127 (2021). https://doi.org/10.3390/app11136127
7. Tursunbayeva, A., Chalutz-Ben Gal, H.: Adoption of artificial intelligence: a TOP framework-based checklist for digital leaders. Bus. Horiz. **67**, 357–368 (2024). https://doi.org/10.1016/j.bushor.2024.04.006
8. Horvat, D., Heimberger, H.: AI readiness: an integrated socio-technical framework. In: Deschamps, F., Pinheiro de Lima, E., Gouvêa da Costa, S.E., Trentin, M.G. (eds.) Proceedings of the 11th International Conference on Production Research–Americas. pp. 548–557. Springer Nature Switzerland, Cham (2023). https://doi.org/10.1007/978-3-031-36121-0_69

9. Mocan, B., Fulea, M., Olaru, M., Buchmüller, M.: From intuitive programming of robotic systems to business sustainability of manufacturing SMEs. Amfiteatru Econ. **18**(41), 215–231 (2016). https://www.econstor.eu/handle/10419/168997
10. Sharp, M., Dadfarnia, M., Sprock, T., Thomas, D.: Procedural guide for system-level impact evaluation of industrial artificial intelligence-driven technologies: application to risk-based investment analysis for condition monitoring systems in manufacturing. J. Manuf. Sci. Eng. **144**, 071008 (2022). https://doi.org/10.1115/1.4053155
11. AI READINESS INDEX. https://www.psdc.org.my/training/airi. Last accessed 24 Nov 2025
12. Fantozzi, I.C., Olhager, J., Johnsson, C., Schiraldi, M.M.: Guiding organizations in the digital era: tools and metrics for success. Int. J. Eng. Bus. Manag. **17**, 18479790241312804 (2025). https://doi.org/10.1177/18479790241312804
13. Asmae, A., Chaimae, A., Mohammed, A.: Bridging the gap between Industry 4.0 readiness and maturity assessment models: an ontology-based approach. Int. J. Adv. Comput. Sci. Appl. **16** (2025). https://doi.org/10.14569/IJACSA.2025.0160245
14. International Telecommunication Union: AI ready–analysis towards a standardized readiness framework. Report, Version 1.0. International Telecommunication Union, Geneva (2024). https://www.itu.int/dms_pub/itu-t/opb/ai4g/T-AI4G-AI4GOOD-2024-2-PDF-E.pdf
15. Alqoud, A., Schaefer, D., Milisavljevic-Syed, J.: Industry 4.0: a systematic review of legacy manufacturing system digital retrofitting. Manuf. Rev. **9**, 32 (2022). https://doi.org/10.1051/mfreview/2022031
16. Windmann, A., Wittenberg, P., Schieseck, M., Niggemann, O.: Artificial Intelligence in Industry 4.0: a review of integration challenges for industrial systems. In: 2024 IEEE 22nd International Conference on Industrial Informatics (INDIN). pp. 1–8 (2024). https://doi.org/10.1109/INDIN58382.2024.10774364
17. Boix-Cots, D., Pardo-Bosch, F., Pujadas, P.: A systematic review on multi-criteria group decision-making methods based on weights: analysis and classification scheme. Inf. Fusion. **96**, 16–36 (2023). https://doi.org/10.1016/j.inffus.2023.03.004
18. Chergui, Z., Jiménez-Martín, A.: On ordinal information-based weighting methods and comparison analyses. Information **15**, 527 (2024). https://doi.org/10.3390/info15090527
19. Dombi, J., Jónás, T.: Learning the weights using attribute order information for multi-criteria decision making tasks. Opsearch **61**, 2379–2409 (2024). https://doi.org/10.1007/s12597-024-00779-9
20. Taherdoost, H.: Analysis of Simple Additive Weighting Method (SAW) as a multiattribute decision-making technique: a step-by-step guide. J. Manag. Sci. Eng. Res. **6**, 21–24 (2023). https://doi.org/10.30564/jmser.v6i1.5400
21. Cremades, L.V., Ponsich, A.: Simple and objective determination of criteria weights for evaluating alternatives when using the Analytic Hierarchy Process. Int. J. Anal. Hierarchy Process. **16** (2025). https://doi.org/10.13033/ijahp.v16i3.1177
22. Casanovas-Rubio, M.D.M., Vinolas, B.: New method for assigning cardinal weights in multicriteria decision-making: the constant weight ratio method. Oper. Res. **24**, 28 (2024). https://doi.org/10.1007/s12351-024-00833-w
23. Vinogradova-Zinkevič, I.: Centroidous method for determining objective weights. Mathematics **12**, 2269 (2024). https://doi.org/10.3390/math12142269
24. Mocan, B., Fulea, M., Brad, E., Brad, S.: State-of-the-art and proposals on reducing energy consumption in the case of industrial robotic systems. In: Proceedings of the 2014 International Conference on Production Research–Regional Conference Africa, Europe and the Middle East, pp. 328–334. Technical University of Cluj-Napoca, Cluj-Napoca (2014). https://publica.fraunhofer.de/entities/mainwork/81318e3e-a97f-4b21-a005-9ef00ebb7ae3
25. Mocan, B., Fulea, M., Brad, S.: Designing a multimodal human-robot interaction interface for an industrial robot. In: Borangiu, T. (ed.) Advances in Robot Design and Intelligent Control.

pp. 255–263. Springer International Publishing, Cham (2016). https://doi.org/10.1007/978-3-319-21290-6_26
26. Mallioris, P., Polychroniadis, C., Moustakas, A., Mitakidis, N., Bechtsis, D.: Holistic maintenance framework for predictive and prescriptive maintenance in Industry 4.0 using artificial intelligence and natural language processing. SSRN Electron. J. (preprint) (2025). https://ssrn.com/abstract=5216697 or https://doi.org/10.2139/ssrn.5216697
27. Alaoui, A.H., Meddaoui, A., Hain, M.: AI-Driven predictive maintenance for Industry 4.0: a systematic review of models, methods, and challenges. https://www.researchsquare.com/article/rs-7618987/v1 (2025). https://doi.org/10.21203/rs.3.rs-7618987/v1
28. Oliveira, C.J.deS.: Artificial intelligence applied to the implementation of a productive reliability framework. In: Iconic Research and Engineering Journals, vol. 8, issue 5, pp. 569–579. IRE J. (Nov 2024). e-ISSN: 2456-8880
29. Mocan, B., et al.: CardioVR-ReTone—Robotic Exoskeleton for upper limb rehabilitation following open heart surgery: design, modelling, and control. Symmetry **14**, 81 (2022). https://doi.org/10.3390/sym14010081
30. Kabashkin, I., Susanin, V.: Unified ecosystem for data sharing and AI-Driven predictive maintenance in aviation. Computers **13**, 318 (2024). https://doi.org/10.3390/computers13120318
31. Zheng, H., Paiva, A.R., Gurciullo, C.S.: Advancing from predictive maintenance to intelligent maintenance with AI and IIoT. San Diego (2020). https://doi.org/10.48550/arXiv.2009.00351
32. Chrissis, M.B., Konrad, M., Shrum, S.: CMMI for development: guidelines for process integration and product improvement, 3rd Edition. Carnegie Mellon University, Software Engineering Institute's Digital Library (2011). Available at: https://www.sei.cmu.edu/library/cmmi-for-development-guidelines-for-process-integration-and-product-improvement-third-edition. Last accessed: 24 Nov 2025
33. Hwang, C.-L., Yoon, K.: Methods for multiple attribute decision making. In: Multiple Attribute Decision Making: Methods and Applications A State-of-the-Art Survey, vol. 186, pp. 58–191. Springer Berlin Heidelberg, Berlin, Heidelberg (1981). https://doi.org/10.1007/978-3-642-48318-9_3
34. Kerschbaumer, S., et al.: VALID: a checklist-based approach for improving validity in psychological research. Adv. Methods Pract. Psychol. Sci. **8**, 25152459241306430 (2025). https://doi.org/10.1177/25152459241306432
35. Lambert, L.S., Newman, D.A.: Construct development and validation in three practical steps: recommendations for reviewers, editors, and authors. Organ. Res. Methods **26**, 574–607 (2023). https://doi.org/10.1177/10944281221115374
36. Marra, M., Nielsen, B.B.: Research methodology: best practices for rigorous, credible, and impactful research: herman Aguinis, Sage Publications, 2024, ISBN 978-1071871942. J. Int. Bus. Stud. **56**, 950–952 (2025). https://doi.org/10.1057/s41267-025-00774-2
37. Kalkbrenner, M.T.: Guidelines and recommendations for writing a rigorous quantitative methods section in counseling and related fields. Prof. Couns. **12**, 217–231 (2022). https://doi.org/10.15241/mtk.12.3.217

Explainability of Artificial Intelligence Models in Production Management Decision-Making: A Conceptual Model and Implications

Kinga Kulawik[1(✉)] and Anna Burduk[2]

[1] Faculty of Computer Science and Telecommunications, Wrocław University of Science and Technology, Wrocław, Poland
263920@student.pwr.edu.pl

[2] Faculty of Mechanical Engineering, Wrocław University of Science and Technology, Wrocław, Poland

Abstract. In the literature on Industry 4.0 and 5.0, the applications of machine learning algorithms for process optimization are widely described; however, few studies consider the perspective of the user who makes operational decisions based on these algorithms. Explainability of artificial intelligence models is understood as the system's ability to present, in a human-understandable way, the premises, mechanisms, and limitations underlying the generated predictions or decision recommendations. In the context of manufacturing management, explainability serves as a bridge between advanced analytical methods and managerial practice, enabling informed use of model outputs under conditions of process variability, time pressure, and accountability for decision outcomes. This article proposes a conceptual model describing the relationships between explanation quality, decision-maker trust, prediction utility, and decision effectiveness. It also presents dimensions for evaluating explainability specific to the production environment and discusses implications for management, auditing, and organizational safety. The work provides a theoretical foundation for further research on XAI metrics and proposes a framework for implementing transparent AI systems in manufacturing enterprises.

Keywords: Explainable Artificial Intelligence · Industrial AI · Manufacturing Process Optimization · Model Transparency · Decision Effectiveness

1 Introduction

The dynamic development of artificial intelligence (AI) technologies has led to their widespread application in the manufacturing sector, encompassing predictive maintenance, scheduling, quality control, inventory management, and material flow optimization. Alongside automation, the importance of decision support systems (DSS) is growing, as they generate recommendations for operators, engineers, and production managers. These users increasingly rely on the outputs of complex machine learning models, which directly affect operational costs and the continuity of plant operations

J. Trojanowska et al. (Eds.): MANUFACTURING 2026, LNME, pp. 96–109, 2026.
https://doi.org/10.1007/978-3-032-21654-0_8

by initiating decisions related to production planning, downtime scheduling, resource allocation, and maintenance interventions [1]. Despite the growing dependence on decision support systems—driven by the increasing complexity of production processes, the need to process large volumes of data, and the pressure to minimize costs and downtime—decisions are still ultimately made by humans, who must assess the reliability and sensibility of the predictions provided [2].

One of the key challenges remains the "black box" phenomenon, meaning the inability to understand how a model arrives at a particular result. In a manufacturing environment, this can increase uncertainty, delay responses, and reduce the effectiveness of actions taken. Oversimplified interpretations of models may lead to their incorrect use and unintended operational consequences. Therefore, explainability becomes a crucial element connecting the technological layer of AI systems with user needs and organizational processes. Explainable AI (XAI) provides tangible benefits for decision-makers in manufacturing: it increases trust in recommendations, enables better risk management, optimizes schedules and resource allocation, allows faster responses to disruptions, and supports employee skill development. Consequently, operational decisions become more informed, efficient, and safe.

This article attempts to systematize the role of XAI from the perspective of production management. The proposed conceptual model identifies key relationships between explanation quality, risk perception, user trust, and decision effectiveness. The goal of the article is to develop a theoretical conceptual model of XAI for manufacturing. The model demonstrates how explainability affects decisions, trust, decision quality, and risk; identifies important dimensions of explanation quality; and proposes directions for designing transparent AI systems in industrial environments. Developing a conceptual XAI model for manufacturing enables managers to make better decisions, increases trust in AI systems, reduces the risk of operational errors, optimizes processes and resources, and supports employee skill development, resulting in more efficient and safer production management.

Section 1 presents a review of research on explainable AI systems in the context of decision support in manufacturing environments, with particular emphasis on the role of the user and the quality of explanations. Section 2 focuses on the development of implementation and management frameworks for explainable AI systems in manufacturing enterprises, including the presentation of the author's conceptual model and its components. Section 3 provides an example of applying the model in the context of operator staffing decisions, illustrating the relationships between explanation quality, user perception, and decision effectiveness. Section 4 discusses the limitations of the proposed approach and its implications for further research. The article concludes with a summary of the main findings and suggests directions for future studies on the explainability of AI systems in manufacturing.

2 Review of Research on Explainable Artificial Intelligence Systems in the Context of Decision Support in Manufacturing

In recent years, the use of artificial intelligence (AI) in production decision support systems has intensified, driven by the increasing complexity of manufacturing processes, greater data availability, and pressure to improve efficiency and resilience. Machine learning models are widely applied in areas such as predictive maintenance, production scheduling, quality control, and workforce allocation, where they generate recommendations that directly influence operational and strategic decisions [3, 4]. However, many of the most accurate models—especially deep learning and ensemble methods—exhibit low transparency, raising concerns regarding trust, accountability, and risk management in industrial environments [5, 6].

In response to the "black box" problem, explainable artificial intelligence (XAI) has emerged as a key research direction aimed at making model behaviors understandable to decision-makers without significantly compromising predictive performance [7]. Early XAI research focused primarily on technical aspects, proposing post hoc explanation techniques such as SHAP (SHapley Additive ExPlanations), LIME (Local Interpretable Model-agnostic Explanations), and feature attribution methods to interpret complex models [8, 9]. Later studies demonstrated that these techniques can effectively identify process parameters that influence production tasks, including fault diagnosis, quality forecasting, and energy consumption optimization [10, 11].

More recent publications extend the discussion beyond technical interpretability, emphasizing the role of XAI in human-AI collaboration and decision-making effectiveness. Empirical studies indicate that explanations improve user understanding, reduce cognitive uncertainty, and increase trust in AI recommendations, particularly in time-critical manufacturing environments [12, 13]. At the same time, many authors warn that poorly designed or unstable explanations can lead to overreliance on AI outputs or, conversely, systematic rejection of model recommendations, undermining potential performance gains [14].

An increasing number of studies highlight that explanation quality should be evaluated in relation to the operational context and user role. Research on intelligent manufacturing shows that explanations relevant to machine operators differ significantly from those required by production managers or process engineers, who focus more on aggregated risk indicators, system-level dependencies, and economic consequences [15, 16]. Consequently, the latest frameworks propose multidimensional criteria for evaluating XAI, including comprehensibility, feasibility, consistency, and timeliness, specifically tailored to industrial decision-making environments [17].

Despite this progress, the literature identifies several open challenges. First, most existing studies are based on laboratory experiments or retrospective case analyses, while large-scale empirical validation in real production environments remains limited. Second, there is no consensus on standardized metrics for evaluating explanation quality and its impact on decision outcomes. Finally, the integration of XAI with organizational management structures, audit procedures, and production management processes is still insufficiently explored [3, 7, 13]. These gaps motivate the development of conceptual models that explicitly link explanation quality, user perception, risk assessment, and decision effectiveness in production systems.

3 Framework for the Implementation and Management of Explainable Artificial Intelligence Systems in Manufacturing Systems

The proposed conceptual XAI model in production management comprises a set of interrelated components that describe how the explainability of artificial intelligence models influences the decision-making process, risk management, organizational culture, and the overall effectiveness of production processes. The model assumes that decisions within an enterprise are not the result of algorithmic outputs alone, but rather the outcome of collaboration between an AI system and a human decision-maker, who ultimately evaluates the available information and takes action. Within this approach, not only the predictive output itself is important, but also how the user understands, interprets, and assesses the credibility of that output.

The model consists of four main blocks: characteristics of the production system, the quality of explanations generated by AI, user perception, and decision effectiveness. The characteristics of the production process include the variability of manufacturing process parameters associated with the execution of individual production orders, the possibility of unplanned downtime (failures), material transport times between manufacturing operations, the number of production units per batch, and the absence of operators possessing key competencies required for specific operations. This approach makes it possible to more accurately capture the factors affecting the functioning of the production system and the role of AI systems in supporting operational decisions. In an environment with such characteristics, the AI system operates as a supportive element rather than a replacement for human decision-makers, which implies that its operation must be transparent and understandable.

The second block of the model encompasses the properties of explanations: comprehensibility, operational adequacy, stability, credibility, and timeliness. Comprehensibility refers to the logical linkage between the explanation and the actual production process; operational adequacy denotes the ability to undertake concrete actions based on the explanation; stability refers to the consistency of explanations under small changes in input data; credibility reflects the alignment of the explanation with both the true behavior of the model and the user's domain expertise; and timeliness concerns the delivery of explanations within a time frame that enables effective response in the production process. These properties influence user perception by shaping trust in the model. Trust is not a one-dimensional construct; it encompasses both confidence in the correctness of predictions and belief in the model's alignment with expert knowledge and the realities of production processes. A user may accept the model's output if it is perceived as consistent with their experience and available data, even without full understanding of the algorithm's technical details.

The third block comprises user perception, which serves as a mediator between explanation quality and decision effectiveness. User perception includes risk assessment, cognitive load, understanding of key relationships, and elements of model interpretation. High-quality explanations reduce perceived risk, enabling faster and more accurate decision-making. Conversely, unclear or operationally inadequate explanations increase concerns about potential model errors, leading to delayed responses or outright rejection of system recommendations.

The fourth block concerns decision effectiveness, which results from human–AI interaction. This effectiveness can be measured in operational terms such as reduced failure rates, shorter downtime, improved production stability, greater process predictability, and lower costs of errors. Process predictability refers to the ability to forecast production performance and plan actions in advance, thereby supporting accurate operational decisions and effective use of AI system recommendations. The model assumes that high-quality explanations lead to higher decision quality by increasing trust, reducing uncertainty, and enabling better interpretation of model outputs. In contrast, low-quality explanations may result in the so-called "black box" effect, in which users either over-rely on the model or, conversely, ignore it entirely.

Based on these four blocks, the model defines the following relationships:

1. Explanation quality → level of trust,
2. Trust → risk assessment and cognitive load,
3. Risk assessment → decision effectiveness,
4. Decision effectiveness → the enterprise's operational performance.

The model has a framework character (Fig. 1) and can be extended with additional elements such as regulatory compliance, the technological maturity of the enterprise, or the level of autonomy of production systems. Its most important feature is that explainability is no longer treated merely as a technical component but becomes an integral part of the management process, exerting a tangible influence on user behavior and organizational outcomes.

3.1 Example of Applying the Conceptual Model–Operator Staffing Decisions

A production system was considered in which orders require operation by workers possessing specific competencies and certifications. Under conditions of variable production plans and employee absences, an artificial intelligence system supports the shift supervisor in making decisions regarding the staffing of workstations.

Block 1–Characteristics of the production process.

The process is characterized by limited availability of operators with key competencies, variability in workstation workloads, and a significant impact of staffing decisions on the quality and pace of order execution. Incorrect staffing decisions may lead to decreased productivity, an increased number of defects, or even the need to stop the production line.

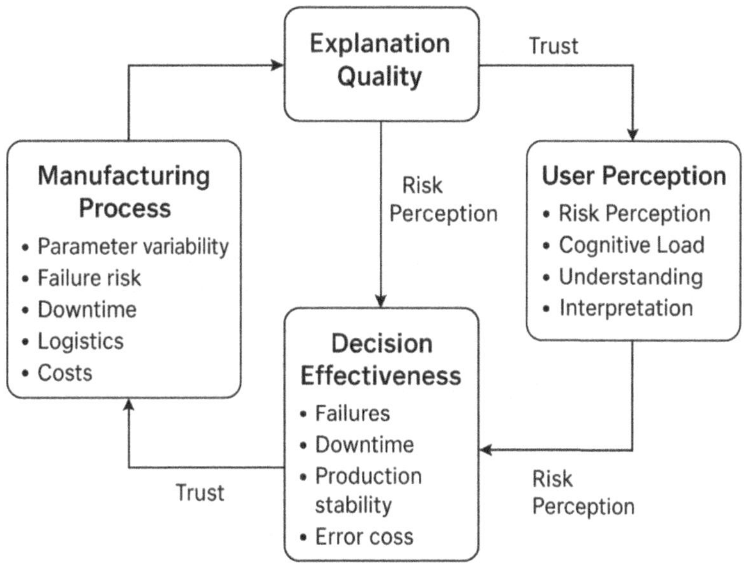

Fig. 1. Conceptual model framework

Block 2–Explanation quality.

The AI system recommends a modification of the production line staffing, indicating that the increased risk of delays and quality defects results from the absence of an operator certified for a critical operation and from the limited experience of the substitute worker at the given workstation. The explanation refers to parameters that can be verified by the decision-maker (competencies, work history, quality data), is stable over time, and is provided at the shift planning stage, ensuring both its operational adequacy and timeliness.

Block 3–User perception.

Thanks to the explanation, the shift supervisor is able to assess the validity of the recommendation and confront it with their own knowledge of the team. Understanding the rationale behind the recommendation reduces the perceived risk of the decision and increases trust in the system, while still leaving room to account for contextual factors such as the current availability and condition of employees.

Block 4–Decision effectiveness.

As a result, a decision is made to modify the staffing configuration or change the sequence of operations, leading to a more stable production flow, a reduction in human errors, and better utilization of team competencies. Decision effectiveness is reflected in increased predictability of order execution and reduced costs resulting from unplanned disruptions.

3.2 Dimensions of Explanation Quality

Explanation quality is one of the key determinants of the effectiveness of XAI systems and constitutes a fundamental element of the proposed model. To properly understand how explanations influence the decision-making process, it is necessary to analyze their individual dimensions.

The first dimension is comprehensibility, which means that the user is able to logically relate the content of the explanation to the reality of the production process. Comprehensibility goes beyond mere textual readability; it also includes appropriate alignment of language, information format, and level of detail with the decision-maker's competencies. In practice, this means that a production line operator should receive explanations referring to material quality parameters or machine settings, whereas a production manager may require more aggregated information related to trends and downtime risk.

The second dimension is operational adequacy, defined as the extent to which the explanation refers to parameters that the user can realistically control. In a production environment, an explanation is meaningful only if it enables action. If a model indicates that a failure is caused by a factor that is not accessible to the decision-maker (e.g. an internal diagnostic indicator hidden within the system), the explanation has limited value. Operational adequacy ensures that explanations are not merely a decoding of the model but provide genuine support for decision-making.

The third dimension is stability of explanations. Stability implies that small changes in input data do not lead to fundamentally different explanations. Models that generate unstable explanations produce chaotic signals, which may result in incorrect decisions. Stability is particularly important in manufacturing, where data are subject to natural variability that does not necessarily indicate changes in the technical condition of equipment.

The fourth dimension is credibility, which encompasses both consistency of the explanation with the data and alignment with the user's expert knowledge. Users evaluate explanations through the lens of experience; if model indications contradict their intuition or process observations, trust in system recommendations is more difficult to establish. Credibility also implies that explanations reflect the actual behavior of the model rather than being artificially generated "justifications."

The final dimension is timeliness. In production environments, decisions must often be made rapidly; therefore, explanations need to be delivered in real time. Delays reduce their operational usefulness and, in extreme cases, may lead to irreversible errors such as equipment damage or loss of an entire production batch.

Each dimension is defined and linked to measurable indicators (Table 1).

Table 1. Operationalization of Explanation Quality

Dimension	Definition	Example Metrics
Comprehensibility	Degree of user understanding	Comprehension test (%), Likert scale
Operational adequacy	Actionability of explanation	Actionability score (0–3)
Stability	Consistency under small input changes	Kendall's Tau of attribution ranks
Credibility	Alignment with expert knowledge	Expert agreement rate (%)
Timeliness	Delivery within operational window	Latency (s), threshold compliance

The selection of metrics follows from the nature of decisions made in production environments and from established research practices in explainable artificial intelligence and decision support systems. **Comprehensibility** is assessed using objective comprehension tests and subjective rating scales, as this combination allows actual understanding of an explanation to be distinguished from the user's declared confidence. **Operational adequacy** is measured through an actionability score, reflecting the fact that in industrial practice the value of an explanation depends on whether it enables a decision-maker to undertake a concrete action. **Stability** is evaluated using Kendall's Tau coefficient, which enables verification of the consistency of feature attributions under small perturbations of input data. **Credibility** is based on agreement with domain expert assessments, reflecting how users validate explanations against process knowledge and experience. **Timeliness** is measured through response latency and compliance with operational time thresholds, as explanations delivered outside the decision window lose their practical value.

3.3 Organizational Implications

The introduction of explainability mechanisms into AI systems has significant implications for organizational structure, employee competencies, and operational procedures in manufacturing enterprises. First and foremost, organizations must perceive themselves not only as users of analytical tools but also as entities responsible for interpreting and monitoring the recommendations generated by the systems. This means that the implementation of XAI should be planned as a cross-functional initiative involving IT, maintenance, quality control, and production planning and control departments. The role of traditional IT teams shifts from being de facto technology installers to partners responsible for the design and quality of delivered explanations. This necessitates the creation of new roles, such as an AI Steward, responsible for auditing explanations, and an Interpretability Analyst, who translates technical model outputs into operational terms.

Another dimension of organizational impact is the change in competency requirements. Implementing XAI requires parallel training initiatives: operators and managers must develop the ability to interpret explanations and assess their relevance; data specialists must be capable of designing explanations suitable for different user groups and monitoring their stability. This requires the development of training programs that integrate process knowledge with the fundamentals of model interpretation. Training should be practical and grounded in operational examples—for instance, analyzing cases in which changing a single machine parameter affected quality metrics, while simultaneously showing how different XAI methods (local vs. global) provide different perspectives.

Organizations should also introduce changes to operational procedures and work regulations. Explanations should be formally incorporated into procedures as part of the documentation of critical decisions, facilitating audits and subsequent incident analyses. For example, in a procedure responding to a predictive maintenance alarm, the step "verify explanations" could become a prerequisite for initiating a service intervention or deciding to stop the production line. This, in turn, requires establishing a trust threshold for explanations and an escalation mechanism for ambiguous explanations. From an HR and personnel policy perspective, transparency in AI operations promotes trust and acceptance of the technology, but only if employees perceive tangible benefits (e.g., fewer failures, fewer routine tasks) and have influence over the interpretation of results.

Finally, organizational implications also involve changes in relationships with AI solution providers and service contracting. Enterprises should require suppliers to provide specifications on how explanations are generated, their format, delivery timing, and guaranteed stability. Service agreements may include provisions on explanation quality (e.g., availability of local explanations for critical predictions) and conditions for transferring supplier audit mechanisms. Such practices facilitate the integration of XAI into governance and reduce the risk of "black box" issues in externally provided systems.

The dimensions of explanation quality can be treated as a set of interrelated parameters describing the usefulness of an explanation in the decision-making process. Each dimension serves a distinct function: comprehensibility and operational adequacy determine the practical applicability of the explanation, stability and credibility shape trust over time, and timeliness determines its operational value. Conceptually, explanation quality can be described as a function of these five dimensions, whose combined effect influences the effectiveness of production decisions.

3.4 Evaluation and Validation Roadmap for Explainable AI in Production Decision-Making

This section proposes a structured evaluation and validation roadmap that translates the conceptual relationships between explanation quality, user perception, and decision effectiveness into measurable and testable constructs. The roadmap is designed to support future empirical research as well as pilot implementations of explainable AI (XAI) systems in real manufacturing environments.

The validation roadmap is organized into three sequential levels: (i) explanation quality validation, (ii) user-centered decision process validation, and (iii) operational performance validation. This layered approach reflects the logic of the conceptual model, in which explanation quality influences user perception, which in turn affects decision effectiveness and operational outcomes.

Level I: Validation of Explanation Quality

The first level focuses on validating the intrinsic quality of explanations generated by the AI system, independent of final production outcomes. The objective is to assess whether explanations meet predefined quality thresholds aligned with the five dimensions proposed in Sect. 2.2: comprehensibility, operational adequacy, stability, credibility, and timeliness. Evaluation at this level should combine expert-based assessment and controlled experimental techniques. Comprehensibility can be measured using structured comprehension tests or scenario-based questionnaires administered to different user groups (e.g., operators, supervisors, production managers). Operational adequacy may be evaluated through actionability analysis, assessing whether explanations reference controllable parameters and lead to clearly identifiable decision options. Stability can be quantified by measuring the consistency of explanation outputs under small perturbations of input data (e.g., using rank correlation metrics such as Kendall's Tau). Credibility should be assessed through expert agreement studies, comparing explanation content with domain expert judgments. Timeliness can be validated by measuring explanation latency relative to predefined operational decision windows. The outcome of Level I validation is a multidimensional explanation quality profile that allows organizations to identify weak dimensions and refine explanation mechanisms before deployment in decision-critical environments.

Level II: Validation of User Perception and Decision-Making Effects

The second level examines how validated explanations influence human decision-makers. This stage operationalizes the mediating role of user perception highlighted in the conceptual model. The primary goal is to verify whether higher explanation quality leads to improved trust calibration, reduced cognitive load, and more accurate risk assessment. Empirical validation at this level may be conducted using controlled experiments, simulation-based decision tasks, or field studies in pilot production settings. Key dependent variables include perceived trust in the AI system, decision confidence, perceived risk, cognitive workload (e.g., measured via NASA-TLX or similar instruments), and decision time. Comparative studies can be designed in which identical decision scenarios are presented with varying levels or formats of explanation quality. Importantly, this level should also evaluate failure modes, such as overreliance on AI recommendations or systematic rejection of model outputs, to assess whether explanations support appropriate trust rather than blind acceptance. The results of Level II validation provide evidence for the hypothesized links between explanation quality, user perception, and decision behavior.

Level III: Validation of Operational and Organizational Impact

The third level focuses on validating the impact of explainable AI on actual production performance and organizational outcomes. This level closes the loop of the conceptual model by linking decision effectiveness to measurable operational indicators. Validation may be conducted through longitudinal case studies, before–after comparisons, or controlled pilot deployments. Relevant performance indicators include production stability, downtime frequency and duration, defect rates, schedule adherence, resource utilization efficiency, and costs associated with operational errors. In addition to quantitative metrics, qualitative indicators—such as auditability of decisions, traceability of responsibility, and acceptance of AI systems by employees—should be incorporated. Crucially, Level III validation should also assess the sustainability of XAI benefits over time. This includes monitoring whether explanation quality remains stable under data drift, whether user trust evolves appropriately with system performance, and whether organizational procedures effectively integrate explanation-based decision support.

Integration and Iterative Improvement

The proposed roadmap is inherently iterative. Findings from higher validation levels may reveal deficiencies in explanation design or organizational integration, necessitating refinement at earlier stages. For example, operational incidents identified at Level III may indicate insufficient operational adequacy or credibility at Level I, prompting redesign of explanation mechanisms. By providing a structured and measurable validation pathway, the roadmap addresses one of the key gaps identified in the literature: the lack of standardized approaches for evaluating explainable AI in real production environments. It also ensures that explainability is treated not as an abstract technical property, but as a verifiable contributor to effective, accountable, and safe decision-making in manufacturing systems.

4 Discussion

The proposed conceptual model and the defined dimensions of explanation quality provide a framework for understanding the impact of XAI, while simultaneously revealing the complexity of the issue and the limitations of current knowledge. First, there is a risk that implementing explanations without concurrently shaping organizational procedures will yield limited benefits. Simply providing technical explanations is insufficient if decision-makers do not understand their operational significance or if the explanations are inconsistent with expert knowledge. Furthermore, low-quality or unstable explanations can introduce misinformation—for example, situations in which local explanations for individual predictions point to completely different factors in closely related observations, potentially leading to contradictory decisions.

A second significant issue is the conflict between transparency and intellectual property protection. In industry, commercial models are often used, the details of which constitute a competitive advantage. Requiring full explainability may conflict with the desire to protect algorithm developers' know-how. An intermediate solution could involve standards and audit mechanisms, in which independent auditors verify the correctness and adequacy of explanations in a secure environment without revealing all internal model mechanisms.

The third area requiring reflection is metrics and validation. The proposed model highlights dimensions of explanation quality and their consequential effects, but practical application requires operationalization of these concepts. Key research questions arising from the model include: which specific measures of comprehensibility, operational adequacy, or stability are relevant in a given production context, how to measure them, and what thresholds constitute "sufficient" explanation quality for different organizational roles. Empirical research is essential here, suggesting field experiments, case studies, and user-centered studies to verify the impact of different forms of explanations on the speed and accuracy of decisions as well as on actual operational performance indicators.

The fourth limitation concerns the influence of the data environment. AI models are only as good as the data on which they are trained; explanations provided under low-quality data conditions may be misleading. Hence, it is necessary to combine XAI practices with data quality control, data drift monitoring, and early-warning mechanisms for prediction quality degradation. In practice, this implies the complementarity of XAI and data governance—without a strong data policy, explanations may merely mask fundamental problems.

Despite these limitations, the proposed approach has both heuristic and practical value. It enables the identification of intervention points where an organization can improve decision effectiveness by adjusting the format of explanations, training personnel, adapting operational procedures, or renegotiating supplier agreements. However, an iterative approach is required: XAI implementation should be tested in controlled scenarios, monitored, and subsequently adjusted based on empirical data.

The proposed conceptual model extends existing research on explainable artificial intelligence in manufacturing by explicitly positioning explainability as a mediator between model output and decision effectiveness. While prior studies primarily focus on the technical interpretability of machine learning models and the classification of XAI methods [5, 12, 15], this work complements the literature by embedding explanation quality within the human decision-making process under real production constraints.

In contrast to existing explainable AI frameworks in manufacturing, which primarily emphasize technical interpretability or assume a direct link between transparency and trust, the proposed framework explicitly positions explainability as a decision-centered, mediating construct between AI model outputs and operational performance. The novelty of the approach lies in decomposing the impact of explainability into distinct stages—explanation quality, user perception, and decision effectiveness—thereby capturing the role of risk perception, cognitive load, and understanding in human–AI interaction. Furthermore, the framework introduces production-specific dimensions of explanation quality, such as operational adequacy and timeliness, which are largely absent from general-purpose XAI models but are critical under real manufacturing constraints. Finally, by linking explanation quality to measurable operational outcomes (e.g., downtime, production stability, and error-related costs) and by proposing a structured evaluation and validation roadmap, the framework extends prior conceptual models toward empirical testability and practical applicability in industrial decision-making contexts.

In contrast to frameworks that assume a direct relationship between transparency and trust [14], the model highlights the role of user perception, including risk assessment and cognitive load, as intermediate factors shaping decision outcomes. This perspective is consistent with findings that poorly designed explanations may either lead to overreliance on AI recommendations or to their systematic rejection [15]. By distinguishing explanation quality from its perceived usefulness, the model addresses an important gap in current XAI research.

Moreover, the identified dimensions of explanation quality partially align with criteria proposed in earlier theoretical and empirical studies [7, 16], but they are extended by incorporating production-specific requirements such as operational adequacy and timeliness. These aspects are often underrepresented in experimental evaluations of XAI, despite being critical in time-sensitive manufacturing environments.

Finally, by discussing organizational roles, procedural integration, and auditing mechanisms, the model contributes to emerging governance-oriented perspectives on industrial AI [17]. Rather than treating explainability as a purely technical feature, the proposed framework emphasizes its role in supporting accountable and effective decision-making in manufacturing systems.

5 Conclusions

The article proposes a conceptual framework that treats the explainability of AI models as an integral component of the decision-making process in manufacturing enterprises. Explainability is not merely a technical aspect but a factor shaping user trust, risk perception, and, ultimately, the effectiveness of operational decisions. The identified dimensions of explanation quality—comprehensibility, operational adequacy, stability, credibility, and timeliness—provide practical criteria that organizations can use when designing, evaluating, and implementing XAI solutions.

For managerial practice, several concrete recommendations emerge from the analysis. First, XAI implementations should be conducted as cross-functional initiatives with clearly defined roles and responsibilities (e.g., AI Steward, Interpretability Analyst). Second, organizations should develop practical training programs that combine process knowledge with the ability to interpret explanations. Third, operational procedures must formally integrate an explanation verification step prior to making critical decisions, along with escalation mechanisms for ambiguous recommendations. Fourth, procurement and contracting policies should include requirements regarding the format, availability, and quality of explanations provided by AI system suppliers.

From a scientific perspective, the article identifies directions for further research. The most urgent tasks include developing operational metrics for explanation quality dimensions, conducting experimental studies on the impact of different forms of explanations on operators' and managers' decisions in real production environments, and developing methodologies for auditing explanations in terms of governance and regulatory compliance. Simultaneously, it is worth investigating the influence of explainability on organizational culture and long-term team competencies, as only a holistic approach can fully leverage the potential of AI in industry.

In conclusion, AI model explainability in the context of production is not only a technical challenge but also an organizational and managerial issue. XAI implementation, if conducted according to the principles presented, can contribute to increased operational efficiency, reduced risk, and the promotion of responsible use of technology in industrial settings.

References

1. Mitchell, E.: Analyzing the impact of AI-Driven technologies on operational efficiency in Industry 4.0. Preprints (2025). https://doi.org/10.20944/preprints202503.2305.v1

2. Soori, M., Ghaleh Jough, F.K., Dastres, R., Arezoo, B.: AI-based decision support systems in Industry 4.0: a review. J. Econ. Technol. (2025). https://doi.org/10.1016/j.ject.2024.08.005
3. Nikiforidis, K., et al.: Enhancing transparency and trust in AI-powered manufacturing: a survey of explainable AI (XAI) applications in smart manufacturing in the era of Industry 4.0/5.0. ICT Express **11**(1) (2025). https://doi.org/10.1016/j.icte.2024.12.001
4. Hoffmann, R., Reich, C.: A systematic literature review on artificial intelligence and explainable artificial intelligence for visual quality assurance in manufacturing. Electronics **12**(22), 4572 (2023). https://doi.org/10.3390/electronics12224572
5. Tzionis, G., et al.: A review of explainable AI methods and their application in manufacturing systems. Discover Appl. Sci. (2025). https://doi.org/10.1007/s42452-025-07908-z
6. Chen, T.-C., Toly, T.: Explainable Artificial Intelligence (XAI) in manufacturing. In Explainable Artificial Intelligence (XAI) in Manufacturing: Methodology, Tools, and Applications (pp. 1–11). Cham: Springer International Publishing (2023). https://doi.org/10.1007/978-3-031-27961-4_1
7. Baum, D., Baum, K., Gros, T.P., Wolf, V.: XAI requirements in smart production processes: a case study. In: Explainable Artificial Intelligence. Springer (2023). https://doi.org/10.1109/RTSI64020.2025.11212395
8. Rottenwalter, G., Tilly, M., Owolabi, V.: Improving industrial injection molding processes with explainable AI for quality classification. In: Proc. IEEE 9th Forum on Research and Technologies for Society and Industry (RTSI). IEEE (2025). https://doi.org/10.1109/RTSI64020.2025.11212395
9. Velmurugan, M., Ouyang, C., Moreira, C., Sindhgatta, R.: Evaluating explainable methods for predictive process analytics: A functionally-grounded approach. arXiv:2008.07258 (2020). https://doi.org/10.48550/arXiv.2012.04218
10. Gawde, S., Patil, S., Kumar, S., Kamat, P., Kotecha, K., Alfarhood, S.: Explainable predictive maintenance of rotating machines using LIME, SHAP, PDP, ICE. IEEE Access (2024). https://doi.org/10.1109/ACCESS.2024.3367110
11. Diaz Marin, G.: Comparative analysis of explainable AI methods for manufacturing defect prediction: a mathematical perspective. In: Artificial Intelligence and Data Science, 2nd edn. Springer (2025). https://doi.org/10.3390/math13152436
12. Sofianidis, G., Rožanec, J.M., Mladenić, D., Kyriazis, D.: A review of explainable artificial intelligence in manufacturing. Trusted Artificial Intelligence in Manufacturing: A Review of the Emerging Wave of Ethical and Human Centric AI Technologies for Smart Production, pp. 93–113 (2021). https://doi.org/10.1561/9781680838770
13. Ahmad, T., Katari, P., Pamidi Venkata, A.K., Ravi, C.S., Shaik, M.: Explainable AI: interpreting deep learning models for decision support. In: Advances in Deep Learning Techniques. Springer (2024)
14. Agostinho, C., et al.: Explainability as the key ingredient for AI adoption in Industry 5.0 settings. Front. Artif. Intell. (2023). https://doi.org/10.3389/frai.2023.1264372
15. Moosavi, S., Farajzadeh-Zanjani, M., Razavi-Far, R., Palade, V., Saif, M.: Explainable AI in manufacturing and industrial cyber–physical systems: a survey. In: Advances in Artificial Intelligence Engineering. Springer (2024). https://doi.org/10.3390/electronics13173497
16. Matejova, M., Paralic, J.: A multi-criteria decision-making approach for the selection of explainable AI methods. Mach. Learn. Knowl. Extr. (2025). https://doi.org/10.3390/make7040158
17. Arunraju, C.: Explainable AI (XAI) for trustworthy and transparent decision-making: a theoretical perspective. World J. Adv. Eng. Technol. Sci. (2025). https://doi.org/10.30574/wjaets.2025.14.3.0106

Applications of Machine Learning Techniques in Asset Management of Engineering Systems: A Bibliometric Analysis

Damjan Maletič(✉), Tilen Medved, and Matjaž Maletič

Faculty of Organizational Sciences, University of Maribor, Kranj, Slovenia
damjan.maletic@um.si

Abstract. This bibliometric analysis systematically maps the scientific landscape of machine learning (ML) applications in asset management for engineering systems. It uses a corpus of 1,158 publications from 2005 to 2025. The study's methodology is based on a comprehensive asset lifecycle perspective (ALC), aligning findings with the strategic clauses of the ISO 55001 standard. The analysis reveals exponential growth in publications since 2018, confirming the field's rapid maturity and relevance. Keyword co-occurrence analysis identified seven thematic clusters, with the intellectual core overwhelmingly concentrated on the operational phase (ISO 55001, Clause 8) and specifically on predictive maintenance (PdM), fault detection, and diagnostics (Clusters 2 and 3). Citation burst analysis reinforces this focus, highlighting systematic reviews on PdM as the most influential literature defining the current research frontier. The study also identifies significant research gaps in the holistic application of ML across the entire ALC. Strategic phases such as planning, acquisition, operation and disposal (including value recovery and end-of-life management) are underrepresented in the core literature. These findings suggest that, although ML provides a highly developed set of tools for technical operations, future research must shift its focus strategically to support the strategic objectives of the ISO 55001 framework. This will facilitate a complete, value-driven ALC management system.

Keywords: Machine Learning · Asset Management · ISO 55001 · Engineering Systems · Bibliometric Analysis

1 Introduction

In modern industrial and infrastructural environments, asset management is crucial for ensuring operational efficiency, reducing costs, extending asset lifecycles, and creating value for organizations [1, 2]. Recently, there has been considerable interest in asset management research in both academia and industry. Similarly, organizations with intensive asset utilization have contributed significantly to advancing knowledge in asset management [3]. Standards help organizations align their practices with common principles, terminology, and requirements, ensuring that asset management encompasses more than the

J. Trojanowska et al. (Eds.): MANUFACTURING 2026, LNME, pp. 110–124, 2026.
https://doi.org/10.1007/978-3-032-21654-0_9

technical maintenance of individual components. According to ISO 55000, asset management is the alignment of organizational objectives with the value derived from assets throughout their lifecycle. This frames asset management as a strategic, value-driven discipline rather than a purely operational activity [4].

In asset management-related fields, a variety of Industry 4.0 (I4.0) technologies are emerging that support highly efficient organizations [5]. The integration of digital technologies across organizational functions is transforming business operations and the delivery of value to customers and stakeholders [6]. Although the term "digital transformation" is widely used, it varies significantly in practice. Essentially, it refers to strategic organizational change and internal restructuring driven by the adoption and use of new digital technologies to transform existing business models, processes, and relationships with external actors [7]. This transformation has important implications for asset management, requiring companies to reassess their approaches across different lifecycle phases. Adopting digital technologies can optimize asset performance, enhance maintenance management, improve decision-making processes, and increase value for the organization and its stakeholders [8].

Despite significant progress in standardization and process-oriented approaches, one of the remaining key challenges is integrating advanced analytical methods, particularly machine learning (ML) and artificial intelligence (AI), into asset management systems operating in real-world environments [9]. This is especially relevant for data-rich systems that rely on sensors, the Internet of Things (IoT), and condition monitoring technologies. These systems can generate large amounts of operational data, but they often lack systematic frameworks for predictive and decision-support applications [10].

The application of ML in asset management is rapidly gaining attention as organizations seek to improve performance, optimize maintenance, and enhance decision-making across engineering systems [11, 12]. Despite the growing body of research in this field, it remains highly interdisciplinary, encompassing areas such as predictive maintenance (PdM), reliability engineering, and digital technologies. This can lead to a fragmented understanding of its developments [13, 14]. Bibliometric analysis is a valuable method for systematically mapping literature, identifying influential works and authors, and uncovering emerging trends and research gaps [15, 16]. While prior studies have examined bibliometric patterns in asset management and digital technologies more broadly, there is still a lack of comprehensive analysis specifically addressing ML applications in engineering asset management. This study aims to address this gap by providing a structured overview of the scientific landscape and highlighting important research directions and opportunities for future exploration.

1.1 Study Focus: ML Applications in Engineering Asset Lifecycle

Effective asset management is essential for the competitiveness, reliability, and sustainability of engineering systems [17, 18], as well as other sectors such as critical infrastructure [19]. As industrial data has proliferated, ML algorithms have become the primary enabler for transitioning from reactive or time-based maintenance to comprehensive, data-driven management optimized across the entire asset lifecycle [11, 20].

This bibliometric analysis primarily focuses on the utilization of ML within manufacturing systems. However, the study's methodology is based on the comprehensive asset lifecycle perspective, as structured by the ISO 55001 [21] standard. Asset management is a universal discipline, and the principles underlying the application of ML for strategic decision-making (e.g., risk reduction, maintenance optimization, and demand forecasting) can be transferred across various engineering systems. Therefore, the following table includes key examples from related fields (e.g., power systems, logistics, and critical rotating equipment) to demonstrate the breadth and maturity of ML techniques available for adoption in the manufacturing sector. This selection provides a holistic view, avoiding an overly narrow or isolated interpretation.

This holistic approach emphasizes that ML serves not only technical operational goals, such as PdM (Clause 8), but also the core strategic objectives of the ISO 55001 standard. These objectives include planning (Clause 6) and actions to address risks and opportunities (Clause 6.1) and maximizing residual value through the principles of sustainable development.

Table 1 presents representative examples of critical ML applications and positions them within this framework, serving as a conceptual model for interpreting the bibliometric analysis's findings. The table illustrates how ML enables the transformation from isolated decisions to comprehensive, data-driven asset management, where optimization is achieved across every phase, from identifying a need to disposal.

Table 1. Machine learning in asset management: A full lifecycle perspective

Asset lifecycle stage	ML application / key function	ISO 55001 alignment	Supporting literature (key reference and focus)
1. Identify need	Demand and capacity forecasting	6.2.1 Asset management objectives: Aligning creation with organizational goals	Muñoz-Villamizar A, Rafavy CY, Casey J (2022) [22]; Rajora, G. L., et al. (2024) [23]: Power systems
2. Create or acquire	Risk assessment & portfolio optimization	6.1 (Balancing C-R-P & opportunities)	Ni, J., Hu, Y., & Zhong, R. Y. (2021) [24]: Price uncertainty; Husna, A. ul, Ghasempoor, A., & Amin, S. H. (2024) [25]: Clustering/Optimization
3. Operate and maintain	PdM optimization & system reliability (ML/DL)	8.1 Operational planning and control: Maximizing performance and reducing operational risk	Zhu, T., Ran, Y., Zhou, X., & Wen, Y. (2024) [26]: Deep Learning for PdM; Shahin, M., Chen, F.F., Hosseinzadeh, A. et al. (2023) [27]: Comparative ML study; Yang, C., Liu, J., Zeng, Y., & Xie, G. (2019) [28]: Real-time monitoring

(*continued*)

Table 1. *(continued)*

Asset lifecycle stage	ML application / key function	ISO 55001 alignment	Supporting literature (key reference and focus)
4. Dispose and/or replace	Value recovery, remanufacturing & recycling optimization	6.2.2 (Sustainability & Life-cycle planning)	Koulinas, G., Paraschos, P., & Koulouriotis, D. (2021) [29]: RL for recycling policies; Cherrington, M., Lu, Z., Xu, Q. et al. (2020) [30]: DL for sustainable DSS

2 Research Methodology

A systematic search was conducted. It was done following the PRISMA 2020 guidelines [31]. The goal was to identify publications related to maintenance and asset management.

2.1 Analytical Tools

Two specialized tools were applied to process the collected data for the subsequent bibliometric analysis.

- Burst Detection: Analysis of sudden increases in the frequency of keyword usage over time (burst detection) was performed using the CiteSpace software tool (https://citespace.podia.com).
- Co-occurrence and Clustering Analysis: Keyword co-occurrence, country collaboration, and clustering analyses were performed using VOSviewer (https://www.vosviewer.com/).

2.2 Search Strategy and Keyword Selection

A comprehensive search was conducted in relevant databases using a broad set of keywords to explore the application of ML techniques in asset management. The search strategy incorporated terms associated with different ML techniques, including "ML," "DL," and "artificial neural networks," to encompass the diverse ML methodologies employed in the field. Additionally, terms such as "asset management," "asset lifecycle," and "life-cycle management" were included to focus on studies related to the entire asset lifecycle, from initial acquisition to decommissioning and replacement. Keywords such as "PdM," "condition monitoring," and "reliability engineering" were also incorporated to ensure the inclusion of research on predictive techniques. The search was also extended to the manufacturing sector, where many asset management strategies are implemented, by including terms such as "manufacturing systems" and "manufacturing assets."

This search strategy was designed to identify foundational research on ML applications in asset management, as well as more specialized studies related to specific phases of the asset lifecycle. There was a particular emphasis on PdM, emerging strategies for risk assessment, portfolio optimization, and asset replacement.

2.3 Scopus Search Query

The comprehensive search strategy outlined above produced the following Scopus search string. It was executed across the title, abstract, and keywords (TITLE-ABS-KEY) fields:

TITLE-ABS-KEY (("machine learning" OR "deep learning" OR "artificial intelligence" OR "artificial neural network*") AND ("asset management" OR "asset life* cycle" OR "life-cycle management" OR "engineering assets" OR "maintenance management" OR "reliability engineering" OR "predictive maintenance" OR "condition monitoring" OR "risk assessment" OR "portfolio optimization" OR "asset replacement strategy") AND ("manufacturing" OR "manufacturing system*" OR "manufacturing asset*"))

The comprehensive filtering process involved several exclusion criteria related to publication date, language, subject area, document type, and source type. Table 2 summarizes the screening and analysis process used in this study. It details the records identified and the number of exclusions applied at each stage.

Table 2. Overview of the document selection process

Step	Stage	Details	Records (n)
Identification of new studies	Identification	Records identified from: Scopus Databases	2,549
	Screening	Records removed before screening:	
		Limited to 2005–2025	2,469
		Limited to English	2,434
		Limited to Engineering area	1,659
		Limited to conference paper, article, book chapter	1,466
		Excluded trade journals	1,460
		Records screened	1,460
		Records excluded	302
	Included	Total studies included in review	**1,158**

The bibliographic data were extracted on November 17, 2025. After the selection process, the final corpus of 1,158 publications was categorized into three types of documents: Articles (545; 47.06%), conference proceedings (482; 41.62%), and book chapters (131; 11.31%).

To ensure the dataset's relevance and consistency, this study focuses on publications classified under engineering, even though some of these articles may also be assigned to other subject areas in Scopus. This choice is justified because the research aims to explore asset management and maintenance practices within industrial and technical systems, topics that are primarily addressed in engineering literature. While other disciplines, such as computer science, mathematics, and decision sciences, may discuss related methods or models, they often lack direct application to practical engineering contexts. Restricting the analysis to engineering enables a more coherent comparison of trends, authorship, and citation patterns and aligns with the methods employed in previous bibliometric studies in this field [9, 14].

3 Results and Discussion

Figure 1 illustrates the chronological analysis of the selected document corpus and reveals a clear and sustained trend of accelerating research interest in the convergence of ML and asset management. The annual publication count remained relatively low and stable until around 2017, confirming the topic's initial status as a niche or emerging area. A significant inflection point occurred around 2018, marking the beginning of an exponential growth phase. This surge reflects the simultaneous industrial adoption of digital transformation, the proliferation of large-scale industrial data, and the increasing maturity of advanced ML techniques. Consequently, the cumulative number of publications demonstrates a steep, almost vertical trajectory in the final years of the analyzed period (2021–2025). This confirms that ML in asset management is now a highly relevant and rapidly maturing field of scholarly inquiry.

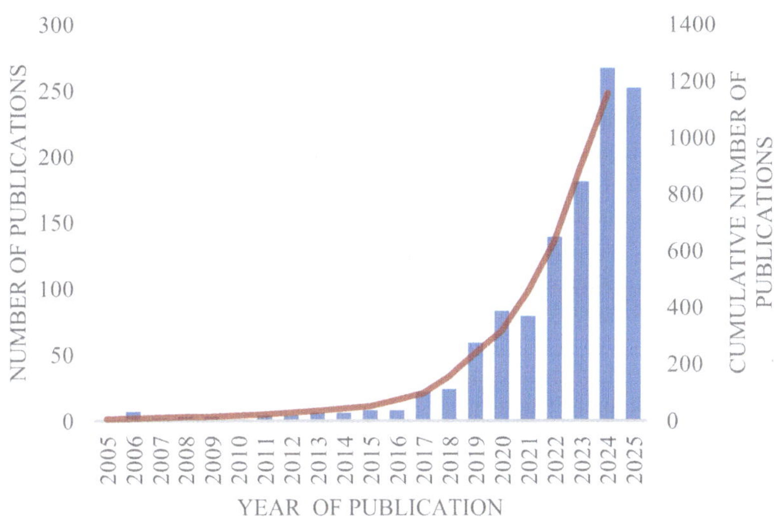

Fig. 1. Annual and cumulative number of publications on machine learning in asset management (2005–2025).

A geographical analysis of the research output reveals a high concentration of scholarly contributions, indicating that a few key nations dominate the field of ML in asset management. As shown in Table 3, the top ten contributing countries account for 938 of the 1,158 documents in the corpus. India emerged as the most prolific country in terms of document volume (n = 266), while the United States exhibited the highest research impact, leading significantly in total citations (n = 6,894). Other major contributors include China (156 documents, 4,368 citations) and the United Kingdom (81 documents, 2,981 citations). Data from the Total Link Strength column suggests that India is a primary driver of production while the United States remains a central, highly influential hub within the global collaborative network.

Tabel 3. Geographical distribution of research contribution (top 10 countries)

Country	Documents	Citations	Total Link Strength
India	266	2,286	111
United States	170	6,894	85
United Kingdom	81	2,981	69
China	156	4,368	65
Germany	99	1,562	38
France	38	572	28
Australia	23	623	25
Italy	62	1,780	25
Spain	34	680	25
Canada	29	700	20
Total (Top 10)	**938**	**22,446**	

Figure 2 provides an additional visualization of the collaborative landscape derived from the bibliometric analysis. This visualization confirms the centrality of the top contributors identified in Table 1, and it reveals distinct co-authorship clusters. The largest cluster, colored red and centered around India, primarily includes other Asian countries, such as China, Saudi Arabia, and Turkey. This suggests a strong regional research alliance in terms of output volume. In contrast, the green cluster (centered around the United States and Germany) and the blue cluster (centered around the United Kingdom) showcase established research partnerships involving European and Anglosphere countries. These partnerships have a high research impact, as evidenced by citation data. These clusters highlight the fragmented yet rapidly forming nature of the global research network on ML in asset management.

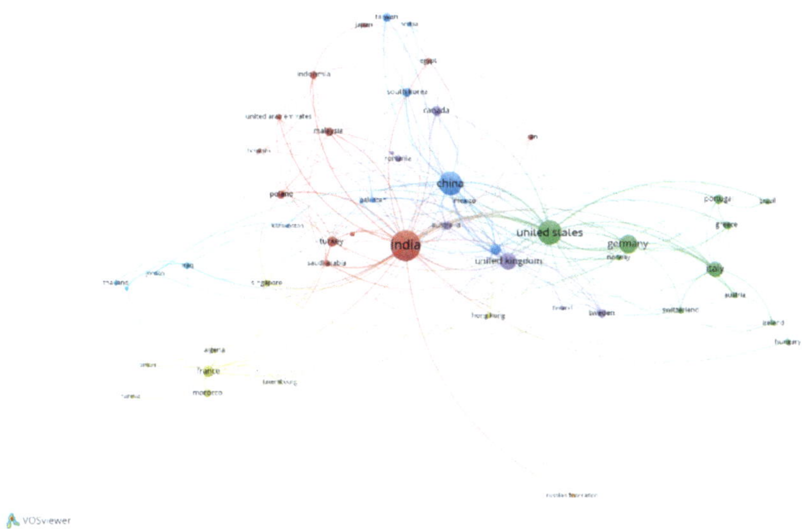

Fig. 2. Country co-authorship network for machine learning in asset management. The size of the nodes (circles) reflects the volume of publications (Documents), and the lines represent co-authorship links. The clusters (colors) indicate groups of countries with strong collaboration ties.

An analysis of the ten most-cited publications (presented in Table 3) reveals the intellectual backbone of ML in the asset management domain. The core research focuses on systematic literature reviews [20] and fundamental PdM methodologies [11, 32]. Although the 2019 article by Carvalho et al. has the highest overall citation count, the calculated Citations Per Year (CPY) metric indicates that the 2020 article by Zonta et al. [11] leads in citation velocity with 152.33 Citations Per Year (CPY), demonstrating its immediate impact. This emphasis on data-driven methods confirms the field's shift toward ML-driven fault detection [33, 34] over traditional methods.

Tabel 4. Top 10 most cited publications

Rank	Authors	Year	Title (as truncated)	Abbreviated title (source)	Cited by (Total)	Citations per year (CPY)
1	Carvalho, T. et al. [20]	2019	A systematic literature review	Computers Ind. Eng	1044	149.14
2	Zonta, T. et al. [11]	2020	Predictive maintenance	Computers Ind. Eng	914	152.33
3	Susto, G.A. et al. [32]	2015	Machine learning for predictive maintenance	IEEE Trans. Ind. Electron	698	63.45

(*continued*)

Tabel 4. *(continued)*

Rank	Authors	Year	Title (as truncated)	Abbreviated title (source)	Cited by (Total)	Citations per year (CPY)
4	Wu, D. et al. [35]	2017	A comparative study	J. Manuf. Sci. Eng	585	65.00
5	Zhang, W. et al. [33]	2019	Data-driven methods	IEEE Syst. J	513	73.29
6	Ayvaz, S. et al. [36]	2021	Predictive maintenance	Expert Syst. Appl	403	80.60
7	Luo, B. et al. [34]	2018	Early fault detection	IEEE Trans. Ind. Electron	275	34.38
8	Essien, A. et al. [37]	2020	A deep learning model	IEEE Trans. Ind. Inform	258	43.00
9	Al-Dulaimi, A. et al. [38]	2019	A multimodal and fusion	Computers Ind	255	36.43
10	Amrutha, M. et al. [39]	2018	A research agenda	5th Int Conf Ind Eng & App (ICIEA)	228	28.50

Note: The publication years used for CPY calculations and temporal analysis adhere to the year provided in the Scopus CSV database export to ensure dataset consistency. Therefore, the year 2018 for Luo et al. is retained, despite the official journal volume print year being 2019

The scholarly publishing landscape further solidifies the field's focus on applied engineering and technology. The International Journal of Advanced Manufacturing Technology is the most prolific publication venue with 53 documents. A significant portion of the corpus is distributed across high-ranking conference proceedings and book series, such as Lecture Notes in Networks and Systems (46 documents) and Procedia CIRP (45 documents), which highlights the rapid dissemination of knowledge through conferences. The prominence of journals such as Computers and Industrial Engineering (20 articles) indicates that the primary area of research is industrial and manufacturing applications.

Citation burst analysis identifies references that received an intense surge in citations over a specific period. This pinpoints the most influential papers currently driving the field's intellectual momentum. Analysis of the top eight references confirms two key aspects of the corpus. First, the research front is focused almost exclusively on ML and DL applications for PdM. Second, the most influential papers are systematic literature reviews (SLRs) and surveys, suggesting that the field is currently in a phase of methodological consolidation and synthesis.

The strongest citation bursts are associated with recent reviews that define the state of the art. The highest burst strength (5.18) was observed for the systematic literature review on ML for PdM by Carvalho (2019) [20], with a burst period from 2022 to 2023. Similarly, Susto's [32] influential paper on ML for PdM (published in 2015) registered a strength of 4.79, indicating a long period of influence from 2015 to 2020.

Crucially, references with burst periods extending into the present and near future (2023–2025) define the current research trajectory. The SLR on PdM in Industry 4.0 by Zonta (2020) [11] exhibits a strong recent burst (3.38) from 2023 to 2025, while the paper on ML in PdM towards Smart Manufacturing by Çınar (2020) [40] also shows a highly active burst (3.22) from 2022 to 2025. This finding reinforces the thematic centrality of Cluster 1 (AI-driven data analytics) and the core concept of PdM as the most active areas of scholarly pursuit.

The structural analysis of the corpus keywords uses a co-occurrence map to delineate the field's intellectual landscape (Fig. 3). A quantitative review confirms that "PdM" (515 occurrences, total link strength 5,096) and "ML" (449 occurrences, total link strength 4,635) are the central concepts for the domain. Frequent terms such as "condition monitoring" (295 occurrences) and "learning systems" (258 occurrences) support this foundation.

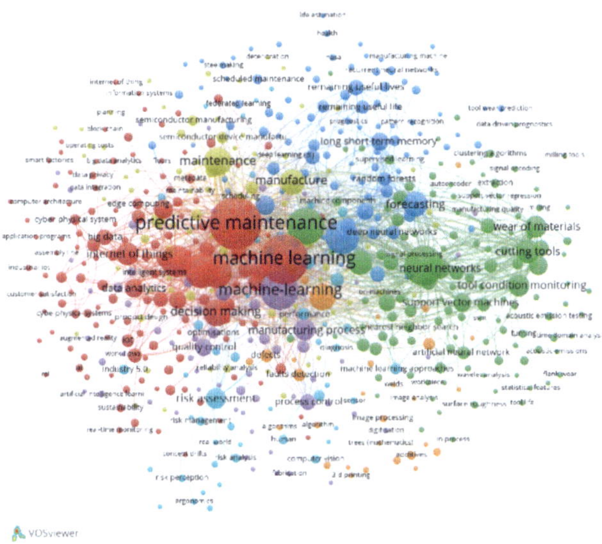

Fig. 3. Keyword co-occurrence network visualization of ML in asset management

The VOSviewer analysis identified seven distinct, interconnected thematic clusters, which are labeled in Table 5. The research focus is organized around four primary domains (Clusters 1–4), which collectively account for the largest share of the literature and form the field's core intellectual structure. The remaining three clusters (Clusters 5–7) represent nascent or peripheral research domains.

Table 5. Identification and labeling of thematic clusters (Keyword co-occurrence analysis)

Cluster	Items	Cluster color	Core concept label	Defining keywords
Cluster 1	146	Red	AI-Driven Data Analytics & Cyber-Physical Systems	AI, Big Data, Cyber-Physical Systems

(continued)

Table 5. *(continued)*

Cluster	Items	Cluster color	Core concept label	Defining keywords
Cluster 2	125	Blue	Signal Processing and Data-Driven Diagnostics	Acoustic Emission, FFT, Data Fusion, ANN
Cluster 3	101	Green	Advanced DL for Prognostics (PHM)	DL, Federated Learning, Explainability, Forecasting
Cluster 4	82	Yellow	Economic and Decision Support Systems	Cost Reduction, Decision Support System, Fuzzy Logic
Cluster 5	63	Purple	Advanced AI Architectures and Intelligent Manufacturing (DRL/GAN)	Deep Reinforcement Learning (DRL), Adversarial Networks, Digital Manufacturing
Cluster 6	47	Orange	Human Factors, Safety, and Industry 5.0	Industry 5.0, Occupational Risks, Human Resource Manag
Cluster 7	30	Brown	Additive Manufacturing and Image-Based Diagnostics	3D Printing, Image Processing, Faults Detection

The visualization further details the internal thematic links (Fig. 3). Cluster 1 (red), the methodological core, centers on the frequent terms "ML" and "PdM" This domain is closely linked to foundational infrastructure technologies, such as the Internet of Things and cyber-physical systems.

Clusters 2 and 3 together define the Prognostics and Health Management (PHM) domain. Cluster 2 (Blue) focuses on traditional ML methods, sensor inputs, and signal processing techniques (e.g., "acoustic emission," "fast Fourier transforms"). Cluster 3 (Advanced DL for Prognostics) introduces more complex models (e.g., "federated learning," "CNN"), focusing on PHM outcomes such as "fault prediction" and the critical aspect of "explainability."

Cluster 4 (yellow) addresses the strategic and financial viability of PdM, covering topics such as cost reduction, investments, and decision support systems.

Clusters 5–7 represent nascent or highly specific research domains. Cluster 5 focuses on advanced AI architectures and intelligent manufacturing. Cluster 6 focuses on human factors, safety, and Industry 5.0. Cluster 7 focuses on niche applications such as additive manufacturing and image-based diagnostics. The strong link between the methodological core (Cluster 1) and the applied clusters (Clusters 2 and 3) highlights the practical, application-oriented nature of the entire research field.

3.1 Core Findings: Validation and Concentration of Research Focus

Integrating the bibliometric findings (Tables 4 and 5) with the asset lifecycle perspective (Table 1) provides a robust framework for evaluating the current intellectual landscape and pinpointing significant research gaps in using ML for engineering asset management.

The results of the keyword co-occurrence analysis clearly show that the field focuses heavily on the operational phase of the asset life cycle (ISO 55001: Clause 8, "Operation"), while the strategic phases ("Planning," "Acquisition," and "Disposal") are significantly underrepresented.

Overwhelming Operational Focus (Stage 3): The largest thematic domains are dedicated to the technical execution of maintenance. Clusters 2 and 3 (Signal Processing and Diagnostics and Advanced DL for Prognostics, respectively) directly support the core functions of PdM, fault detection, and condition monitoring, as detailed in Table 1 (Stage 3).

Technical Foundation: The dominance of Cluster 1 (AI-driven data analytics and cyber-physical systems) confirms that foundational research focuses on establishing the necessary infrastructure (IoT and big data) and methodology (AI and ML) required to enable data-driven operational decisions across the entire system.

Support for Strategic Clauses: The bibliometric results demonstrate that ML research extends beyond simple technical operations and supports strategic management clauses, albeit often indirectly.

Risk and Economic Alignment (Clause 6.1): Cluster 4 (Economic and Decision Support Systems) supports the ISO 55001 requirement to manage risks and opportunities, facilitating the balance between cost, risk, and performance.

Human Factors Integration: Including Cluster 6 (Human Factors, Safety, and Industry 5.0) broadens the strategic scope by addressing the socio-technical aspects of future industrial environments.

Advanced Applications: The presence of Cluster 5, which focuses on advanced AI architectures and intelligent manufacturing, indicates a forward-looking trend of integrating sophisticated AI models, such as deep reinforcement learning (DRL), into highly optimized, self-regulating manufacturing processes.

3.2 Major Research Gaps in the Asset Lifecycle

The analysis reveals clear gaps in holistic support for the full ALC framework in the literature:

- Underrepresented planning and acquisition (stages 1 and 2): The research focus on demand and capacity forecasting or market trend analysis (Table 1, Stage 1) is not defined by a dedicated cluster.
- Underrepresented disposal phase and value recovery (stage 4): The most significant gap is found in the final stage of the ALC concerning disposal decision-making and value recovery. Research on utilizing ML for disposal strategies, such as remanufacturing, recycling policy optimization, and maximizing asset residual value (Table 1, Stage 4) is minimal. This crucial strategic stage is only peripherally addressed by niche topics, such as Cluster 7 (additive manufacturing and image-based diagnostics). This gap is particularly notable as it aligns with the ISO 55001 emphasis on sustainability and climate change, which requires assets to be managed as part of a circular economy. This suggests a lack of focus on sustainable, long-term asset management goals beyond operational longevity.

4 Conclusion

The bibliometric analysis confirms that the field is driven primarily by the technical imperatives of PdM, which provides a robust and highly developed set of methodologies and applications for the operation and maintenance phase of the asset lifecycle. However, to transition to a comprehensive, strategic approach to asset management aligned with ISO 55001, future research must shift its focus to the underrepresented areas of initial planning and acquisition, as well as the final phases of an asset's disposal. The strong link between methodological and application clusters underscores the actionable nature of existing research and positions the field for strategic expansion. This study provides the first evidence-based roadmap demonstrating that future ML value lies in solving complex strategic challenges of asset planning and disposal. Closing these gaps is crucial to realizing the full potential of digital transformation and achieving holistic, value-driven asset management that aligns with global sustainability goals.

References

1. Maletič, D., Marques de Almeida, N., Gomišček, B., Maletič, M.: Understanding motives for and barriers to implementing asset management system: an empirical study for engineered physical assets. Prod. Plann. Control **34**, 1497–1512 (2023). https://doi.org/10.1080/09537287.2022.2026672
2. Almeida, N., Trindade, M., Komljenovic, D., Finger, M.: A conceptual construct on value for infrastructure asset management. Utilities Policy **75**, 101354 (2022). https://doi.org/10.1016/j.jup.2022.101354
3. Trindade, M., Almeida, N., Finger, M., Ferreira, D.: Design and development of a value-based decision making process for asset intensive organizations. In: Mathew, J., Lim, C.W., Ma, L., et al. (eds.) Asset Intelligence through Integration and Interoperability and Contemporary Vibration Engineering Technologies, pp. 605–623. Springer International Publishing, Cham (2019)
4. ISO 55000: Asset management—Vocabulary, overview and principles. International Organization for Standardization (ISO), Geneva, Switzerland (2024)
5. Sandengen, O.C., Estensen, L.A., Rødseth, H., Schjølberg, P.: High performance manufacturing-An innovative contribution towards Industry 4.0. Atlantis Press, pp. 14–20 (2016)
6. Maletič, D., Grabowska, M., Maletič, M.: Drivers and barriers of digital transformation in asset management. Manage. Prod. Eng. Rev. **14**, 118–126 (2023). https://doi.org/10.24425/mper.2023.145370
7. Plekhanov, D., Franke, H., Netland, T.H.: Digital transformation: a review and research agenda. Eur. Manag. J. **41**, 821–844 (2023). https://doi.org/10.1016/j.emj.2022.09.007
8. GFMAM: Digital transformation in maintenance and asset management (2024)
9. Maletič, D., Trojanowska, J., Maletič, M.: Adoption of digital technologies in asset management: a bibliometric and text analysis. Struct. Infrastruct. Eng. (2025) [in press]. https://doi.org/10.1080/15732479.2025.2598801
10. Rosati, R., Romeo, L., Cecchini, G., et al.: From knowledge-based to big data analytic model: a novel IoT and machine learning based decision support system for predictive maintenance in Industry 4.0. J. Intell. Manuf. **34**, 107–121 (2023). https://doi.org/10.1007/s10845-022-01960-x

11. Zonta, T., da Costa, C.A., da Rosa Righi, R., et al.: Predictive maintenance in the Industry 4.0: a systematic literature review. Comput. Ind. Eng. **150**, 106889 (2020). https://doi.org/10.1016/j.cie.2020.106889
12. Oluwasegun, A., Jung, J.-C.: The application of machine learning for the prognostics and health management of control element drive system. Nucl. Eng. Technol. **52**, 2262–2273 (2020). https://doi.org/10.1016/j.net.2020.03.028
13. Rampini, L., Moretti, N., Re Cecconi, F., Dejaco, M.C.: Digital asset management enabling technologies: a bibliometric analysis (2020)
14. Silva, R.F.D., Martha De Souza, G.F.: Mapping the literature on asset management: a bibliometric analysis. JSCIRES **10**, 27–36 (2021). https://doi.org/10.5530/jscires.10.1.4
15. Muhuri, P.K., Shukla, A.K., Abraham, A.: Industry 4.0: a bibliometric analysis and detailed overview. Eng. Appl. Artif. Intell. **78**, 218–235 (2019). https://doi.org/10.1016/j.engappai.2018.11.007
16. Manzoor, B., Othman, I., Pomares, J.C.: Digital technologies in the Architecture, Engineering and Construction (AEC) Industry—A bibliometric—Qualitative literature review of research activities. Int. J. Environ. Res. Public Health **18**, 6135 (2021). https://doi.org/10.3390/ijerph18116135
17. Alsyouf, I., Alsuwaidi, M., Hamdan, S., Shamsuzzaman, M.: Impact of ISO 55000 on organisational performance: evidence from certified UAE firms. Total Q. Manage. Bus. Excellence 1–19 (2018). https://doi.org/10.1080/14783363.2018.1537750
18. Maletič, D., Maletič, M., Al-Najjar, B., Gomišček, B.: An analysis of physical asset management core practices and their influence on operational performance. Sustainability **12**, 9097 (2020). https://doi.org/10.3390/su12219097
19. Maletič, D., Trojanowska, J., Lorber, M., Maletič, M.: Identification and prioritization of barriers to the implementation of an asset management system in the healthcare sector using a Delphi-AHP approach. J. Health Organ. Manag. **39**, 535–560 (2025). https://doi.org/10.1108/JHOM-12-2024-0513
20. Carvalho, T.P., Soares, F.A.A.M.N., Vita, R., et al.: A systematic literature review of machine learning methods applied to predictive maintenance. Comput. Ind. Eng. **137**, 106024 (2019). https://doi.org/10.1016/j.cie.2019.106024
21. ISO 55001 (2024) Asset management—Management systems—Requirements. International Organization for Standardization (ISO), Geneva, Switzerland
22. Muñoz-Villamizar, A., Rafavy, C.Y., Casey, J.: Machine learning and optimization-based modeling for asset management: a case study. Int. J. Product. Perform. Manag. **71**, 1149–1163 (2020). https://doi.org/10.1108/IJPPM-05-2020-0206
23. Rajora, G.L., Sanz-Bobi, M.A., Tjernberg, L.B., Urrea Cabus, J.E.: A review of asset management using artificial intelligence-based machine learning models: Applications for the electric power and energy system. IET Gener. Transm. Distrib. **18**, 2155–2170 (2024). https://doi.org/10.1049/gtd2.13183
24. Ni, J., Hu, Y., Zhong, R.Y.: A hybrid machine learning method for procurement risk assessment of non-ferrous metals for manufacturing firms. Int. J. Comput. Integr. Manuf. **35**, 1028–1042 (2022). https://doi.org/10.1080/0951192X.2021.1901315
25. Husna A ul, Ghasempoor A, Amin SH (2024) A proposed framework for supplier selection and order allocation using machine learning clustering and optimization techniques. J. Data Inf. Manag. **6**, 235–254. https://doi.org/10.1007/s42488-024-00127-y
26. Zhu, T., Ran, Y., Zhou, X., Wen, Y.: A survey on Intelligent Predictive Maintenance (IPdM) in the Era of fully connected intelligence. IEEE Commun. Surv. Tutorials (2025). https://doi.org/10.1109/COMST.2025.3567802
27. Shahin, M., Chen, F.F., Hosseinzadeh, A., Zand, N.: Using machine learning and deep learning algorithms for downtime minimization in manufacturing systems: an early failure detection

diagnostic service. Int. J. Adv. Manuf. Technol. **128**, 3857–3883 (2023). https://doi.org/10.1007/s00170-023-12020-w
28. Yang, C., Liu, J., Zeng, Y., Xie, G.: Real-time condition monitoring and fault detection of components based on machine-learning reconstruction model. Renew. Energy **133**, 433–441 (2019). https://doi.org/10.1016/j.renene.2018.10.062
29. Koulinas, G., Paraschos, P., Koulouriotis, D.: A machine learning-based framework for data mining and optimization of a production system. Procedia Manuf. **55**, 431–438 (2021). https://doi.org/10.1016/j.promfg.2021.10.059
30. Cherrington, M., Lu, Z.(Joan)., Xu, Q., et al.: Deep learning decision support for sustainable asset management. In: Ball, A., Gelman, L., Rao, B.K.N. (eds.) Advances in Asset Management and Condition Monitoring. Springer International Publishing, Cham, pp. 537–547 (2020)
31. Page, M.J., Moher, D., Bossuyt, P.M., et al.: PRISMA 2020 explanation and elaboration: updated guidance and exemplars for reporting systematic reviews. BMJ **372**, n160 (2021). https://doi.org/10.1136/bmj.n160
32. Susto, G.A., Schirru, A., Pampuri, S., et al.: Machine learning for predictive maintenance: a multiple classifier approach. IEEE Trans. Industr. Inf. **11**, 812–820 (2015). https://doi.org/10.1109/TII.2014.2349359
33. Zhang, W., Yang, D., Wang, H.: Data-Driven methods for predictive maintenance of industrial equipment: a survey. IEEE Syst. J. **13**, 2213–2227 (2019). https://doi.org/10.1109/JSYST.2019.2905565
34. Luo, B., Wang, H., Liu, H., et al.: Early fault detection of machine tools based on deep learning and dynamic identification. IEEE Trans. Industr. Electron. **66**, 509–518 (2019). https://doi.org/10.1109/TIE.2018.2807414
35. Wu, D., Jennings, C., Terpenny, J., et al.: A comparative study on machine learning algorithms for smart manufacturing: tool wear prediction using random forests. J. Manuf. Sci. Eng. (2017). https://doi.org/10.1115/1.4036350
36. Ayvaz, S., Alpay, K.: Predictive maintenance system for production lines in manufacturing: a machine learning approach using IoT data in real-time. Expert Syst. Appl. **173**, 114598 (2021). https://doi.org/10.1016/j.eswa.2021.114598
37. Essien, A., Giannetti, C.: A Deep learning model for smart manufacturing using convolutional LSTM neural network autoencoders. IEEE Trans. Industr. Inf. **16**, 6069–6078 (2020). https://doi.org/10.1109/TII.2020.2967556
38. Al-Dulaimi, A., Zabihi, S., Asif, A., Mohammadi, A.: A multimodal and hybrid deep neural network model for remaining useful life estimation. Comput. Ind. **108**, 186–196 (2019). https://doi.org/10.1016/j.compind.2019.02.004
39. Amruthnath, N., Gupta, T.: A research study on unsupervised machine learning algorithms for early fault detection in predictive maintenance. In: 2018 5th International Conference on Industrial Engineering and Applications (ICIEA). pp. 355–361 (2018)
40. Çınar, Z.M., Abdussalam Nuhu, A., Zeeshan, Q., et al.: Machine learning in predictive maintenance towards sustainable smart manufacturing in Industry 4.0. Sustainability **12**, 8211. https://doi.org/10.3390/su12198211

Time Series Forecasting Under Irregular Seasonality and Anomalies

Patrycja Guzanek[1(✉)], Mykola Karpenko[2], and Anna Borucka[3]

[1] Doctoral School of Military, University of Technology, Warsaw, Poland
patrycja.guzanek@student.wat.edu.pl
[2] Vilnius Gediminas Technical University, Vilnius, Lithuania
[3] Military University of Technology, Warsaw, Poland

Abstract. The study examines challenges in forecasting time series with sudden peaks and large fluctuations, common in logistics and supply data of the cosmetics industry. Such irregularities violate key assumptions of classical models, including stationarity and predictable seasonality.

Daily delivery data from selected cosmetics suppliers were analyzed using three models: ARIMA, ETS, and TBATS, applied to both raw and preprocessed datasets. Preprocessing included outlier adjustment via decomposition-based cleaning and winsorization.

Results showed strong weekly seasonality across all suppliers and, in some cases, additional monthly effects. Data cleaning improved forecast accuracy, reducing the mean absolute error (MAE) by over 50%. ARIMA performed best on cleaned data, while TBATS handled multiple seasonalities most effectively. The findings highlight the importance of preprocessing for reliable forecasting in volatile environments, though excessive smoothing can obscure meaningful irregular patterns.

Keywords: Time Series Forecasting · Optimization · ARIMA · ETS · TBATS · Seasonality · Data Cleaning · Supply Chain Management

1 Introduction

1.1 Research Description

Forecasting in the cosmetics industry is challenging due to irregular daily delivery patterns caused by promotions, seasonal peaks, and logistical disruptions. Such variability violates the assumptions of classical models like ARIMA and ETS, reducing forecast accuracy and efficiency.

This study evaluates ARIMA, ETS, and TBATS models using daily delivery data from a cosmetics warehouse and examines how data cleaning affects forecast performance. After identifying structural components and applying an outlier-removal procedure, results show that even basic preprocessing markedly improves accuracy and interpretability, helping balance robustness and responsiveness in short-term forecasting.

J. Trojanowska et al. (Eds.): MANUFACTURING 2026, LNME, pp. 125–140, 2026.
https://doi.org/10.1007/978-3-032-21654-0_10

1.2 Literature Review

Process optimization models constitute a key tool in the analysis and improvement of production, logistics, and service systems. Their primary objective is to identify the most efficient solutions through cost minimization, performance maximization, or enhancement of other operational efficiency indicators, while taking into account defined constraints. In practice, various mathematical and computational approaches are employed, such as linear, nonlinear, dynamic, and stochastic programming, as well as heuristic and metaheuristic methods. An effective optimization model is optimization based on multi-criteria decision trees [1, 2]. The application of optimization models enables the support of decision-making processes [3], enhances the competitiveness of enterprises, and promotes the rational utilization of available resources.

Forecasting under irregular and dynamic conditions is a key challenge across multiple sectors of the economy. Industries such as retail, transport, energy, and healthcare generate vast amounts of operational data characterized by irregular demand patterns, nonlinear seasonality, and short-term anomalies [4, 5]. Accurately modeling such variability is essential for efficient resource allocation, production scheduling, and supply chain management [6]. In these contexts, time series models remain the foundation of quantitative forecasting, providing interpretable and computationally efficient tools for operational decision-making [7].

The ARIMA (*AutoRegressive Integrated Moving Average*) family of models is one of the most widely applied forecasting approaches in practice. It has been successfully used in predicting short-term demand in retail [8, 9] and manufacturing [10], energy consumption in public utilities [11], and transport flows [12, 13] and logistics systems [14]. Owing to its autoregressive and moving-average structure, ARIMA performs well when short-term dependencies dominate, though its accuracy declines under volatile conditions or when sudden disruptions occur [15]. Seasonal extensions such as SARIMA have proven effective for weekly or monthly forecasting cycles, especially in production planning and distribution scheduling [16].

The ETS (*Error, Trend, Seasonality*) model, based on exponential smoothing, has been extensively used in sales [17] and inventory forecasting [18], healthcare sector predictions [19], and energy load management [20]. Its decomposition into level, trend, and seasonal components makes it well suited for operational planning [21], particularly where smooth seasonal dynamics are present [22]. Despite its flexibility, ETS is sensitive to anomalies and tends to over- or underreact when extreme observations distort smoothing parameters [23].

More advanced models such as TBATS (*Trigonometric, Box-Cox transformation, ARIMA errors, Trend, and Seasonal components*) have been developed to handle multiple overlapping seasonalities, for instance, in e-commerce [24], transportation networks [25], or electricity markets [26], where data exhibit both daily and weekly cycles. TBATS combines trigonometric functions, ARMA terms, and Box-Cox transformations to improve robustness against irregular periodicity [27].

Across these applications, an increasingly recognized aspect of forecasting is data preprocessing [28]. Procedures such as anomaly detection, decomposition-based filtering, and winsorization are used to stabilize variance and isolate meaningful structure from noise. Studies consistently show that cleaning time series before modeling enhances forecast reliability in complex operational environments [29]. For industries such as cosmetics, where demand fluctuates with marketing campaigns and seasonal peaks, combining robust models with preprocessing is essential for accurate and interpretable forecasting.

2 Results

2.1 Research Problem

The analysis uses daily delivery time series from three cosmetics suppliers. Supplier A (mass market) shows high volatility and frequent restocking. Supplier B (premium) delivers less often but very regularly. Supplier C (personal care and wellness) shows moderate regularity with strong promotional and e-commerce effects. The data reflect real logistics processes shaped by fluctuating demand, replenishment cycles, and short-term promotions. Initial inspection revealed clear irregularities, multiple seasonalities, and high day-to-day variability, indicating both systematic seasonal effects and event-driven fluctuations.

The study focuses on identifying and quantifying recurring seasonal patterns in irregular data. Before choosing forecasting models, it was necessary to determine whether variability came from systematic seasonality or random noise. A nonparametric seasonality test was therefore applied to assess monthly and weekly effects.

2.2 Assessment of Seasonal Patterns Using the Kruskal-Wallis Test

To verify whether the observed variability in deliveries is systematic or random, the seasonality of the analyzed time series was examined using the Kruskal-Wallis test. This non-parametric test is robust to deviations from normality and suitable for data with irregular fluctuations [30]. Two types of seasonality were tested: monthly and weekly, with a significance level of $\alpha = 0.05$ (Tables 1 and 5).

Table 1. Results of the Kruskal-Wallis test for delivery time series.

Supplier	Type of seasonality	p-value
Supplier A	Monthly	3.717148×10^{-7}
Supplier A	Weekly	9.918455×10^{-7}
Supplier B	Monthly	3.402569×10^{-1}
Supplier B	Weekly	1.989614×10^{-30}
Supplier C	Monthly	6.842062×10^{-2}
Supplier C	Weekly	3.732317×10^{-28}

The results show that all three suppliers exhibit at least one statistically significant seasonal pattern. Supplier A displays both monthly and weekly effects ($p < 0.001$), reflecting the steady rhythm of high-turnover operations. Supplier B shows a strong weekly pattern ($p < 0.001$) but no significant monthly seasonality ($p > 0.05$), while Supplier C has weak monthly ($p \approx 0.07$) and strong weekly effects ($p < 0.001$) linked to promotional cycles.

Overall, a dominant weekly pattern is observed across all suppliers, consistent with cosmetics distribution processes, where deliveries peak on weekdays and drop on weekends. Figure 1 shows the monthly delivery volumes for each supplier.

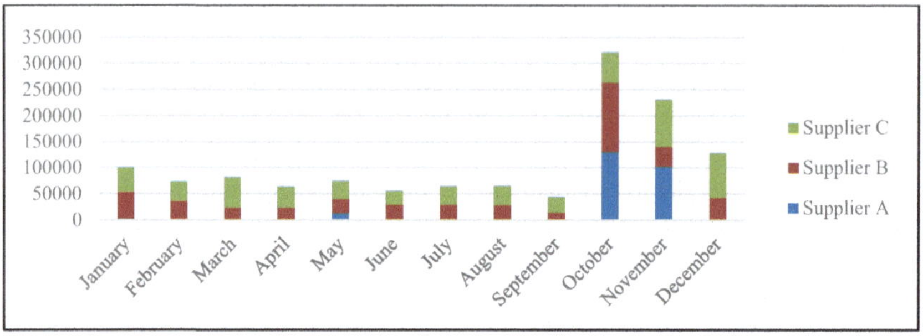

Fig. 1. Monthly delivery volumes by supplier.

The highest delivery activity occurs in October and November, reflecting pre-holiday stock accumulation. Supplier C maintains the largest share of total deliveries, while Suppliers A and B record similar but lower volumes. This pattern confirms strong seasonality driven by demand peaks in late autumn and early winter. Figure 2 shows the distribution of daily deliveries across weekdays for comparison between suppliers.

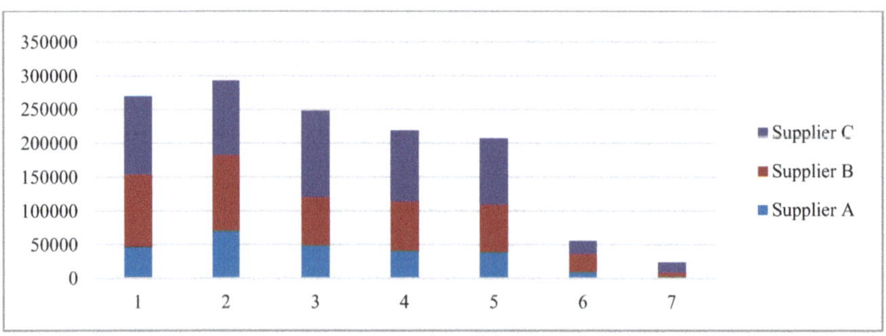

Fig. 2. Total number of product units delivered per supplier by weekday.

Deliveries are concentrated on weekdays, peaking between Monday and Wednesday (days 1–3) and declining toward the weekend, with minimal activity on days 6–7. This pattern is consistent across all suppliers, reflecting standard logistics cycles without weekend dispatching. Supplier C again records the highest delivery volumes, while Suppliers A and B show similar weekday patterns at lower levels.

2.3 Time Series Decomposition Using the MSTL Method

To illustrate the internal structure of the analyzed data and better understand the interaction between multiple seasonal cycles, one representative supplier (Supplier C) was selected for detailed examination. In the first stage of the analysis, the time series was visually decomposed using the MSTL (Multiple Seasonal-Trend decomposition using Loess) method. This approach allows for the simultaneous extraction of two distinct cyclical components-weekly and monthly seasonality-while isolating the long-term trend and the random residual component (Fig. 3).

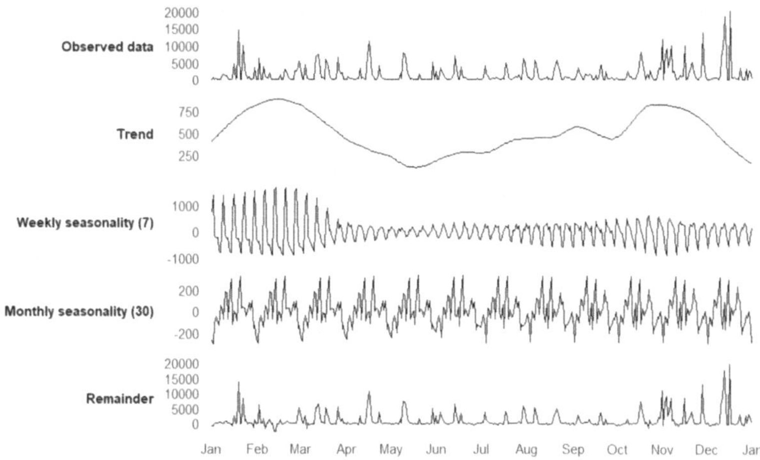

Fig. 3. MSTL decomposition-Supplier C (weekly and monthly seasonality).

The upper panel shows raw daily delivery volumes with clear peaks in January-February and November-December, driven by post-holiday restocking and pre-holiday demand. The trend reveals two main peaks: early spring and September to December, indicating a biannual delivery cycle. Weekly patterns show regular weekday highs and weekend lows, while monthly effects reflect end-of-month stock adjustments. The remainder contains isolated promotional spikes, highlighting the influence of event-driven fluctuations and the need for preprocessing to avoid distorted seasonality estimates.

2.4 Forecasting Models and Evaluation Procedure

Given the data characteristics, several seasonal forecasting models were applied to compare how they capture trend and seasonality under irregular fluctuations. ARIMA and its seasonal form SARIMA model autoregressive and moving-average effects but handle sudden peaks poorly. ETS, based on exponential smoothing, works well with clear structures but is sensitive to outliers. TBATS, using Box-Cox transformation, ARMA terms, and trigonometric components, manages complex and multiple seasonalities but can still be influenced by abrupt anomalies. Models were trained on historical data, with the last 14 days reserved for testing. Accuracy was assessed using AIC and standard error metrics (MAE, RMSE, WAPE, sMAPE, MASE).

ARIMA Model

As the first forecasting approach, the ARIMA (AutoRegressive Integrated Moving Average) model was estimated. The parameters were selected using automatic information-criterion optimization, resulting in the specification ARIMA(0,1,2), expressed as:

$$(1-B)y_t = \left(1 - 0.596B - 0.3654B^2\right)\varepsilon_t \tag{1}$$

where B denotes the backshift operator and ε_t represents the random error term.

The model includes two moving-average coefficients, both statistically significant, with low standard errors (0.057 and 0.0582). This indicates the presence of short-term dependencies within the time series, suggesting that current delivery volumes are partly influenced by random shocks occurring one or two days earlier. In practical terms, the ARIMA(0,1,2) model captured a short-run inertia effect but did not detect a persistent long-term trend (Figs. 4 and 5).

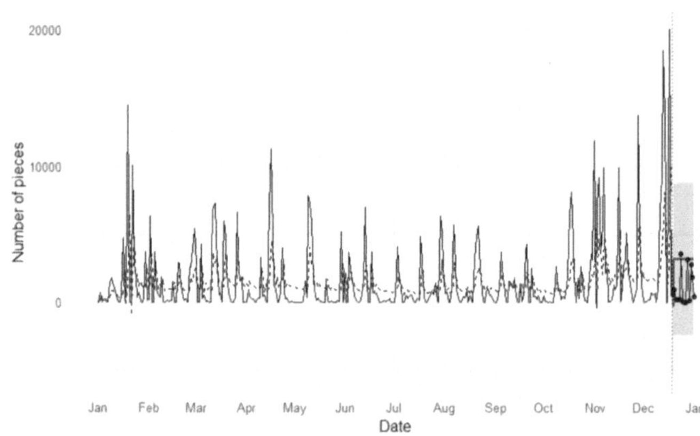

Fig. 4. Fitted ARIMA(0,1,2) model and forecast.

The ARIMA(0,1,2) model captures short-term dependencies through its moving-average structure, reflecting limited inertia in daily delivery volumes. While it follows moderate fluctuations reasonably well, it remains unable to anticipate abrupt peaks and sudden drops, treating extreme deviations primarily as random noise. This confirms the known limitations of ARIMA in highly volatile and event-driven environments.

The model's information criteria values were AIC = 6534.9 and BIC = 6546.5, confirming an adequate but not optimal fit relative to the more flexible approaches applied later in the analysis.

ETS Model

The second forecasting approach applied was the ETS (Error, Trend, Seasonality) model, based on exponential smoothing. The best-performing structure was an additive model with both trend and seasonality components, represented by the following equation:

$$y_t = l_{t-1} + s_{t-m} + \varepsilon_t \tag{2}$$

where l_t denotes the level (trend), s_t the seasonal component, and ε_t. The random error term.

The estimated smoothing parameters were: level $\alpha = 0.1933$, seasonality $\gamma = 0.0001$, and the initial level $l_0 = 621.3.$. The information criteria reached values of AIC = 7627.4 and BIC = 7666.1.

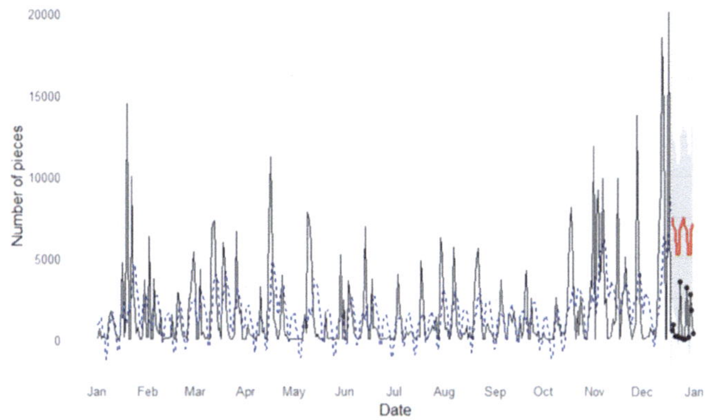

Fig. 5. Fitted ETS model and forecast.

The ETS model, characterized by a relatively low level-smoothing coefficient (α), responds moderately to new information, leading to a smoother representation of short-term fluctuations. The extremely low value of the seasonal smoothing parameter (γ) suggests that the seasonal component remains stable and does not change significantly over time.

Described model provides a stable representation of the average delivery level and recurring seasonal pattern but shows limited adaptability to rapid changes. The low level-smoothing parameter results in delayed reactions to recent shifts, leading to systematic overestimation during periods of declining activity. Consequently, ETS is more suitable for descriptive or medium-term forecasting than for short-term prediction under irregular conditions.

TBATS Model

The third forecasting approach employed was the TBATS (Trigonometric, Box-Cox, ARMA, Trend, and Seasonal) model, designed to handle multiple and complex seasonalities. The best-fitting specification was identified as TBATS(1, {2,2}, -, {<7,3 >}), with the corresponding parameter estimates summarized in Table 2.

Table 2. Estimated parameters of the TBATS(1, {2,2}, -, {<7,3 >}) model.

Component	Parameter
α	0.0215
γ_1, γ_2	0.0000034, 0.0000071
AR(1), AR(2)	0.176, 0.325
MA(1), MA(2)	0.175, – 0.338

The model's information criterion value was AIC = 7606.6.

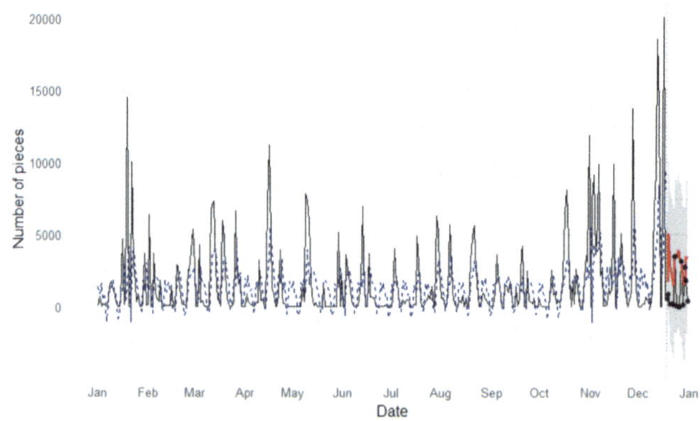

Fig. 6. Fitted TBATS model and forecast.

The TBATS model, characterized by very low smoothing parameters (α and γ), acts as a strongly smoothed filter, responding slowly to abrupt trend changes but effectively reducing noise in high-frequency daily data. The inclusion of an ARMA(2,2) component captures short-term autocorrelations between consecutive observations, improving local fit precision. The trigonometric seasonal component with a 7-day cycle successfully models the weekly delivery rhythm identified in earlier analyses.

The model presented on Fig. 6 effectively captures the dominant weekly seasonality while suppressing high-frequency noise through strong smoothing. The inclusion of ARMA components improves local fit by modeling short-term autocorrelations, resulting in a more stable forecast than ARIMA. However, very low smoothing parameters imply a delayed response to sudden structural changes, limiting sensitivity to abrupt demand shocks.

Model Results and Data Preprocessing

The results obtained for all three forecasting models are summarized in Table 3. The performance metrics include the Mean Absolute Error (MAE), Root Mean Square Error (RMSE), Weighted Absolute Percentage Error (WAPE), Symmetric Mean Absolute Percentage Error (sMAPE), and Mean Absolute Scaled Error (MASE).

Table 3. Forecast accuracy and information criteria for all models.

Rank	Model	MAE	RMSE	WAPE	sMAPE	MASE	AIC
1	ARIMA	2105.020	2410.842	211.287	131.978	0.989	6534.9
2	TBATS	2121.863	2481.859	212.977	126.942	0.997	7606.6
3	ETS	5554.414	5756.861	557.512	154.604	2.609	7627.4

Among the analyzed models, ARIMA(0,1,2) achieved the lowest forecast error values in most metrics (MAE, RMSE, WAPE, MASE), followed closely by TBATS, which obtained a slightly lower sMAPE while the ETS model performed considerably worse. The results confirm that models incorporating autoregressive and moving average components (ARIMA and TBATS) are better suited to handling short-term dependencies and irregular fluctuations in delivery volumes.

Although AIC values are reported for all models, comparisons across fundamentally different model classes (e.g., ARIMA, ETS, and TBATS) should be interpreted with caution and treated as indicative rather than definitive.

The TBATS model is theoretically more flexible in representing multiple seasonalities; however, its higher AIC value suggests a slightly weaker overall fit compared to ARIMA for this particular dataset.

Despite the models' ability to capture general patterns in the data, the forecasting errors remain relatively high across all approaches. This outcome reflects the strong volatility and presence of extreme outliers in the original time series.

Therefore, in the next stage of analysis, the data were preprocessed to remove anomalous observations, under the assumption that they result from isolated logistical events or short-term promotional campaigns. Such anomalies distort the estimation of trend and seasonality, causing models to fit extreme points rather than the underlying structure of the series.

To address this issue, a two-step cleaning procedure was applied. First, outliers were identified based on the deviation of observed values from the trend and seasonal components extracted through MSTL decomposition. Observations deviating by more than approximately 2.5 standard deviations from the residual distribution were treated as anomalies and replaced with fitted values from the decomposition model.

In the second step, the winsorization technique was employed to further limit extreme values to the 2nd-98th percentile range of the distribution. The 2.5 standard deviation threshold was selected as a commonly used compromise between sensitivity and robustness in anomaly detection, while the 2nd–98th percentile winsorization was applied to limit extreme values without excessively distorting the central distribution. A more extensive sensitivity analysis is left for future research.

As a result, the time series was cleared of anomalies and short-term shocks, yielding a more stable and representative dataset for subsequent forecasting. The cleaned series was then re-modeled using the same three forecasting methods to assess the improvement in predictive accuracy.

2.5 Forecasting Results After Data Cleaning

ARIMA Model after the data cleaning

After removing outliers and stabilizing the time series, the best-fitting specification was obtained using a seasonal ARIMA model with both short-term and weekly autoregressive structures. The optimal model was identified as ARIMA(2,0,0)(2,0,0) [7], expressed as:

$$\left(1 - 0.5216B + 0.1048B^2\right)\left(1 - 0.0924B^7 - 0.2499B^{14}\right)(y_t - 1033.97) = \varepsilon_t \quad (3)$$

where B is the backshift operator and ε_t denotes the random error term.

The presence of moderate daily autoregression of order two (AR(2)) effectively captures short-term dependencies and the 7-day cyclical behavior observed in the data. Parameter estimates indicate that current deliveries depend strongly on the previous day's values ($\phi_1 = 0.52$), are slightly adjusted by the influence of two days earlier ($\phi_2 = -0.10$), and are moderately correlated with deliveries from the previous week ($\Phi_1 = 0.09$) and two weeks prior ($\Phi_2 = 0.25$). This confirms a clear weekly seasonality and short-term inertia in delivery patterns.

Importantly, the absence of differencing ($d = 0$) suggests that the cleaned time series is stationary, which improves model stability and interpretability.

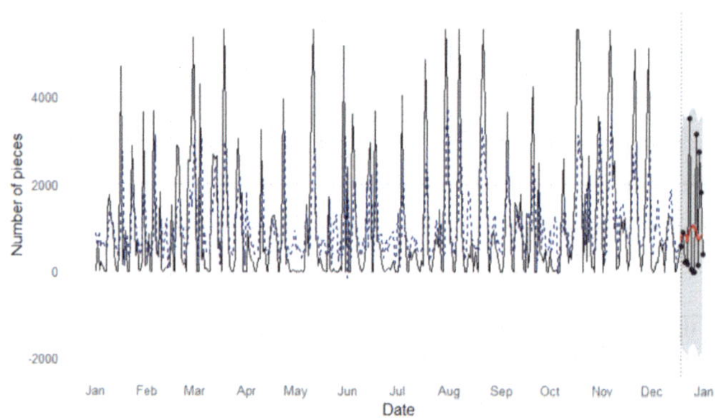

Fig. 7. Fitted seasonal ARIMA model and forecast for the cleaned data.

The fitted model and forecast are presented in Fig. 7. The model accurately captures the regular weekly structure and short-term dependencies present in the cleaned series. The removal of extreme fluctuations improves stationarity and model stability, allowing recurring operational patterns to be identified more clearly.

The information criteria indicate a strong improvement in model fit compared to the unprocessed series: AIC = 6007.69, AICc = 6007.94, and BIC = 6030.88, confirming that data cleaning significantly enhanced the predictive quality of the ARIMA framework.

ETS Model after the data cleaning

For the cleaned dataset, the best-fitting exponential smoothing structure was the additive model with no trend and additive seasonality, denoted as ETS(A, N, A). The estimated smoothing parameters were $\alpha = 0.5009$, $\gamma = 0.0001$, and the initial level was $l_0 = 595.4$. The information criteria were AIC = 7121.9 and BIC = 7160.6.

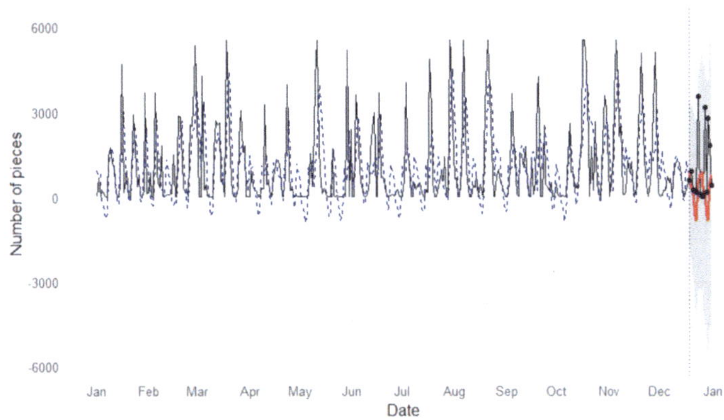

Fig. 8. Fitted ETS model and forecast for the cleaned data.

The relatively high value of α indicates that the model reacts quickly to new observations, effectively adapting to recent shifts in delivery levels. Conversely, the extremely low seasonal smoothing coefficient (γ) implies that the seasonal component remains nearly constant over time, reflecting a stable and recurring weekly pattern.

As shown in Fig. 8, the model produces a stable short-term forecast that reflects the dominant weekly seasonality of the cleaned data. Higher responsiveness to recent observations improves adaptation compared to the unprocessed series, while the nearly constant seasonal component indicates a persistent and regular cycle. Despite improved stability, the model remains less flexible than ARIMA and TBATS under changing demand conditions.

TBATS Model after the data cleaning

For the cleaned series, the best-fitting specification was the TBATS(1, {1,2}, 0.885, {<7,3 >}) model. The estimated parameters are presented in Table 4.

The model achieved an AIC = 7058.21, indicating an improved fit compared to the unprocessed data.

The TBATS model, characterized by very low smoothing parameters (α, β, γ), behaves as a highly smoothed filter, responding slowly to rapid changes in trend but effectively reducing noise in the daily observations. The inclusion of an ARMA(1,2) component captures short-term autocorrelations, while the trigonometric seasonal term with a 7-day period accurately reflects the weekly delivery cycle.

Table 4. Estimated parameters of the TBATS model after data cleaning.

Component	Parameter
α	−0.0444
β	0.0048
γ_1, γ_2	0.00020, −0.00021
Damping parameter	0.885
AR(1)	0.537
MA(1), MA(2)	−0.051, −0.051

A visual inspection of the fitted values (not shown) confirmed that the model successfully reproduces the dominant seasonal structure, while substantially reducing random fluctuations. The results indicate the presence of strong weekly seasonality and moderate short-term dependencies. The TBATS model performs well in short-term forecasting under stable seasonal conditions, providing greater flexibility than ETS and more effective noise suppression. However, due to the low α value, its response to sudden structural shifts remains relatively slow.

3 Discussion

The comparison of forecasting performance before and after data cleaning highlights the substantial impact of preprocessing on model behavior and accuracy.

Figure 9 on the left side presents the results for the ARIMA, ETS, and TBATS models based on the original, unprocessed dataset, while the right side shows the same models fitted to the cleaned series.

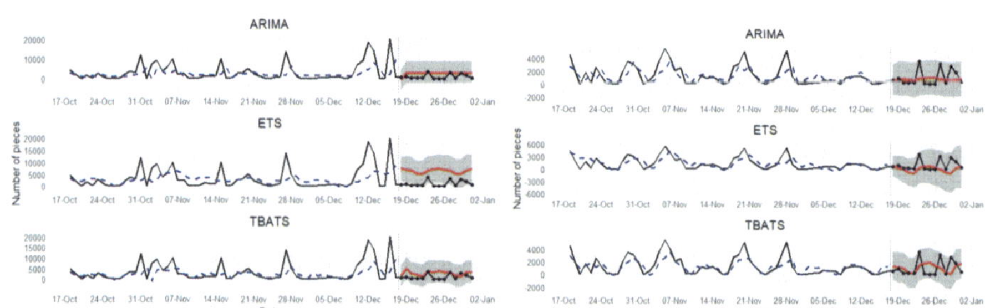

Fig. 9. Forecast comparison for ARIMA, ETS, and TBATS models before (left) ande after (right) data cleaning.

As illustrated in Fig. 9, all models trained on the original data exhibit visible overestimation and wide confidence intervals during the test period. The irregular spikes and extreme fluctuations distort the identification of seasonal patterns, leading to inflated variance and lower predictive precision. After applying the cleaning procedure, the forecasts became notably smoother and more consistent. The ARIMA(2,0,0)(2,0,0)[7] model provided the best overall fit, capturing both short-term dependencies and clear weekly seasonality. The TBATS model achieved comparable accuracy, maintaining flexibility in representing multiple seasonal cycles, while the ETS(A,N,A) model showed stable but less adaptive behavior.

Table 5. Forecasting errors and information criteria before and after data cleaning.

Stage	Rank	Model	MAE	RMSE	WAPE	sMAPE	MASE	AIC
Before cleaning	1	ARIMA	2105.020	2410.842	211.287	131.978	0.989	6534.9
	2	TBATS	2121.863	2481.859	212.977	126.942	0.997	7606.6
	3	ETS	5554.414	5756.861	557.512	154.604	2.609	7627.4
After cleaning	1	ARIMA	998.745	1236.499	100.247	113.108	0.791	6007.7
	2	TBATS	1129.809	1387.907	113.402	106.333	0.895	7058.2
	3	ETS	1219.086	1643.404	122.363	150.662	0.966	7121.9

The table confirms that data cleaning significantly improved the predictive accuracy of all models. The ARIMA model achieved the greatest reduction in error, with MAE and RMSE decreasing by more than 50%, and AIC improving from 6534.9 to 6007.7.TBATS also benefited from preprocessing, especially in terms of RMSE and sMAPE, confirming its robustness in handling multi-seasonal data once outliers were suppressed. The ETS model, though initially weakest, gained stability and smoother residuals after cleaning, showing that even simple exponential smoothing can perform well under stable conditions.

Despite these improvements, data cleaning involves a trade-off. Eliminating outliers or irregular observations can lead to a loss of meaningful information and excessive smoothing of the series. Such operations may distort the true structure of the data and reduce the model's sensitivity to genuine events such as supply interruptions or demand shocks.

In practice, the cleaning process should therefore be viewed as a balance between improving model stability and preserving representativeness. While the cleaned data improve forecast accuracy and interpretability, real anomalies often carry operational significance. Thus, data preprocessing should not aim for complete regularity but should retain the natural variability that reflects the underlying business dynamics.

While data cleaning substantially improves forecast accuracy, it inevitably introduces a trade-off between stability and sensitivity. Excessive smoothing may suppress not only noise but also operationally meaningful anomalies, such as promotion-driven demand surges or temporary supply disruptions. Therefore, preprocessing should be applied selectively and interpreted in conjunction with domain knowledge.

4 Conclusion

This study assessed the performance of three forecasting models-ARIMA, ETS, and TBATS-applied to irregular time series of daily cosmetics deliveries. The analysis was conducted in two stages: modeling the original data and re-estimating models after outlier removal and winsorization.

Results showed that simple preprocessing significantly improved accuracy and interpretability, reducing forecast errors by around 50% for the best-performing model and improving information criteria. Irregular spikes and noise were major sources of inaccuracy. Among the models, ARIMA(2,0,0)(2,0,0) [7] offered the best balance between robustness and responsiveness, TBATS captured complex seasonality most effectively, and ETS produced smoother forecasts on cleaned data.

These findings emphasize the importance of preprocessing in forecasting workflows. In logistics and retail, irregularities often stem from short-term promotions or disruptions; addressing them enhances forecast reliability but excessive smoothing may remove meaningful patterns [31].

Future research could test hybrid or machine learning models-such as Prophet, XGBoost, or neural networks-on preprocessed datasets to explore the trade-off between interpretability, complexity, and accuracy in dynamic environments. However, such approaches typically require extensive feature engineering, longer time horizons, and careful validation, which fall outside the scope of the present study focused on interpretable classical models for operational forecasting.

References

1. Deptuła, A., et al.: Application of the methodology of multi-valued logic trees with weighting factors in the optimization of a proportional valve. Axioms **12**, 8 (2023). https://doi.org/10.3390/axioms12010008
2. Stosiak, M., et al.: Analysis of the impact of vibrations on a micro-hydraulic valve using a modified induction algorithm. Machines **11**, 184 (2023). https://doi.org/10.3390/machines11020184
3. Deptuła, A., Augustynowicz, A., Stosiak, M., Towarnicki, K., Karpenko, M.: The concept of using an expert system and multi-valued logic trees to assess the energy consumption of an electric car in selected driving cycles. Energies **15**(13) (2022). https://doi.org/10.3390/en15134631
4. Wang, R., Gong, C., et al.: Supply chain anomaly detection and prediction models based on large-scale time series data. In: Proceedings of the ACM International Conference on Data Engineering and Intelligent Systems (DEIS 2025), pp. 1–12. ACM, New York (2025)
5. Piechowski, M., Meller, A.: Use of KPI indicators to optimize production process performance–case study. In: Mehmood, R., Alves, V., Praça, I., et al. (eds.) Distributed Computing and Artificial Intelligence, Special Sessions I, 20th International Conference (DCAI 2023). Lecture Notes in Networks and Systems, vol. 741, pp. 3–12. Springer, Cham (2023)
6. Wang, Q.: A hybrid transformer-ARIMA model for forecasting global supply chain disruptions using multimodal data. Int. J. Adv. Comput. Sci. Appl. **16**(1) (2025)
7. Loska, A., Paszkowski, W., Waszkowski, R.: The scenario approach to the concept of maintenance of technical systems of urban engineering. In: Trojanowska, J., et al. (eds.) Advances in Manufacturing IV: Volume 2–Production Engineering: Digitalization, Sustainability and

Industry Applications. Lecture Notes in Mechanical Engineering, pp. 84–97. Springer, Cham (2024)
8. Vaz, C.B., et al.: Predicting retail store transaction patterns: a comparison of ARIMA and machine learning models. In: International Conference on Optimization, Learning Algorithms and Applications (OLAA 2024), pp. 1–12. Springer Nature, Cham (2024)
9. Borucka, A.: Seasonal methods of demand forecasting in the supply chain as support for the company's sustainable growth. Sustainability **15**(9), 7399 (2023). https://doi.org/10.3390/su15509739
10. Fatima, S.S.W., Rahimi, A.: A review of time-series forecasting algorithms for industrial manufacturing systems. Machines **12**(6), 380 (2024)
11. Serrano, A.L.M., et al.: Statistical comparison of time series models for forecasting Brazilian monthly energy demand using economic, industrial, and climatic exogenous variables. Appl. Sci. **14**(13), 5846 (2024)
12. Verma, R., Pandey, S., Rup, S.: Traffic flow prediction using LSTM, ARIMA & SARIMA model for intelligent transportation system. In: 2024 International Conference on Intelligent Computing and Sustainable Innovations in Technology (IC-SIT 2024). IEEE (2024)
13. Ziółkowski, J., Zieja, M., Oszczypała, M.: Forecasting of the traffic flow distribution in the transport network. Transport Means: Proceedings of the International Conference (2019)
14. Mbonyinshuti, F., et al.: Health supply chain forecasting: a comparison of ARIMA and LSTM time series models for demand prediction of medicines. Acta Logistica **11**(2), 269–280 (2024)
15. Thota, V.R.: Comparative study of time series forecasting for trucking shipments: Evaluating ARIMA, LSTM, and Hybrid ARIMA-LSTM Models. MS Thesis. State University of New York at Binghamton (2025)
16. Anh, N.T.N., Anh, N.N., Thang, T.N., Solanki, V.K., González Crespo, R., Quang Dat, N.: Online SARIMA applied for short-term electricity load forecasting. Appl. Intell. **54**, 1003-1019 (2024)
17. Ma, Y., He, L., Zheng, J.: STL-DCSInformer-ETS: a hybrid model for medium- and long-term sales forecasting of fast-moving consumer goods. Appl. Sci. **15**(3) (2025)
18. Omri, O., et al.: On the inventory performance of hierarchical forecasting approaches: case of a vaccine supply chain. J. Oper. Res. Soc. 1–27 (2025)
19. Shaibani, M.J., Fazaeli, A.A.: Robust prediction of healthcare inflation rate with statistical and AI methods in Iran. J. Appl. Math. **2024**(1), 1193134 (2024)
20. Algarni, A., et al.: Predictive energy management for Docker containers in cloud computing: a time series analysis approach. IEEE Access **12**, 52524–52538 (2024)
21. Rojek, I., Jasiulewicz-Kaczmarek, M., Piszcz, A., Galas, K., Mikołajewski, D.: Review of the 6G-Based supply chain management within Industry 4.0/5.0 Paradigm. Electronics **13**(13), 2624 (2024)
22. Oh, J., Seong, B.: Forecasting with a combined model of ETS and ARIMA. Commun. Statist. Appl. Methods **31**(1), 143–154 (2024)
23. Agaj, T., et al.: Using ARIMA and ETS models for forecasting water level changes for sustainable environmental management. Sci. Rep. **14**(1), 22444 (2024)
24. Yao, Y.: An investigation into the performance of time series models in predicting US E-commerce data. In: 2025 International Conference on Financial Risk and Investment Management (ICFRIM 2025). Atlantis Press (2025)
25. Patel, M., et al.: Enhancing accuracy in hourly passenger flow forecasting for urban transit using TBATS boosting. Modelling **6**(2), 32 (2025)
26. Gellert, A., et al.: Estimating electricity consumption at city-level through advanced machine learning methods. Connect. Sci. **36**(1), 2313852 (2024)
27. Duran, H.E., Elburz, Z., Değerli Çifçi, B.: The future of European regional inequalities: box-cox transformed ARMA process trend smoothing (BATS) forecasting. Growth Chang. **56**(2), e70031 (2025)

28. Antosz, K., Jasiulewicz-Kaczmarek, M., Machado, J., Relich, M.: Application of principle component analysis and logistic regression to support six sigma implementation in maintenance. Eksploatacja i Niezawodność–Maintenanced Reliab. **25**(4) (2023)
29. Chudo, S.B.: Modeling and forecasting time-series data with multiple seasonal periods using periodograms. Econometrics **13**(2), 14 (2025)
30. Gładysz, P., Merkisz, J., Borucka, A.: Reliability of unmanned aerial vehicles in the context of selected factors. Eksploatacja i Niezawodność-Maintenance Reliab. (2026). https://doi.org/10.17531/ein/210312
31. Guzanek, P., Sobecki, G., Bawoł, P., Borucka, A.: Wiarygodne prognozy kluczem do usprawnienia realizacji zamówień. Gospodarka Materialowa & Logistyka pp. 31–39 (2023). https://doi.org/10.33226/1231-2037.2023.1.4

A Multi-Case Document Analysis of Global AI Tools for Data-Driven Circular Manufacturing

Thi Bao Chau Huynh and Paulina Golińska-Dawson(✉)

Institute of Logistics, Poznan University of Technology, Poznan, Poland
paulina.golinska@put.poznan.pl

Abstract. Manufacturers are under increasing pressure to improve resource efficiency and reduce waste in order to support circular manufacturing practices. Artificial intelligence (AI) has been widely adopted to facilitate this change; however, there is limited empirical evidence on the effectiveness of industrial digital tools. This paper presents a document-centric analysis of three representative industrial platforms, focusing on their data acquisition methods, AI-based functionalities, operational support and outcomes. The results demonstrate that these platforms address various stages of the manufacturing process. One category supports asset reliability through continuous monitoring and predictive maintenance; another enhances production line performance by reducing scrap and auxiliary resource consumption; and a third improves quality assurance through AI-enabled visual inspection, resulting in reduced rework, scrap, and emissions. The findings suggest that circular manufacturing is facilitated by the combined use of industrial connectivity, line-level analytics, and specialized AI applications rather than by a single, comprehensive solution. This study provides practical insights into a field of research that has predominantly been conceptual.

Keywords: Circular manufacturing · industrial AI tools · Internet of Things (IoT) analytics · predictive maintenance · quality inspection · multi-case analysis · resource efficiency

1 Introduction

The linear model of extraction, production, consumption, and disposal is no longer viable as it depends on the assumption of unlimited resources [1]. Therefore, the circular economy (CE) has gained increasing attention because it seeks to keep materials and products at their highest value, as long as possible, and reduce environmental impacts [2]. Manufacturers are now being urged to support this transition, since circular manufacturing can decouple industrial growth from environmental harm [3, 4]. However, firms still require clearer guidance on how to implement circularity in practice. [5] suggest that CE can improve sustainability at multiple levels of the economy, from individual firms to regional systems. Its potential increases when supported by digital technologies, especially AI. Recent studies show that AI, Digital Twins (DT) and IoT infrastructures strengthen process monitoring and decision making in Industry 4.0 and 5.0 [6]. In this

J. Trojanowska et al. (Eds.): MANUFACTURING 2026, LNME, pp. 141–153, 2026.
https://doi.org/10.1007/978-3-032-21654-0_11

context, DTs rely on real-time sensor data to produce virtual representations of physical assets [7]. Edge and cloud computing reduce latency in these systems and IoT enables continuous observation of machine conditions [8]. In addition, the DT concept makes it possible to accurately check the condition of the equipment. Predictive maintenance (PdM) is becoming more common in industry, which uses inspections or diagnostics to guess how long important parts will last [9]. AI, through integration with technologies such as DT and IoT, supports circular manufacturing by improving efficiency, enabling predictive maintenance, and more sustainable resource management.

The academic literature has examined the role of AI in recycling and circular processes, but much of this work remains conceptual. [10] show the scope of AI applications through large-scale text analysis, but do not evaluate industrial tools. Similarly, [11] propose frameworks for the adoption of AI in SMEs without assessing a specific platform. Furthermore, [12] outline how AI can support prediction, inspection, and sorting in circular systems, but does not analyze operational technologies. Furthermore, although [13] demonstrate the use of ThingWorx and Kepware in the implementation of the industrial Internet of Things (IIoT) for factory optimization and logistics management, the study does not evaluate how this tool supports circular manufacturing functions such as component recovery, lifecycle extension, material reuse, or compares its performance with other AI platforms in a multicase framework. Taken together, there is still no systematic multi-case analysis that evaluates industrial AI platforms within a circular manufacturing context. Published research has not shown how documented AI tools support circular functions such as data acquisition, quality control, process optimization, and sustainability. To fill this gap, the purpose of this paper is to examine how leading industrial AI platforms operationalize data-driven circular manufacturing. The study analyzes MindSphere, PTC ThingWorx, and IBM Maximo Visual Inspection based on their technical documentation and case evidence. It investigates their capabilities for data integration, analytics, and circular outcomes. The study addresses the following research questions (RQ):

RQ1: How do the selected AI tools operationalize data-driven circular manufacturing in practice?
RQ2: Which circular manufacturing functions, such as data collection, quality assurance, process optimization, and sustainability, are most strongly supported by each tool?
RQ3: What strengths and limitations can be identified across these tools in terms of readiness, scalability, and alignment with circular manufacturing principles.

The reminder of the paper is organized as follows. Section 2 presents the theoretical background of this research. Section 3, describes the research methodology. The results are presented in Sect. 4. The discussion of the results and the final conclusions is stated in Sect. 5.

2 Theoretical Background - Analytical Framework for This Study

2.1 Circular Manufacturing and the Current Challenges

Companies are increasingly adopting sustainability-oriented frameworks, and within this context, the circular economy (CE) has emerged as a fundamental approach to reduce waste and improving resource efficiency [14]. Research on Circular Manufacturing (CM) has expanded significantly in recent years, as scholars seek to understand how production systems can retain the value of products and materials while minimizing environmental impacts. However, barriers such as organizational inertia, budgetary constraints, and limited understanding of the principles of circular economy continue to hinder broader implementation [15]. Moreover, an effective method for achieving the triple bottom line (3BL) in enterprises is the establishment of closed-loop supply chains (CLSCs) that encompass the entire life cycle [16].

Digitalization has become a central enabler of this transformation. In this regard, the convergence of digital technologies with the circular economy and life cycle assessment can significantly advance sustainability practices by improving resource efficiency and improving the accuracy of environmental impact evaluations in multiple sectors [17]. However, despite the growing importance of CM, much of the existing research remains largely conceptual and rarely examines how firms operationalize circular strategies through specific technological tools. This gap constrains a practical understanding of how circularity can be effectively integrated into everyday manufacturing operations.

2.2 AI and Data-Driven Technologies in Manufacturing

Recent advancements in AI particularly in Machine Learning (ML), demonstrate significant promise to revolutionize the manufacturing sector by utilizing sophisticated analytics tools to process the extensive volumes of manufacturing data, referred to as Big Data [18].

Despite the significant interest in DT in several fields, the literature suggests that the idea remains theoretically ambiguous, as definitions of digital transformation have not achieved conceptual stability, underscoring the continuous need for revision as new technologies arise and current ones become obsolete [19]. [20] investigated four DT categories within the framework of the circular economy and identified thirty-one applications in eight industries. IoT was the most widely implemented technology, succeeded by big data analytics, AI, and blockchain [21]. However, the utilization of these technologies for sustainability objectives remains constrained and predominantly in the experimental phase. Prominent uses include IoT for data acquisition, AI for predictive maintenance, big data analytics for demand forecasting, and blockchain for traceability. The beneficial impact of AI on resource efficiency is enhanced under rigorous environmental restrictions, suggesting that companies subjected to elevated pollution governance and carbon emission restrictions obtain more advantages from AI technologies [21]. In the automotive industry, data from On-Board Diagnostics (OBD) systems offer ongoing insights into vehicle condition and performance, facilitating data-driven diagnostics that prolong component lifespan and minimize unnecessary replacements. [22] introduces a hybrid Predictive Maintenance (PdM) model that integrates Long Short-Term Memory (LSTM) networks with K-means clustering to analyze unlabeled time-series data and precisely forecast engine health. Together, these studies confirm that AI-supported systems improve efficiency, reliability, and resource use, thus supporting the transition towards circular manufacturing.

2.3 Analytical Framework for This Study

The analytical framework utilized in this study is based on three recurring themes identified in the reviewed literature. The first is about the basics of circular manufacturing, which focuses on saving resources by reusing, remanufacturing and extending their life. These strategies continue to be difficult to implement due to operational and organizational limitations [14]. The second theme is about AI-powered manufacturing intelligence. DT, IoT data collection, and predictive maintenance are all ways to keep an eye on assets, predict failures, and make better use of resources [20]. The third theme highlights the role of data-driven loops that connect operational activity to continuous information flows. This loop involves data acquisition, analytical processing, operational intervention, and feedback into future decisions, enabling improvements in efficiency and material preservation.

Together, these three themes form an integrated analytical framework that links circular manufacturing principles with concrete digital and analytical capabilities. The framework provides a structured lens for examining how industrial AI platforms operationalize circular strategies at different levels of manufacturing systems. This structure is illustrated in Fig. 1. The study concentrates on three industrial AI platforms, within this framework:

- MindSphere, rebranded As Insights Hub, is an industrial IoT platform that integrates sensor data acquisition, cloud analytics, and AI-driven applications to support predictive maintenance, asset monitoring, and data-driven optimization.
- ThingWorx, is an industrial IoT and digital thread platform that connects machines, sensors, and enterprise systems to enable real-time monitoring, prescriptive analytics, and digital twin-based optimization across manufacturing operations.
- Maximo Visual Inspection, is a computer vision platform that applies deep-learning models to image and video inputs for automated defect detection, quality assessment, and component classification in industrial production environments.

These tools are mature, widely used solutions that combine IoT connectivity, AI-based analytics, and industrial applications. There is a lot of publicly available technical documentation and case material to back them up. As a whole, they provide a good empirical basis to look at how data-driven platforms put circular manufacturing principles into practice in different areas of business, such as monitoring, optimization, and quality control.

3 Methodology

The study applies a qualitative multi-case design to examine how industrial AI platforms support data-driven circular manufacturing. This approach is suitable for analyzing contemporary technological systems, following [23] case study principles. Document analysis was selected as the main method because operational industrial data are proprietary, whereas technical documentation provides reliable evidence for platform capabilities [24].

In this study, three platforms were selected for examination: MindSphere (Case A), ThingWorx (Case B), and Maximo Visual Inspection (Case C). These systems were

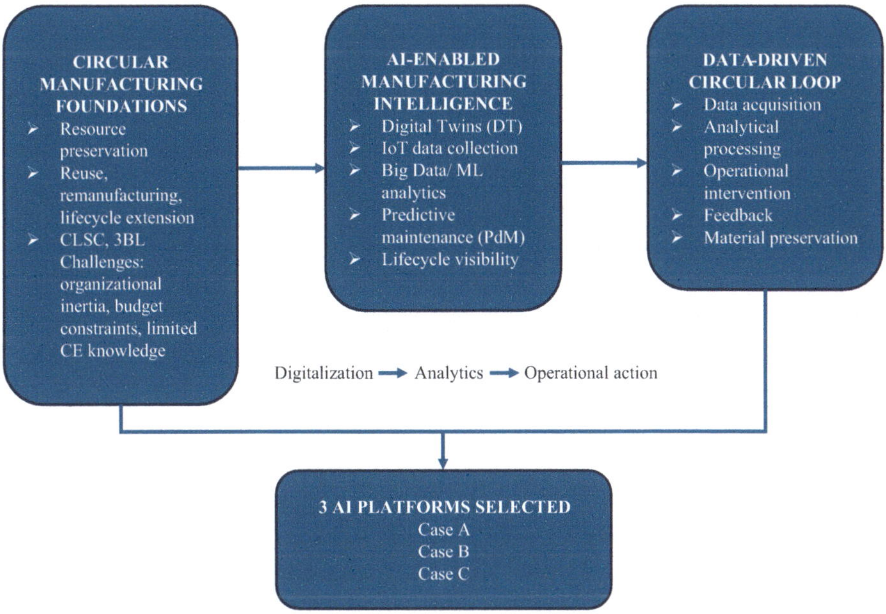

Fig. 1. Analytical framework applied to three industrial AI platforms. Source: Own elaboration.

chosen because they have a strong global presence, provide rich publicly available documentation, and are closely connected to the technological requirements of modern IoT-based and data-driven manufacturing environments.

At the same time, the study recognizes that publicly available documents differ in scope, depth, and level of aggregation between platforms. The selected cases represent different levels of industrial application, ranging from cross-industry asset platforms to line-level and product-level implementations. As a result, the reported performance indicators reflect distinct operational contexts rather than directly comparable metrics. Therefore, the analysis treats quantitative values as documented evidence of functional impact within specific contexts, not as standardized benchmarks across tools. This positioning allows the study to examine how each platform operationalizes circular manufacturing functions while acknowledging the inherent limitations and potential bias of vendor-provided case material.

Data were drawn from technical white papers, product manuals, validated case studies, and peer-reviewed publications. Only authoritative and publicly available documents were included. Vendor white papers and official documentation were used to identify the architecture and core functionalities. The product manuals and technical specifications supported the analysis of the data acquisition and analytics capabilities. Validated industrial case studies provided documented performance indicators such as reductions in downtime, scrap, rework, or resource use. Peer-reviewed publications were included to contextualize the platforms within existing research on AI and circular manufacturing.

A deductive coding procedure was used. Coding categories were derived from the analytical framework and covered data acquisition, analytics, operational functions, circular outcomes, and business value. The structured Excel matrix served as the coding tool, allowing consistent extraction and systematic comparison between cases. The quantitative indicators were coded together with their documented industrial context to avoid direct performance comparison between platforms.

Cross-case analysis followed [23] pattern matching logic. The patterns were identified first and then compared between cases to reveal similarities and differences in how each platform operationalizes circular manufacturing functions. Validity was enhanced through triangulation of document types, transparent coding steps, and replication logic across the three cases.

4 Results

4.1 Analysis of the Platforms

This section presents the findings for three platforms (Case A, Case B, and Case C). Each case is examined across four dimensions: data acquisition, AI analytics, operational functions, and contributions to circular manufacturing. All business value indicators are derived from the reviewed documents. A cross-case analysis follows the individual case descriptions. The reported quantitative indicators originate from tool-specific industrial implementations and are therefore interpreted as context-dependent rather than universally comparable. This interpretation is consistent with prior circular economy assessments, which show that performance metrics are sensitive to methodological choices and operational settings [25].

Table 1 summarizes the coded evidence for Case A, Table 2 for Case B, and Table 3 presents the corresponding results for Case C.

Case A: MindSphere has been rebranded as Insights Hub as part of a broader digital industrial ecosystem strategy. The platform retains its core IIoT and analytics architecture, including asset connectivity, cloud-based data processing, and predictive maintenance capabilities [26]. The platform is deployed at the asset and plant levels across distributed sites. The reported performance indicators reflect system-level operational trends rather than line-specific outcomes. This context explains the emphasis on asset reliability and operational stability rather than direct material or scrap-related metrics [27]. The platform functions as an IIoT and analytics solution for industrial assets. The white paper Four Pillars of the Industrial IoT and the IndustryWeek survey reported in [27] provide evidence of how firms use operational data to reduce downtime and operating costs. This platform supports circular manufacturing primarily at the asset level. Reductions in downtime and operating costs extend asset life and reduce premature equipment replacement. The results indicate a closed operational loop based on continuous monitoring, analytics, and condition-based maintenance. This loop aligns with circular manufacturing principles that emphasize resource retention through improved asset reliability. The reviewed documents focus on operational efficiency rather than material flows. Accordingly, the platform serves as a data foundation for circular strategies but does not directly manage material recovery or reuse processes. This distinction

Table 1. Coded evidence for Case A. Source: own elaboration.

Element	Key content	Business value
Data acquisition	Collection of equipment and machine data through IIoT, OPC UA and gateways	The IndustryWeek survey reports that most firms use operational data to measure downtime, cost, and productivity
AI and analytics	Anomaly detection and predictive analytics based on real-time data	Predictive maintenance reduces unexpected failures and improves maintenance planning
Operational functions	Asset monitoring, early warnings, and energy optimization in factories	Firms report lower downtime, lower operating cost, and higher output
Circular contribution	Extension of the useful life of the asset and reduction of waste from unplanned failures	54% of companies reduce downtime. 52% reduce operating cost. 46% increase productivity. 46% gain better asset visibility. 41% improve operational transparency. 30% increase in revenue
Business value	Support for connectivity, control, digitalization, and augmentation	Firms achieve lower maintenance costs and fewer unplanned stops

contrasts with the following cases, which address scrap reduction, quality losses, and emissions more directly.

Case B Includes a large-scale deployment across multiple production lines covering diaper, hygiene, and aerosol manufacturing [28]. The values therefore reflect improvements within a defined organizational and operational context. Differences in production scale, product type, and baseline performance limit direct comparison with other cases in this study. The Case B data from [28] presents the deployment of ThingWorx and Kepware across fifty production lines. Real-time data supports monitoring of scrap, OEE, safety and paper use. The case shows how IoT data reduces material waste and improves the use of auxiliary resources.

Platform Case B supports circular manufacturing at the production line level. Scrap was reduced by 23%, corresponding to approximately two million preserved products. This outcome demonstrates a direct link between the utilization of real-time data and material preservation. The reduction of five tons of paper per year extends the circular impact to auxiliary resources, while savings of 1,500 labor hours represent a secondary benefit resulting from the elimination of manual reporting. Case B demonstrates a complete data loop in which data collection enables detection, detection enables intervention, and intervention reduces waste while improving efficiency. Improvements in overall equipment effectiveness (OEE), quality, and safety further confirm the value of IoT technologies to enable circular practices in production environments. Overall, the

Table 2. Coded evidence for Case B. Source: own elaboration.

Element	Key content	Business value
Data acquisition	Real-time data from PLCs and sensors across about fifty lines	Data provides insight into scrap, downtime and OEE for diaper, aerosol and hygiene lines
AI and analytics	Real-time analysis and early quality warnings	The system detects trends in scrap earlier than manual checks
Operational functions	Shop-floor dashboards and digital forms	Multiple lines are monitored on one screen and paper reports are removed
Circular contribution	Reduction of scrap and paper use and an increase in process transparency	Scrap on the diaper line drops 23%. The reduction equals two million products and about USD 250,000. Paper use drops five tons per year. The shift saves 1,500 working hours
Business value	Gains in OEE and safety	OEE rises 3% in the hygiene plant. Aerosol performance rises 2%. Safety incidents dropped 60%. QC pass rate rises from 75% to 98%

case illustrates how data-driven approaches support both primary material flows and supporting operational activities.

Case C Is derived from a documented industrial application in which computer vision was used to automate quality inspection activities within the production process [29]. Accordingly, the reported results reflect the impact of artificial intelligence at the quality inspection stage rather than system-wide manufacturing performance. Platform in Case C applies deep learning techniques for automated quality control by processing images in near real time to identify defective products. The documented case in [30] reports significant reductions in scrap, rework, and associated emissions.

Platform C supports circular manufacturing at the product level. The model identifies defects with high accuracy, resulting in a 20% reduction in scrap and a 30% reduction in rework. These improvements prevent the unnecessary consumption of materials and energy. An annual reduction of 25 tons of CO_2 demonstrates clear environmental benefits. The tool operates at a critical decision point in the production process, where firms determine whether a product should be accepted, repaired, or discarded. Unlike previously analyzed platforms (Case A & Case B), platform C does not function across the entire factory, but instead targets a single production stage. Despite its limited scope, the effects on material flows and emissions are substantial. This finding indicates that circular manufacturing benefits not only from comprehensive IoT platforms but also from specialized AI modules tailored to specific operational decisions.

Table 3. Coded evidence for Case C

Element	Key content	Business value
Data acquisition	Image and video capture from industrial cameras	The system processes 2,000 products per hour with a 10 ms inference time
AI and analytics	CNN-based detection and localization of defects	Precision reaches 92%. Recall reaches 89%. IoU reaches 0.83
Operational functions	Automated quality checks and product classification	Manual inspection is reduced and QC becomes consistent
Circular contribution	Reduction of scrap and rework and lower emissions	Scrap drops 20%. Rework drops 30%. CO_2 decreases 25 tons per year. Energy per inspection is 0.05 kWh
Business value	Lower production and labor cost	Scrap savings reach USD 100,000 per year. Labor savings reach USD 50,000 per year. Compliance remains at 100%

4.2 Cross-Case Analysis

The cross-case analysis emphasizes the comparison of the functional roles and underlying mechanisms of the platforms, rather than their quantitative performance results. This approach acknowledges that the platforms function at varying analytical tiers, encompassing asset management, line-level optimization, and product-level evaluation. This focus aligns with methodological recommendations on cross-case analysis, which stresses the importance of identifying similarities and contrasts among instances to facilitate theoretical interpretation rather than direct numeric comparison [31].

The three cases present different levels of intervention in circular manufacturing, as Platform A operates at the asset level, Platform B operates at the line level, and Platform C operates at the product level. Each level creates distinct circular outcomes. Table 4 provides a functional cross-case comparison.

Cases A–C show that circular manufacturing depends on various analytical layers. Platform A strengthens asset reliability, while Platform B reduces material losses and auxiliary resource consumption. Platform C minimizes scrap, rework, and emissions through highly accurate inspections. No single tool can address the entire circular value chain. These findings align with the identified research gap, as prior studies largely describe the role of AI in circular manufacturing at a conceptual level. In contrast, the presented cases illustrate how these roles are implemented in practice, albeit with varying scopes and depths. The results underscore the need for an ecosystem perspective on AI tools, positioning this study as an initial comparative analysis of real industrial platforms that support data-driven circular manufacturing.

Table 4. Cross-case comparison. Source: own elaboration.

Aspect	Platform A	Platform B	Platform C
Data scope	Asset data	Line-level operational data	Image data
AI role	Prediction & monitoring	Real-time analysis	Defect detection
Circular lever	Asset life extension	Scrap and paper reduction	Scrap and rework reduction
Evidence strength	Industry survey	Multi-line case evidence	Single-line deep evidence

5 Discussion and Conclusions

Analyzing the three cases shows that circular manufacturing does not depend on one tool. Rather, it requires a combination of tools because each one addresses a different analytical layer within a circular structure that spans the entire production system. Thus, circularity emerges from an ecosystem of AI-enabled tools rather than from a single platform. To scale circular manufacturing, firms must integrate an IoT platform to stabilize equipment performance, a line-level analytics system to reduce material waste, and a specialized AI module to manage critical decision points, such as quality inspection. This integrated approach generates quantifiable circular outcomes and aligns with the structure of industrial operations. It also reflects manufacturing environment realities, where circularity is achieved through coordinated approach rather than isolated technologies.

These findings further confirm that none of the analyzed AI platform supports the entire circular manufacturing. Each tool contributes to circular outcomes through a specific functional role at the asset, line, or product level. Therefore, circular manufacturing depends on an ecosystem of complementary AI tools rather than isolated solutions. While the analyzed tools support complementary circular functions, differences in system architectures, data models, and interoperability standards constrain their integration. Previous research on smart manufacturing systems suggests that integrating multiple vendors often necessitates additional middleware or standardized interfaces, which can hinder the seamless exchange of data across platforms [32].

This study does not evaluate vendor marketing claims or attempt to independently validate reported performance metrics. Quantitative indicators are interpreted as documented outcomes within specific industrial contexts. The analysis focuses on how documented functionalities illustrate the operational mechanisms through which industrial AI tools can support circular manufacturing. All three tools are proprietary commercial platforms. A direct evaluative ranking would risk analytical bias. Therefore, a functional comparison approach is adopted to maintain neutrality and align with qualitative research standards for industrial technologies.

The study aims to examine how three industrial AI tools support data-driven circular manufacturing and to clarify the circular functions and limitations of each tool. Previous studies describe the role of AI in circularity at a conceptual level. This study extends existing knowledge by examining documented industrial platforms that are already deployed in manufacturing environments. The analysis focuses on operational roles and documented functionalities rather than on validating absolute performance outcomes.

This study has several limitations, as it uses publicly available documents, so the level of detail differs between cases. This research does not access internal operational data, so it cannot verify the original performance metrics. Reported quantitative indicators are therefore interpreted as context-specific and illustrative rather than as generalizable benchmarks. The scope is limited to three well-documented tools, so it does not cover the full industrial AI landscape.

Future studies can expand in three directions. First, more AI tools from wider industrial ecosystems can be analyzed to assess circularity across sectors. Second, interviews and operational data can be combined to increase reliability and validate reported metrics. Third, new assessment models can be developed to evaluate circular outcomes throughout the product life cycles and to study the impact of using multiple AI tools within a single value chain. These directions would support a more integrated understanding of how AI-enabled systems jointly contribute to circular manufacturing.

Acknowledgments. The authors gratefully acknowledge the support of the Poznan University of Technology, Institute of Logistics for providing an academic environment conducive to this research. The authors further thank the reviewers for their constructive comments and valuable suggestions, which significantly contributed to improving the clarity, structure, and academic quality of the paper.

References

1. Kara, S., Hauschild, M., Sutherland, J., McAloone, T.: Closed-loop systems to circular economy: a pathway to environmental sustainability? CIRP Ann. **71**, 505–528 (2022). https://doi.org/10.1016/j.cirp.2022.05.008
2. Bjørnbet, M.M., Skaar, C., Fet, A.M., Schulte, K.Ø.: Circular economy in manufacturing companies: a review of case study literature. J. Clean. Prod. **294**, 126268 (2021). https://doi.org/10.1016/j.jclepro.2021.126268
3. Sakao, T., Bocken, N., Nasr, N., Umeda, Y.: Implementing circular economy activities in manufacturing for environmental sustainability. CIRP Ann. **73**, 457–481 (2024). https://doi.org/10.1016/j.cirp.2024.06.002
4. Fohl, F., Rajčić, V., Ungureanu, V., et al.: Circular Manufacturing. In: Bragança, L., Griffiths, P., Askar, R., et al. (eds.) Circular Economy Design and Management in the Built Environment: A Critical Review of the State of the Art. Springer Nature Switzerland, Cham, pp. 69–91 (2025)
5. Acerbi, F., Forterre, D.A., Taisch, M.: Role of artificial intelligence in circular manufacturing: a systematic literature review. IFAC-Pap **54**, 367–372 (2021). https://doi.org/10.1016/j.ifacol.2021.08.040
6. Rojek, I., Mikołajewski, D., Dostatni, E., et al.: Personalization of AI-Based digital twins to optimize adaptation in industrial design and manufacturing—Review. Appl. Sci. **15**, 8525 (2025). https://doi.org/10.3390/app15158525
7. Bellavista, P., Bicocchi, N., Fogli, M., et al.: Requirements and design patterns for adaptive, autonomous, and context-aware digital twins in industry 4.0 digital factories. Comput. Ind. **149**, 03918 (2023). https://doi.org/10.1016/j.compind.2023.103918

8. Botín-Sanabria, D.M., Mihaita, A.-S., Peimbert-García, R.E., et al.: Digital twin technology challenges and applications: a comprehensive review. Remote Sens. **14**, 1335 (2022). https://doi.org/10.3390/rs14061335
9. Abd Wahab, N.H., Hasikin, K., Wee Lai, K., et al.: Systematic review of predictive maintenance and digital twin technologies challenges, opportunities, and best practices. PeerJ Comput. Sci. **10**, e1943 (2024). https://doi.org/10.7717/peerj-cs.1943
10. Kim, Y., Ye, W., Kumar, R., et al.: Unlocking the potential of remanufacturing through machine learning and data-driven models—A survey. Algorithms **17**, 562 (2024). https://doi.org/10.3390/a17120562
11. Corti, D., Matteri, D., Masiero, S., et al.: Definition of a solution space to guide AI adoption in manufacturing SMEs. In: IFAC-PapersOnLine. Elsevier B.V., pp. 2298–2303 (2025)
12. Raut, S., Hossain, N.U.I., Kouhizadeh, M., Fazio, S.A.: Application of artificial intelligence in circular economy: a critical analysis of the current research. Sustain. Futur. **9**, 100784 (2025). https://doi.org/10.1016/j.sftr.2025.100784
13. Doshi, S., Ghatage, P., Borole, R.: IIoT for Factory optimization and logistics management using PTC Thingworx and Kepware (2023)
14. Mansouri, A., Bellat, A., Bennis, I., et al.: Enhancing circular economy in reconditioned spare parts through artificial intelligence and genetic algorithms. Results Eng. **28**, 107122 (2025). https://doi.org/10.1016/j.rineng.2025.107122
15. Bourdin, S., Jacquet, N.: Closing the loop at the local scale: Investigating the drivers of and barriers to the implementation of the circular economy in cities and regions. Ecol. Econ. **231**, 108542 (2025). https://doi.org/10.1016/j.ecolecon.2025.108542
16. Ramanathan, U., He, Q., Subramanian, N., et al.: Collaborative closed-loop supply chain framework for sustainable manufacturing: evidence from the Indian packaging industry. Technol. Forecast. Soc. Change **191**, 122489 (2023). https://doi.org/10.1016/j.techfore.2023.122489
17. Hariyani, D., Hariyani, P., Mishra, S., Kumar Sharma, M.: Leveraging digital technologies for advancing circular economy practices and enhancing life cycle analysis: a systematic literature review. Waste Manag. Bull. **2**, 69–83 (2024). https://doi.org/10.1016/j.wmb.2024.06.007
18. Arinez, J.F., Chang, Q., Gao, R.X., et al.: Artificial intelligence in advanced manufacturing: current status and future outlook. J. Manuf. Sci. Eng. (2020). https://doi.org/10.1115/1.4047855
19. Abdelrahman, M., Macatulad, E., Lei, B., et al.: What is a Digital Twin anyway? Deriving the definition for the built environment from over 15,000 scientific publications. Build. Environ. **274**, 112748 (2025). https://doi.org/10.1016/j.buildenv.2025.112748
20. Schöggl, J.-P., Rusch, M., Stumpf, L., Baumgartner, R.J.: Implementation of digital technologies for a circular economy and sustainability management in the manufacturing sector. Sustain. Prod. Consum. **35**, 401–420 (2023). https://doi.org/10.1016/j.spc.2022.11.012
21. Zhao, L., Xu, J., Zhang, B., Lu, J.: Leveraging AI to enhance firms' resource efficiency: ecological modernization theory and resource-based view perspectives. Int. J. Prod. Econ. (2025). https://doi.org/10.1016/j.ijpe.2025.109723
22. Errezgouny, A., Chater, Y., González, C.D.B., Cherkaoui, A.: An integrated deep learning approach for predictive vehicle maintenance. Decis Anal J **16**, 100597 (2025). https://doi.org/10.1016/j.dajour.2025.100597
23. Yin, R.K.: Case study research and applications: design and methods, Sixth edition. SAGE, Los Angeles London New Delhi Singapore Washington DC Melbourne (2018)
24. Bowen, G.A.: Document analysis as a qualitative research method. Qual. Res. J. **9**, 27–40 (2009). https://doi.org/10.3316/QRJ0902027

25. Pakuła, N., Łapniewska, Z., Dutra, C.J.C.: Comparative measurements of circular economy performance among European countries: reviewing approaches and limitations. J. Environ. Manage. **375**, 124414 (2025). https://doi.org/10.1016/j.jenvman.2025.124414
26. Insights Hub industrial IoT | Siemens Software. In: Siemens Digit. Ind. Softw. https://plm.sw.siemens.com/en-US/insights-hub/. Accessed 14 Jan 2026
27. Siemens's White paper: Siemens technical document template A4 portrait (2023)
28. PTC Case Studies: Evyap Digitally Transforms with DVM IoT Solutions | PTC (2023). https://www.ptc.com/en/case-studies/evyap-dvm-iot-solutions-reduce-scrap-improve-oee. Accessed 19 Nov 2025
29. Prolifics: AI-Powered quality assurance with IBM maximo-prolifics (2025). https://prolifics.com/usa/resource-center/case-studies/ai-powered-quality-assurance. Accessed 19 Nov 2025
30. IBM Prolifics Case Study: IBM MVI's case study (2025)
31. Ridder, H.-G.: The theory contribution of case study research designs. Bus. Res. **10**, 281–305 (2017). https://doi.org/10.1007/s40685-017-0045-z
32. Zeid, A., Sundaram, S., Moghaddam, M., et al.: Interoperability in smart manufacturing: research challenges. Machines (2019). https://doi.org/10.3390/machines7020021

Artificial Intelligence in Sustainable Product Design and Manufacturing: A Bibliometric Analysis

Weronika Marchewka(✉) and Paulina Golińska-Dawson

Faculty of Management Engineering, Poznan University of Technology, Poznan, Poland
weronika.marchewka@doctorate.put.poznan.pl

Abstract. Artificial intelligence plays an increasingly important role in supporting sustainable product design, eco-innovation, and circular manufacturing. Although research in this area is expanding, its development dynamics and thematic orientation have not been systematically examined yet. This study fills this gap by providing a bibliometric overview of how AI-related technologies intersect with sustainability-driven design and manufacturing concepts. A structured search strategy was applied to publications from 2015–2025 using the Biblioshiny tool. The data set was built using keywords related to artificial intelligence, eco-design and sustainable or circular manufacturing. The analysis included annual publication trends, keyword co-occurrence mapping, and thematic clustering. The findings showed that the field is evolving towards a strong integration of digital and sustainability perspectives. AI-driven tools and data-centric methods. These are becoming increasingly important in the development of sustainable or circular manufacturing and product development. As the area is rapidly expanding, future research should analyse publication patterns after 2024 to determine whether the sharp surge indicates a stable long-term shift or a short-term acceleration and also what implications this may have for the field's further development.

Keywords: Smart Manufacturing · Eco-Design · Circular Manufacturing · Artificial Intelligence · Sustainability

1 Introduction

Today's modern manufacturing should depend on production systems that use material processing technologies in a way that minimises energy consumption, greenhouse gas emissions, waste generation, and the use of non-renewable resources [1]. The manufacturing industry operates in a dynamic and constantly changing environment, shaped by increasing uncertainty and complexity caused by financial crises, pandemics, or geopolitical conflicts [2–4]. The literature emphasises that sustainable development is not only a response to global environmental challenges, but also strengthens organisational resilience, competitiveness, and the ability of companies to adapt to changing conditions [2]. Sustainable practices also help firms deal with supply chain disruptions,

J. Trojanowska et al. (Eds.): MANUFACTURING 2026, LNME, pp. 154–167, 2026.
https://doi.org/10.1007/978-3-032-21654-0_12

rising energy costs, climate regulations, and growing customer expectations, as modern consumers increasingly expect products to be environmentally friendly [5]. As a result, manufacturing companies that are engaged in environmental protection can at the same time improve their reputation, build customer loyalty, and increase competitive advantage by implementing sustainability-orientated practices, such as circular economy models [6]. The literature also notes that sustainable development within modern industrial policy includes decarbonisation, reducing environmental impact, developing green technologies and circular models, building new competencies, and integrating social and environmental perspectives [4].

As sustainability pressures grow, the design of the product becomes crucial since most environmental impacts are determined at the very early stage of a lifecycle. Regulations and eco-innovations strongly shape sustainable production [1], forming the foundation of eco-design. Design practices are rapidly changing under sustainability and circular economy requirements [7]. Environmental criteria are still addressed too late, even though design decisions define materials, technologies, and functionality [8]. Eco-design should therefore be embedded in routine engineering and aligned with circular models based on extended product life, reuse, remanufacturing, and recycling [9]. To achieve these goals, advanced data-driven tools to support complex design and decision making. Digital transformation enhances sustainability through better energy efficiency, waste reduction, and responsible resource use [10]. However, knowledge of the role remains fragmented: few studies combine advanced AI with solid sustainability principles, and most focus on isolated technologies rather than integrated sustainable product design [11].

This paper aims to fill this gap by providing a bibliometric overview of how AI-related technologies intersect with sustainability-driven design and circular manufacturing concepts. The main contribution is a systematic analysis that integrates research on artificial intelligence, sustainable design, and sustainable or circular manufacturing. This approach reveals key trends, uncovers thematic gaps, and outlines future directions of development within this interdisciplinary field. The remainder of the paper is organised as follows: Sect. 2 outlines the theoretical background and identifies research gaps. Section 3 presents the research problem and methodology, while Sect. 4 reports the results of the bibliometric analysis. Section 5 provides the discussion and Sect. 6 offers the final conclusions.

2 Theoretical Background

2.1 The Role and Potential of AI in Product Design and Circular Manufacturing

Digital transformation remains a key pillar of Industry 4.0 because technologies such as AI, IoT, big data, and automation significantly improve operational efficiency, support sustainable development, and enable the creation of intelligent, connected production systems [12]. Today, we also a transition from Industry 4.0 to Industry 5.0, which affects how sustainability goals are understood, especially by highlighting the role of humans in production systems [3, 13]. Recent research [14] shows that Industry 5.0 increasingly focuses on human-centred AI solutions that enhance smart manufacturing while supporting sustainability goals. Technologies such as additive manufacturing, IoT and AI make production more flexible, data-driven, and energy-efficient, while also improving workers' well-being and enabling customised product design. These systems, supported by advanced connectivity, allow companies to combine intelligent automation with stronger human involvement.

Among the digital technologies that transform modern production systems, artificial intelligence plays a particularly important role because it can analyse complex data and optimise product and process performance. AI includes a wide range of technologies, such as machine learning, neural networks, and expert systems, which automate tasks that traditionally require human intelligence. These technologies help solve modern production problems such as optimising resource use, reducing downtime through predictive maintenance, and improving product quality by automatically detecting defects [15].

As noted by Kumar and Shahin [16], AI supports optimisation in modern product design and manufacturing by improving resource management, reducing waste, and increasing energy efficiency through machine learning,, deep learning and computer vision. These technologies also enable circular design by integrating LCA methods and optimising material use, which helps reduce the carbon footprint. AI is also an important tool for planning and controlling production, quality inspection, and implementing Industry 5.0 solutions. For example, machine learning models can monitor tool wear, optimise tool usage, and shift maintenance strategies from preventive to predictive, reducing downtime [17]. In sustainable product design, AI, big data, and machine learning reduce development time, lower costs, improve resource efficiency, and limit environmental impact [18].

The concept that strengthens the role of AI-based tools in supporting sustainable product design and manufacturing is the Digital Product Passport (DPP). According to Regulation (EU) 2024/1781 [19], this concept was presented as a data-driven instrument to enhance product transparency, lifecycle management, and circularity. As it is mentioned in the literature, DDP enable data-driven interconnections between industrial ecosystems and create favourable conditions for AI-supported optimization of circular manufacturing [20]. Current DPP frameworks remain largely static, which highlights the need for AI-supported solutions that allow dynamic lifecycle data updated to better support circular economy objectives [21].

2.2 AI in Eco-Design and LCA

AI not only improving production efficiency, but is increasingly used in the early stages of product development, especially in eco-design and life-cycle assessment. It enables the integration of environmental criteria from the very beginning of the design process. According to Ashkbous et al. [22], AI is most commonly used in recycling-orientated design strategies and material selection, supporting activities such as evaluating sustainable alternatives, choosing clean or recovered materials, and designing for disassembly. Most solutions are based on neural networks, while generative models and large-language models increasingly support automated searching and analysing LCA knowledge bases. However, researchers emphasise the need to further develop tools that combine AI with a holistic life-cycle perspective.

LCA is a key element of eco-design, which opens many opportunities for AI-based solutions. Designers often struggle to use environmental data effectively because of its complexity, time limitations, or lack of expertise. In this context, AI tools are especially valuable when LCA data are incomplete or difficult to interpret [23]. AI can automate the analysis of extensive LCA documentation, extract key information quickly, and support interpretation. Furthermore, trained machine learning models can predict environmental impacts even before the product is manufactured [23–25].

2.3 Research Gap

Despite growing interest in the use of artificial intelligence in supporting sustainable development within the manufacturing industry, the current state of knowledge remains fragmented and lacks coherence. As noted by Gohr et al. [11], only a small number of studies effectively combine advanced AI applications with a deeper understanding of sustainability principles. This means that most existing publications focus on isolated technological aspects without clearly linking them with the goals and practices of sustainable product design and production.

At the same time, an increasing number of studies point to the potential of AI in circular economy applications. Raut et al. [26] emphasise that recent work highlights emerging AI trends in this domain and introduces frameworks designed to guide the implementation of AI-based tools for building technical capabilities that enhance circular performance. However, although such conceptual models exist, the field still lacks a comprehensive perspective showing how AI supports circularity from both design and manufacturing points of view. Furthermore, as observed by Moghrabi et al. [10], the steady rise in the number of publications motivates researchers to analyse the growing range of technologies and assess their relevance to sustainable production. At the same time, this rapid growth reveals the lack of a structured and synthesised overview of how research directions in this area are evolving.

A major challenge remains the lack of studies that present a complete and current picture of AI applications in the manufacturing sector. Tarasiuk and Halicka [15] point out, that the literature still does not offer a comprehensive understanding of the current state of research on AI in manufacturing. As a result, there is a shortage of analyses that integrate design, production, and environmental perspectives into one coherent framework.

Previous bibliometric reviews, which typically examine artificial intelligence or sustainability [27, 28] in manufacturing as separate research streams. This study provides an integrated analysis of AI in the context of the eco-design and manufacturing. The bibliometric approach is used not only to map publication trends but also to identify the thematic structure and temporal evolution of research linking digital technologies with the principles of sustainability and circular economy.

3 Research Problem and Methodology

The analysis systematises the existing body of knowledge by examining publications indexed in the Scopus database and mapping their thematic, temporal, and conceptual structure. Research problems with respect to AI-related issues in sustainable product design and manufacturing include:

- RQ1: Which countries lead research on the integration of AI tools in eco-design and circular manufacturing, and how are their publications linked to specific research topics and keyword networks?
- RQ2: Which keywords appear most frequently in this literature and what thematic clusters do they form?
- RQ3: Which topics in the analysed field are the most important and well-developed, and which are more specialised or still emerging?

The methodology of this research is presented in Fig. 1, and it follows:

Step1: Identification

In the first stage, the scope of the study was defined and a comprehensive search strategy was developed. Relevant keywords were identified and combined using Boolean operators to capture the intersection of artificial intelligence, sustainable product design, and sustainable or circular manufacturing. Scopus was selected as the primary database because of its wide coverage of engineering, environmental and industrial research. The final search query returned 270 publications, all of which were imported for further analysis. The set query is indicated below.

TITLE-ABS-KEY (("*artificial intelligence*" OR "*machine learning*" OR "*deep learning*" OR "*data analytics*" OR "*data-driven*" OR "*AI-driven*" OR "*AI-assisted*" OR "*intelligent system**") AND ("*ecodesign*" OR "*eco-design*" OR "*sustainable design*" OR "*sustainable product design*" OR "*design for environment*" OR "*design for sustainability*" OR "*design for circularity*" OR "*sustainable product development*" OR "*eco-innovation*" OR "*eco-efficient design*" OR "*product design*") AND ("*sustainable manufacturing*" OR "*green manufacturing*" OR "*circular manufacturing*" OR "*circular production*" OR "*resource-efficient manufacturing*" OR "*closed-loop manufacturing*" OR "*remanufactur**" OR "*recycl**" OR "*circular economy*" OR "*circular value chain*" OR "*smart manufacturing*")) AND PUBYEAR > 2014 AND PUBYEAR < 2026 AND (LIMIT-TO (SUBJAREA, "ENGI") OR LIMIT-TO (SUBJAREA, "ENVI") OR LIMIT-TO (SUBJAREA, "BUSI") OR LIMIT-TO (SUBJAREA, "ECON")) AND (LIMIT-TO (DOCTYPE, "ar") OR LIMIT-TO (DOCTYPE, "cp") OR LIMIT-TO (DOCTYPE, "re") OR LIMIT-TO (DOCTYPE, "cr") OR LIMIT-TO (DOCTYPE, "sh")) AND (LIMIT-TO (LANGUAGE, "English")).

Step 2: Data Preparation

In the preparation stage, the bibliographic information was structured and standardised. Author names, countries, keywords, publication years, abstracts, and journal information were extracted and harmonised. Keywords were inspected for variations and merged where appropriate to ensure coherence in the co-occurrence analysis. The cleaned data set was then imported into the Biblioshiny tool environment for quantitative evaluation.

Step 3: Bibliometric Analysis

The analysis consisted of three complementary components:

- Descriptive analysis: This included examining annual scientific production, identifying the most productive countries, and analysing the most frequent author keywords.
- Network analysis**:** Keyword co-occurrence networks were generated to identify conceptual linkages and thematic clusters, a Sankey diagram was used to visualise the flow between countries, abstract keywords and major research topics, cluster detection was applied to reveal dominant and interconnected themes.
- Thematic mapping: A thematic map was constructed based on centrality and density indicators to classify topics into motor, basic, niche, and emerging themes. This visualisation helped uncover the structural organisation and maturity of the research area.
- *Step 4: Interpretation and Synthesis*

In the final stage, the results of the descriptive, network, and thematic analyses were integrated and interpreted. The discussion focused on the main publication's trends,

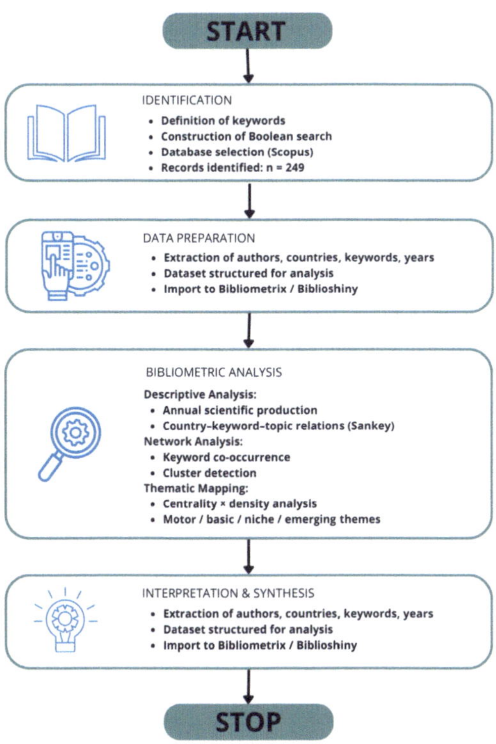

Fig. 1. A process diagram. Source: own elaboration.

the structure of thematic clusters, the relationship between digital technologies and sustainability concepts, and the evolution of the research field over time. These findings were used to identify existing research gaps and outline potential directions for future studies.

4 Results

This section presents the results of a comprehensive bibliometric analysis examining how AI-related technologies intersect with sustainability-driven design and circular or sustainable manufacturing. The findings reveal dominant research trends, emerging thematic gaps, and key directions for future development within this interdisciplinary landscape. The chart (see Fig. 2) shows the annual number of scientific publications in the studied research area from 2015 to 2025[1]. A clear upward trend can be observed, although the growth happens in stages: It starts slowly and becomes much more dynamic in the later years.

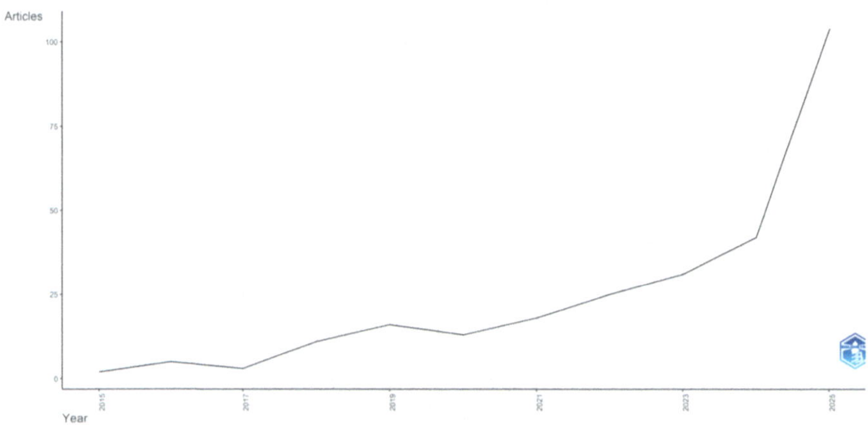

Fig. 2. Annual Scientific Production

Between 2015 and 2018, the number of publications was quite low, ranging from 2 to 11 per year. This suggests that the field was still in an early stage of development and attracted limited global interest. An analysis of the most cited documents indicates that the highest-impact publications originated in 2018. For example, Kusiak [29] discussed future trends in manufacturing based on smart solutions, such as simulation, sensors, and cyber-physical systems. In contrast, Bressanelli et al. [30] examined how digital technologies can enable the transition toward a circular economy (CE). Starting in 2019, the number of papers increases and stabilizes at 16 per year, indicating that the topic is becoming more noticeable in the scientific community. A stronger growth appears from

[1] The data for 2025 include publications indexed in the Scopus database up to the end of 2025.and do not account for articles indexed in 2025 but added to the database after the beginning of 2026.

2021 onward. Notably, the majority of the most cited publications originate from 2022 and 2023 [31]. The number of publications rises from 18 in 2021 to 25 in 2022 and 31 in 2023, reaching about 42 in 2024. This period, after 2024, reflects more intensive research activity and a growing maturity of the field. The most significant increase occurs in 2025, when the number of publications jumps to about 104, more than double the previous year. This sharp rise shows rapid expansion of the topic, likely linked to growing research interest, new projects, and broader relevance across scientific disciplines.

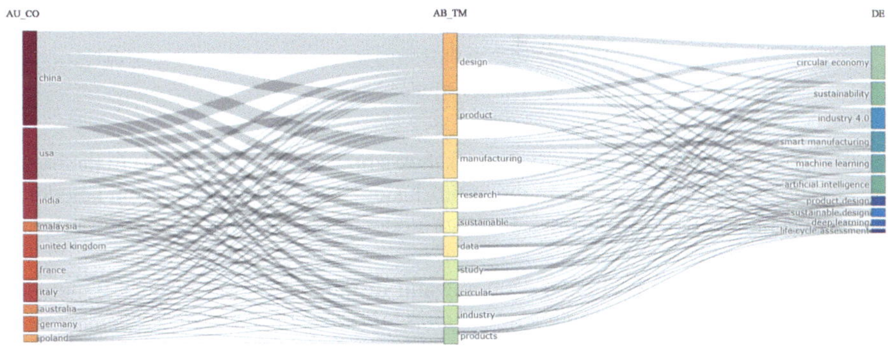

Fig. 3. The Sankey diagram

The Sankey diagram (Fig. 3) shows the links between the authors' countries of origin, the most common words used in the abstracts, and the main research topics. It helps to understand how research starts in different countries, uses similar keywords, and ends in shared topics. On the left side of the diagram, the countries of authors' affiliations are presented. The largest number of publications come from China and the USA. These are followed by India, Malaysia, the United Kingdom, France, Italy, Australia, Germany, and Poland. This distribution confirms the international character of the research, with strong contributions from European countries. The middle column displays the most common abstract keywords, including *design, manufacturing, product, sustainable, research, industry, and data.* These terms represent the core concepts of the analysed studies and indicate a strong focus on product development, manufacturing processes, and sustainability-related issues. On the right side, the main research topics are shown. The strongest connections lead to themes such as *circular economy, sustainability, industry 4.0, smart manufacturing, artificial intelligence, machine learning, sustainable manufacturing, sustainable design product design, or deep learning.* Overall, the diagram highlights that current research increasingly integrates digital technologies with sustainable and circular economy approaches. These observations directly address the research question RQ1.

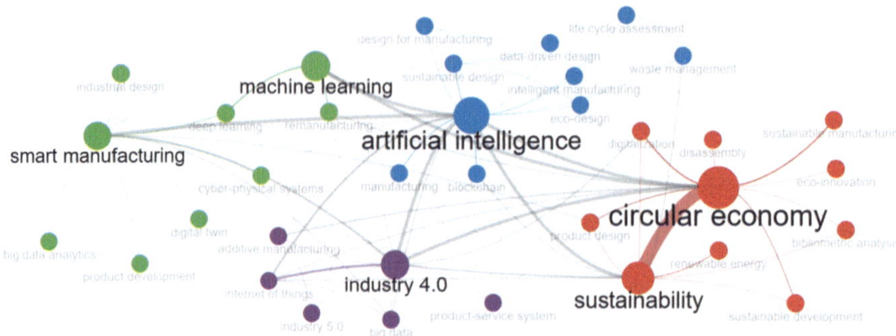

Fig. 4. Co-occurrence Network

The co-occurrence network map (Fig. 4) shows how the main keywords in the literature are connected and grouped into thematic clusters. Larger nodes represent keywords that appear more often, while links between nodes show how frequently these terms occur together.

The strongest cluster is circular economy and sustainability (red cluster). It includes keywords such as *circular economy, sustainability, sustainable development, eco-innovation, sustainable manufacturing, renewable energy, product design, bibliometric analysis, digitalisation,* and *disassembly*. This cluster indicates that many studies focus on environmental strategies, circular practices, and sustainable industrial development.

Another important cluster is artificial intelligence (blue cluster). It contains terms such as *artificial intelligence, manufacturing, blockchain, sustainable design, design for manufacturing, data-driven design, intelligent manufacturing, eco-design, lifecycle assessment,* and *waste management.* This shows that AI is mainly studied in relation to data-based decision-making and manufacturing processes, especially to designing.

The Industry 4.0 cluster (purple cluster) groups keywords related to digital transformation, including *Industry 4.0, industry 5.0, Internet of Things, Product Service System, Additive Manufacturing,* and *big data.* These terms are loosely related to AI and manufacturing.

The machine learning and smart manufacturing cluster (green cluster) includes keywords such as *machine learning, deep learning, smart manufacturing, remanufacturing, cyber-physical systems, product development, industrial design, cyber-physical systems, digital twin, and big data analytics.* This cluster connects strongly with both AI and sustainability-related topics, showing that machine learning is often used as a tool to support sustainable and circular solutions.

Overall, the map highlights two loosely connected research streams: digital technologies (e.g. artificial intelligence, machine learning, Industry 4.0) and sustainability-oriented concepts (e.g., circular economy, sustainable design, and eco-innovation). Strong links between these clusters confirm that current research increasingly integrates technological innovation with sustainability goals, directly addressing the research question RQ2.

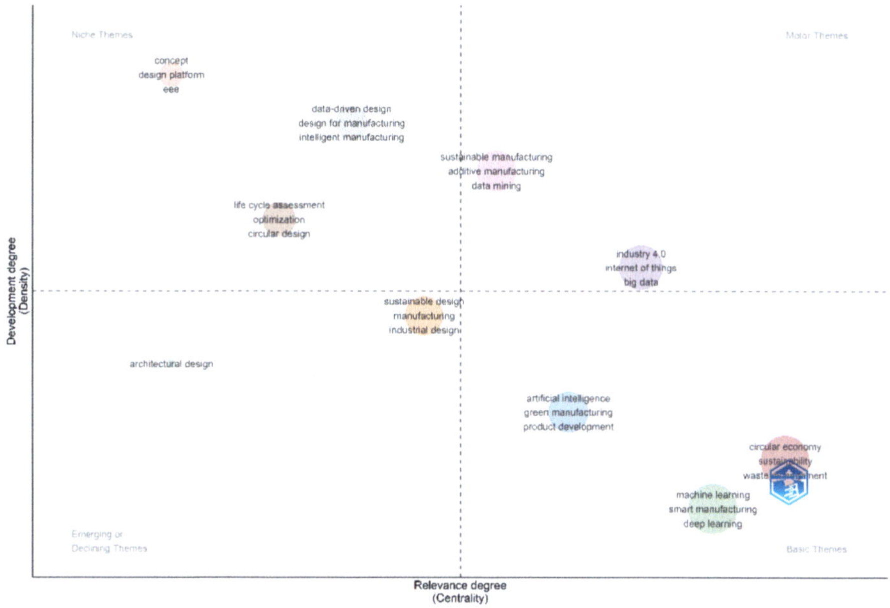

Fig. 5. Thematic map

The thematic map (see Fig. 5) presents the main research themes according to their relevance (centrality) and level of development (density).

Motor Themes (upper-right quadrant) include *sustainable manufacturing, additive manufacturing, data mining, Industry 4.0, Internet of Things,* and *big data.* These themes are both highly developed and strongly connected with other topics, indicating that data-driven and digital manufacturing solutions play a central role in the field and actively drive its development. One of the most cited studies [32], presents that the successful development of Smart Product-Service Systems requires an integrated perspective combining technological advancement, human-machine collaboration, and innovative value creation aligned with sustainability goals. Basic Themes (lower-right quadrant) are dominated by *circular economy, sustainability,* and *waste managemen*t, closely linked with *machine learning, smart manufacturing,* and *deep learning.* Their high relevance but lower density suggests that these topics are fundamental for the field, widely used across studies, yet still evolving and expanding toward more integrated frameworks. There are also concepts connected to *artificial intelligence, green manufacturing,* and *product development.* Bag et al. [31] demonstrate an approach that was classified within this cluster. Their findings indicate that institutional pressures foster eco-innovation and green supply chain practices in SMEs, enabling circular economy capability and selectively improving performance through big data-driven supply chains. Niche Themes (upper-left quadrant) include more specialized and well-developed topics such as *life cycle assessment, optimization, circular design, data-driven design, design for manufacturing, and intelligent manufacturing.* These themes show strong internal coherence but weaker links to the core research streams. Emerging or Declining Themes (lower-left quadrant) consist of less developed and less connected topics, such as architectural design, industrial design, and sustainable design. Their position suggests either early-stage research directions or themes with decreasing influence in the broader research landscape. Overall, the map highlights a strong interaction between digital technologies and sustainability-oriented concepts and illustrates the ongoing evolution of the research field, directly addressing research question RQ3.

5 Discussion

Product design is a key enabler of circular and sustainable manufacturing, as it determines whether products can be reused, repaired, refurbished, or remanufactured. The transition from linear to circular business models increases the complexity of the design and requires early consideration of multiple use and recovery scenarios [33, 34]. The analysed literature shows that these challenges are increasingly addressed through AI-supported and data-driven decision-making, which supports product evaluation, recovery planning, and sustainable manufacturing outcomes.

The bibliometric results indicate a strong growth of research on artificial intelligence in sustainable product design and circular manufacturing, particularly after 2021 with a clear increase in 2024. The Sankey (see Fig. 3) diagram reveals that most publications originate from countries actively engaged in digital and industrial transformation, mainly China, the United States, and India, with additional contributions from Europe. Research themes consistently focus on design, manufacturing, and sustainability, confirming the global convergence of research priorities. These findings requested the RQ1.

The keyword co-occurrence network highlights two loosely connected research streams: digital technologies (e.g., artificial intelligence, machine learning, Industry 4.0) and sustainability-orientated concepts (e.g., circular economy and sustainable manufacturing). The thematic map further shows that the circular economy and sustainable manufacturing act as the core driving themes, while digital technologies remain central but are still being developed. Niche topics such as life cycle assessment and waste management are well established but less influential. These findings requested RQ2 and RQ3. These findings are consistent with prior literature review studies addressing selected aspects of artificial intelligence, digital technology design, and product development within the context of sustainable and circular manufacturing [2, 3, 14, 26, 35].

6 Conclusions

This study presents a bibliometric overview of how AI supports sustainable product design and circular manufacturing. The results show a strong and accelerating growth of the field, with a sharp increase in publications from 2024, indicating rising maturity and global relevance. Research activity is led by China, the United States and India, with consistent contributions from Europe. Across regions, studies connect digital technologies, such as AI, machine learning and Industry 4.0, with sustainability themes including circular economy, sustainable design and low-impact manufacturing. Thematic mapping reveals that sustainability-related topics form the core of the field, while digital technologies remain central but continue to evolve. Overall, the findings indicate that the integration of AI, sustainability and circular economy is becoming a key direction in industrial transformation and will continue to shape future research on sustainable and circular product development.

This research is subject to certain limitations arising from potential bias in the search parameters and the defined time scope, which may have led to the exclusion of recent relevant publications. To mitigate this limitation, the findings were triangulated with the most recent studies available beyond the primary dataset.

Future research should re-examine publication trends beyond 2025 to determine whether the observed sharp increase in research output reflects a long-term structural shift or a temporary acceleration. Furthermore, future research will focus on the development of a decision support framework that integrates artificial intelligence tools to guide product design, manufacturing, and the application of circular value retention processes after the product usage.

Acknowledgements. This study was supported by the internal grant from the Poznan University of Technology, Institute of Logistics, number 0812/SBAD/4238.

References

1. Hermawan, A.N., Masudin, I., Zulfikarijah, F., Restuputri, D.P., Shariff, S.S.R.: The effect of sustainable manufacturing on environmental performance through government regulation and eco-innovation. Int. J. Ind. Eng. Oper. Manage. **6**, 299–325 (2023). https://doi.org/10.1108/IJIEOM-04-2023-0039
2. Patalas-Maliszewska, J., Łosyk, H.: Changes in sustainable development in manufacturing in cases of unexpected occurrences—A systematic review. Sustainability **16**, 717 (2024). https://doi.org/10.3390/su16020717
3. Patalas-Maliszewska, J., Szmołda, M., Łosyk, H.: Integrating Artificial Intelligence into the supply chain in order to enhance sustainable production—A systematic literature review. Sustainability **16**, 7110 (2024). https://doi.org/10.3390/su16167110
4. Kastelli, I., Mamica, L., Lee, K.: New perspectives and issues in industrial policy for sustainable development: from developmental and entrepreneurial to environmental state. Rev. Evol. Polit. Econ. **4**, 1–25 (2023). https://doi.org/10.1007/s43253-023-00100-2
5. Dennison, M.S., Kumar, M.B., Jebabalan, S.K.: Realization of circular economy principles in manufacturing: obstacles, advancements, and routes to achieve a sustainable industry transformation. Discov. Sustain. **5**, 438 (2024). https://doi.org/10.1007/s43621-024-00689-2
6. Hidalgo-Crespo, J., Golinska-Dawson, P., Riel, A.: Product-as-a-service transition for original equipment manufacturers: challenges, performance metrics, and design guidelines–The case of electrical and electronic equipment. Cleaner Environ. Syst. **19**, 100330 (2025). https://doi.org/10.1016/j.cesys.2025.100330
7. Favi, C., Marconi, M.: Product Eco-Design in the era of the circular economy. Sustainability **17**, 213 (2024). https://doi.org/10.3390/su17010213
8. Peigné, S., Ben Rejeb, H., Monnier, E., Zwolinski, P.: Navigating the eco-design paradox: criteria and methods for sustainable eco-innovation assessment in early development stages. Sustainability **16**, 2071 (2024). https://doi.org/10.3390/su16052071
9. Werner-Lewandowska, K., Golinska-Dawson, P., Mierzwiak, R.: Enablers and barriers in building the circular supply chain through remanufacturing-Grey DEMATEL approach. Int. J. Prod. Econ. **284**, 109617 (2025). https://doi.org/10.1016/j.ijpe.2025.109617

10. Moghrabi, I.A.R., Bhat, S.A., Szczuko, P., AlKhaled, R.A., Dar, M.A.: Digital transformation and its influence on sustainable manufacturing and business practices. Sustainability **15**, 3010 (2023). https://doi.org/10.3390/su15043010
11. Gohr, C., et al.: Artificial intelligence in sustainable development research. Nat. Sustain. **8**, 970–978 (2025). https://doi.org/10.1038/s41893-025-01598-6
12. Sartono, Y., Siti Astuti, E., Wilopo, W., Noerman, T.: Sustainable digital transformation: its impact on perceived value and adoption intention of Industry 4.0 in moderating effects of uncertainty avoidance. F1000Res. **13**, 821 (2024). https://doi.org/10.12688/f1000research.152228.2
13. Rame, R., Purwanto, P., Sudarno, S.: Industry 5.0 and sustainability: an overview of emerging trends and challenges for a green future. Innovation Green Dev. **3**, 100173 (2024). https://doi.org/10.1016/j.igd.2024.100173
14. Martini, B., Bellisario, D., Coletti, P.: Human-centered and sustainable artificial intelligence in Industry 5.0: challenges and perspectives. Sustainability **16**, 5448 (2024). https://doi.org/10.3390/su16135448
15. Tarasiuk, W., Halicka, K.: Artificial intelligence in manufacturing–Systematic literature review. SPSUTOM. **2025**, 678–701 (2025). https://doi.org/10.29119/1641-3466.2025.215.39
16. Tiwari, M.K., Sinha, N., Kumar, S., Rai, R., Mukhopadhyay, S.K.: A Petri net based approach to determine the disassembly strategy of a product. Int. J. Prod. Res. 40–45 (2002)
17. Powell, D.J.: Artificial intelligence in lean manufacturing: digitalization with a human touch? IJLSS **15**, 719–729 (2024). https://doi.org/10.1108/IJLSS-05-2024-256
18. Lin, H., Deng, X., Yu, J., Jiang, X., Zhang, D.: A study of sustainable product design evaluation based on the analytic hierarchy process and deep residual networks. Sustainability **15**, 14538 (2023). https://doi.org/10.3390/su151914538
19. Regulation (EU) 2024/1781 of the European Parliament and of the Council of 13 June 2024 establishing a framework for the setting of ecodesign requirements for sustainable products, amending Directive (EU) 2020/1828 and Regulation (EU) 2023/1542 and repealing Directive 2009/125/EC (Text with EEA relevance) (2024)
20. Langley, D.J., Rosca, E., Angelopoulos, M., Kamminga, O., Hooijer, C.: Orchestrating a smart circular economy: guiding principles for digital product passports. J. Bus. Res. **169**, 114259 (2023). https://doi.org/10.1016/j.jbusres.2023.114259
21. Wan, P.K.F., Jiang, S.: Enabling a dynamic information flow in digital product passports during product use phase: a literature review and proposed framework. Sustain. Prod. Consumption **54**, 362–374 (2025). https://doi.org/10.1016/j.spc.2025.01.014
22. Ashkbous, M., Ghorbani, E., Keivanpour, S.: Artificial intelligence for eco-design: a systematic review. Adv. Eng. Inform. **69**, 103989 (2026). https://doi.org/10.1016/j.aei.2025.103989
23. Choudhary, D., Qaiser, F.H., Choudhary, A., Fernandes, K.: A model for managing returns in a circular economy context: a case study from the Indian electronics industry. Int. J. Prod. Econ. **249**, 108505 (2022). https://doi.org/10.1016/j.ijpe.2022.108505
24. Goridkov, N., Wang, Y., Goucher-Lambert, K.: Empowering designers to create life cycle informed products: heuristics for extracting insights from LCA reports. Des. Sci. **11**, e27 (2025). https://doi.org/10.1017/dsj.2025.10014
25. Hidalgo-Crespo, J., Riel, A., Sakao, T., Duflou, J.R.: Design platform concept for product-as-a-service by electrical and electronic equipment manufacturers. Procedia CIRP. **130**, 1176–1181 (2024). https://doi.org/10.1016/j.procir.2024.10.224
26. Raut, S., Hossain, N.U.I., Kouhizadeh, M., Fazio, S.A.: Application of artificial intelligence in circular economy: a critical analysis of the current research. Sustain. Futur. **9**, 100784 (2025). https://doi.org/10.1016/j.sftr.2025.100784

27. Ellili, N.O.D.: Bibliometric analysis of sustainability papers: evidence from environment. Dev. Sustain. Environ. Dev. Sustain. **26**, 8183–8209 (2023). https://doi.org/10.1007/s10668-023-03067-6
28. Arora, S.: Bibliometric study on ESG, SDG Goals, Green Finance, and Startups Using VOS Viewer and Gephi: In: Tripathi, V., Rastogi, S., and Singh, N. (eds.) Advances in Human Resources Management and Organizational Development. pp. 401–432. IGI Global (2024). https://doi.org/10.4018/979-8-3693-6617-2.ch019
29. Kusiak, A.: Smart manufacturing. Int. J. Prod. Res. **56**, 508–517 (2018). https://doi.org/10.1080/00207543.2017.1351644
30. Bressanelli, G., Adrodegari, F., Perona, M., Saccani, N.: Exploring how usage-focused business models enable circular economy through digital technologies. Sustainability **10**, 639 (2018). https://doi.org/10.3390/su10030639
31. Bag, S.: Big data and predictive analysis is key to superior supply chain performance: a South African experience. IJISSCM **10**, 66–84 (2017). https://doi.org/10.4018/IJISSCM.2017040104
32. Zheng, P., Wang, Z., Chen, C.-H., Pheng Khoo, L.: A survey of smart product-service systems: key aspects, challenges and future perspectives. Adv. Eng. Inform. **42**, 100973 (2019). https://doi.org/10.1016/j.aei.2019.100973
33. Kurilova-Palisaitiene, J., Golinska-Dawson, P., Gorroño-Albizu, L., Duberg, J.V.: Turning challenges into opportunities when adopting remanufacturing by original equipment manufacturers. Cleaner Logistics Supply Chain. **17**, 100265 (2025). https://doi.org/10.1016/j.clscn.2025.100265
34. Hidalgo-Crespo, J., Riel, A., Golinska-Dawson, P., Peeters, J.R., Werner-Lewandowska, K., Duflou, J.R.: Facilitating circularity: challenges and design guidelines of Product-as-a-Service (PaaS) business model offers for electrical and electronic equipment. Procedia CIRP. **128**, 567–572 (2024). https://doi.org/10.1016/j.procir.2024.03.037
35. Di Vaio, A., Palladino, R., Hassan, R., Escobar, O.: Artificial intelligence and business models in the sustainable development goals perspective: a systematic literature review. J. Bus. Res. **121**, 283–314 (2020). https://doi.org/10.1016/j.jbusres.2020.08.019

Artificial Intelligence in Engineering Design, Case Study

Wojciech Babirecki[1(✉)], Katarzyna Skrzypek[1], Karol Dąbrowski[2], Małgorzata Śliwa[1], and Wojciech Śmierski[1]

[1] Institute of Mechanical Engineering, University of Zielona Góra, Zielona Góra, Poland
w.babirecki@iim.uz.zgora.pl

[2] Institute of Management and Quality Sciences, University of Zielona Góra, Zielona Góra, Poland

Abstract. The ongoing digitalisation of engineering processes, combined with the rapid development of artificial intelligence (AI), is transforming contemporary approaches to mechanical design. This article presents generative design as a methodology that employs optimisation algorithms and machine learning techniques to autonomously generate alternative structural solutions. The study demonstrates the design process of a vacuum gripper for a Universal Robots UR10E manipulator, developed in Autodesk Fusion 360 using the Generative Design module. The results are analysed in terms of efficiency and applicability within mechanical engineering. In the article, analyses were performed using FEM simulations. The final design was fabricated using FDM 3D printing technology to enable prototype visualisation.

Keywords: Generative design · artificial intelligence · CAD/CAE · mechanical engineering · additive manufacturing

1 Introduction

The digitalisation of engineering processes, accompanied by the dynamic development of artificial intelligence (AI) methods, is significantly transforming contemporary approaches to mechanical design [12]. With increasing computational capabilities and the growing integration of CAD/CAE/CAM tools with machine learning algorithms, a clear paradigm shift is observed—from classical, deterministic geometric modelling towards systems based on the automatic exploration of design solution spaces [19]. A key manifestation of this transformation is generative design, a modern methodology that employs optimisation and heuristic algorithms to autonomously generate structural variants according to predefined design objectives, geometric constraints and boundary conditions [1, 3, 5]. Unlike traditional design methods, in which the engineer defines the geometry and model parameters, generative design shifts the user's role towards specifying criteria, while the computational system independently proposes solutions with a high degree of functional and structural optimisation.

J. Trojanowska et al. (Eds.): MANUFACTURING 2026, LNME, pp. 168–180, 2026.
https://doi.org/10.1007/978-3-032-21654-0_13

The importance of this technology in mechanical engineering stems from its ability to perform intelligent optimisation of structures characterised by reduced mass, increased stiffness, or improved adaptation to operational conditions. This article presents the design of a vacuum gripper intended for a Universal Robots UR10E robot. The design process was carried out in Autodesk Fusion using the Generative Design module. The publication outlines the theoretical foundations, the modelling workflow, and the analysis of the results in the context of efficiency and potential applications in modern robotic systems. Using a real industrial case, the study examines the feasibility and advantages of applying this approach in low-volume or single-piece production. To support visualisation, a physical prototype of the gripper was also produced. The model was manufactured using FDM 3D printing technology at full scale. This comprehensive approach provides tangible benefits for designers and R&D engineers, particularly during functional testing. It enables rapid generation of successive prototype iterations and facilitates the creation of demonstrative solutions suitable for immediate presentation to an end-client.

1.1 Research Scope and Evaluation Criteria

The application of generative design techniques in the development of lightweight robotic end-effectors has gained increasing attention in recent years, particularly in the context of additive manufacturing and low-volume production. Within this framework, the following research questions are addressed in order to assess the applicability and limitations of generative design workflows for a vacuum gripper used in robotic manipulation tasks.

RQ1: Is it possible to obtain a functionally equivalent vacuum gripper component with a reduced mass by applying generative design techniques, while maintaining the load-bearing capacity and stiffness required for robotic manipulation tasks?
RQ2: Is the use of generative algorithms in engineering design an effective tool for designing machine parts for single-unit production?

The evaluation criteria adopted in this work include compliance with predefined functional requirements, acceptable stress and displacement levels verified through numerical simulations, and a quantitative comparison of component mass relative to the reference design. The assessment is based on simulation-driven verification, which is considered sufficient for the comparative purpose of this analysis, while experimental validation is identified as a potential direction for future studies.

2 Literature Review

Generative design is a rapidly developing field that integrates structural optimisation, artificial intelligence methods and additive manufacturing technologies. Technical literature emphasises that many generative methods are fundamentally based on genetic algorithms. These algorithms, inspired by mechanisms of biological evolution, enable the iterative improvement of a population of solutions through operations such as selection, crossover and mutation. Neural networks, in turn, support the system's learning process by predicting structural behaviour based on simulation data and previous design iterations [10, 14, 15]. Examples of algorithms enabling multi-objective optimisation include NSGA-II, NS-DSDSC and MMOO [2, 11]. Autodesk, the developer of the software used in this study, has published a series of technical resources introducing the Generative Design technology, covering both theoretical foundations and practical industrial applications [3, 5, 6]. Particularly valuable are analyses of implementations within the aerospace sector. Airbus, in collaboration with Autodesk, developed a bionic cabin partition, achieving a mass reduction of approximately 45% [1, 8]. Similarly, General Motors adopted a generative approach to the design of automotive components, obtaining structures that were simultaneously lighter (−40%) and stronger (+20%) for a seat bracket [7].

In the context of Design for Additive Manufacturing (DfAM), increasing attention is given to the integration of generative design with additive processes, emphasising their compatibility and the significant potential they offer for industrial applications [4].

The literature on mass reduction in mechanical structures highlights a clear trend towards the use of advanced optimisation techniques for designing lightweight and energy-efficient systems. As supporting evidence, Shu et al. present an example of applying a GAN-based model trained on 4,045 three-dimensional aircraft models [18], while further studies demonstrate significant mass reduction in the design of automotive body structures [13].

3 Basics of Generative Design and Tools

Generative design represents a modern approach to the development of engineering structures, in which the designer's role is focused on defining design objectives, boundary conditions, geometric constraints and material parameters, while the generation of solutions is performed automatically using artificial intelligence and optimisation algorithms [1, 3, 5]. The computational system evaluates criteria such as mass minimisation, stiffness maximisation or displacement limitation, and subsequently produces multiple alternative design variants that fulfil the specified requirements. Generative design should be distinguished from other computer-aided design methodologies, such as parametric modelling or topology optimisation [9]. In parametric modelling, the user defines geometric relationships and dimensional dependencies, enabling rapid modification of the model but not allowing for autonomous generation of novel structural concepts. Topology optimisation, on the other hand, enables material reduction within an existing structure based on the analysis of stress distribution and loading conditions; however, the resulting shapes often require further processing and adaptation to manufacturing constraints. Generative design integrates elements of both approaches, extending them with the capability for independent exploration and evaluation of diverse structural concepts.

The algorithms employed within Generative Design rely on iterative optimisation methods, machine learning techniques and numerical analysis. The system generates and assesses successive design variants based on finite element simulations, material data and applied loading conditions. This process is adaptive, meaning that the obtained results are used to progressively refine the models in subsequent iterations [17]. A conceptual overview of the generative design workflow is presented in Fig. 1.

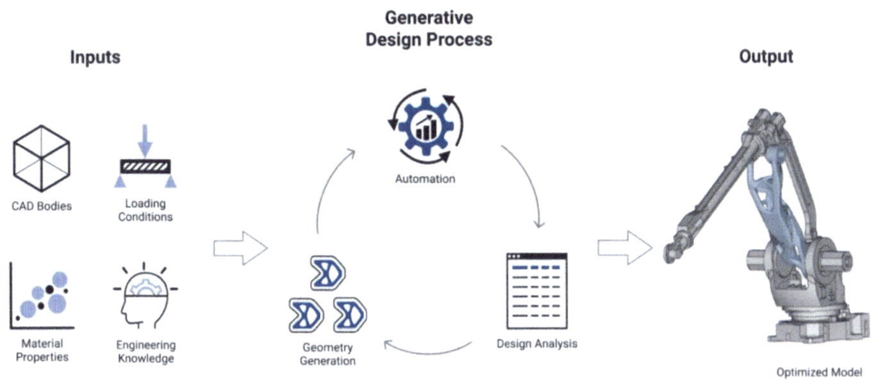

Fig. 1. General scheme of the generative design process, source [16]

The development of generative design technologies has led to the emergence of numerous commercial and open-source software solutions that integrate artificial intelligence algorithms with traditional CAD/CAE environments. Among these, Autodesk Fusion occupies a particularly prominent position as a reference tool in industrial applications [3, 5]. The software provides a dedicated Generative Design module that combines 3D modelling capabilities, structural analysis and design optimisation within a single, integrated environment. The user defines boundary conditions, materials and manufacturing constraints, while the system automatically generates and evaluates multiple design variants, taking into account mechanical, cost-related and manufacturability criteria.

Platforms that incorporate generative design concepts include, for example:

- Simens NX,
- SolidWorks,
- PTC Creo,
- Altair Inspire,
- nTypology.

These systems differ in their functional scope, their level of integration with FEA environments, and their capabilities for exporting results to additive manufacturing technologies. Siemens NX and PTC Creo are characterised, among other features, by strong integration with CAE modules, enabling comprehensive structural and fluid-flow analyses. Altair Inspire and nTopology, in turn, focus on topology optimisation and the preparation of models for 3D printing, offering high flexibility in defining design constraints and manufacturing conditions [3, 16]. Ultimately, the development of generative design tools is progressing in parallel with advancements in additive manufacturing, which enables the direct fabrication of geometries with a high degree of structural complexity.

4 Case Study

To verify the applicability of generative design combined with additive manufacturing technologies in single-piece and low-volume production, an attempt was made to design a vacuum gripper intended for integration with a Universal Robots UR10E manipulator. In one of the ongoing industrial projects, focused on optimising a production line, the need arose to develop a dedicated pneumatic gripper for transferring small boxes between workstations. The mass of the handled box is 2 kg, while the robotic arm used in the application is the UR10E model, featuring a payload capacity of 10 kg.

4.1 Gripper: Conventional Design Version

Using Autodesk Inventor software, a baseline version of the gripper was developed for manufacturing using conventional technologies. The model was constructed from designed sheet-metal components, a machined sleeve and commercially available structural profiles. The individual elements were processed accordingly and subsequently assembled by welding and threaded fasteners. The gripper components were made from aluminium alloy (AW-2017A) with a density of 2.82 g/cm^3. The assembled gripper, including all fastening elements, weighs approximately 1.38 kg.

The following assumptions were adopted for the design calculations:

- g = 9.81 m/s^2
- design acceleration: $a = 2.0$ m/s^2
- number of suction cups: 6
- vacuum level: 0.8 bar ≈ 80 000 Pa
- safety factor (SF): 2.0
- load mass: m = 2.0 kg

During the basic design calculations, the following values were determined:

- gravitational force (weight): $F_g = m \cdot g = 2.0 \cdot 9.81 = 19.62$ N,
- inertial force (due to acceleration): $F_{in} = m \cdot a = 2.0 \cdot 2.0 = 4.00$ N,
- total stopping force (static + inertial): $F_{tot} = 19.62 + 4.0 = 23.62$ N,
- design force with safety factor SF = 2: $F_{obl} = F_{tot} \cdot 2 = 47.24$ N
- force carried by a single suction cup (6 units): $Fs = 47.24/6 = 7.8733$ N.

The baseline gripper design is shown in Fig. 2.

4.2 Gripper: Design Based on the Generative Design Approach

The sequence of actions undertaken to generate an alternative design using the Generative Design module is as follows:

- importing the 3D model created in Autodesk Inventor into the Fusion environment. This operation is shown in Fig. 3:

- disabling the components that will not be included in the optimisation process, leaving only the parts intended to be retained in the final design. This operation is shown in Fig. 4:

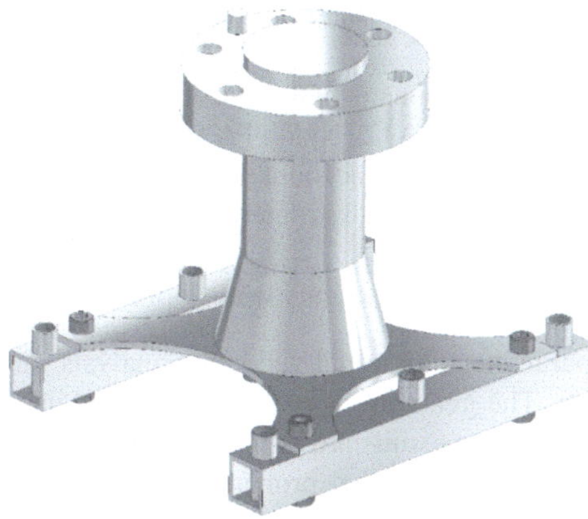

Fig. 2. Gripper: baseline version (source: own work)

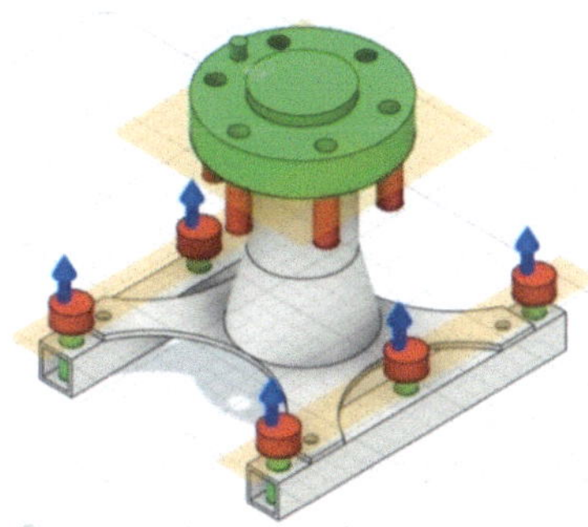

Fig. 3. The gripper model imported into Autodesk Fusion with structural components isolated for optimisation (source: own work)

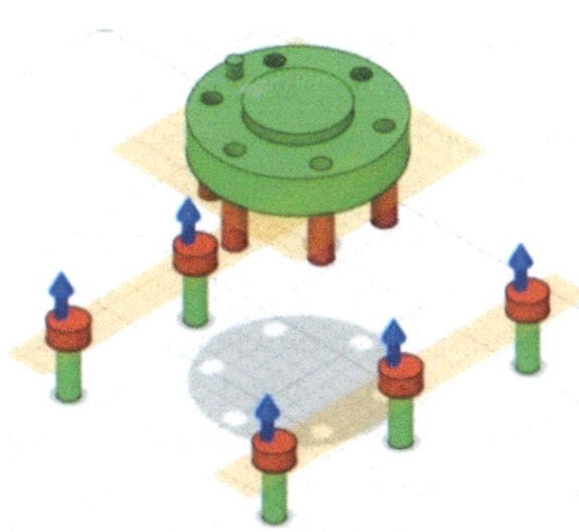

Fig. 4. Geometry retained from the original model (source: own work)

- defining obstacle geometry, i.e. the regions that the generative system must avoid during the shape-generation process,

- specifying the design objectives – mass minimisation was adopted, with a safety factor of 2,
- defining manufacturing conditions – both unrestricted and additive manufacturing scenarios were selected, using PA12 (Nylon PA 603-CF) as the material,
- the computational model takes into account the force of gravity; the specified force value assigned to a single suction cup was determined by rounding the value calculated in Sect. 4.1.
- defining the constraints – fixed constraints were applied to selected support regions.

After entering all parameters, an initial validation analysis was performed to allow for potential corrections. The generative process is executed using Autodesk's cloud-based computing resources. The results are presented as a set of design variants, which are compared based on their mechanical properties and visualisation of their shape-generation iterations. The generated solutions for the analysed example are shown in Fig. 5. The output data can be exported to a CSV format for further examination in a spreadsheet.

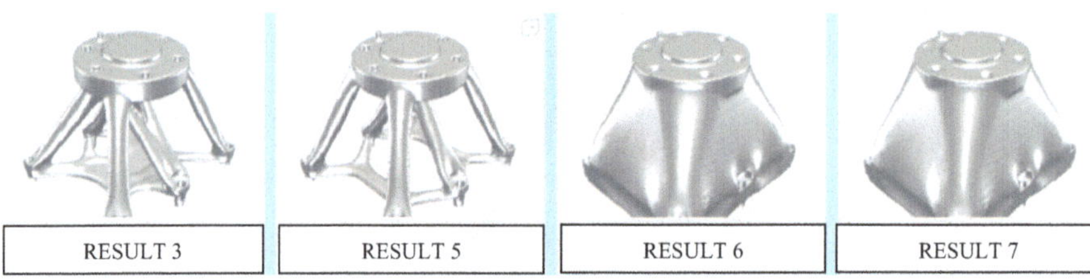

Fig. 5. Solutions generated by the Generative Design module (source: own work)

The system returns several solution variants, all characterised by a similar mass, as this optimisation criterion was selected. It is the designer who decides which of the generated options will be adopted as the final solution. The choice should take into account factors such as geometry, overall appearance, manufacturability requirements and individual preferences arising from engineering experience.

After analysing the obtained results, the variant labelled "Result 3" was selected for further evaluation. The selection was based on a skeletal structure that contributes to reduced air resistance during rapid robot movements; therefore, variants 6 and 7 were rejected. In variant 3, the solid surface between the suction cup sockets allows for the addition of further mounting points for additional suction cups in subsequent testing phases. This was the reason for rejecting variant 5. The adopted solution, along with a prototype manufactured using FDM 3D printing technology, is shown in Fig. 6. The prototype was printed on a BambuLab P1 Combo printer equipped with an AMS filament management system. The printing process used a layer height of 0.2 mm and 100% infill.

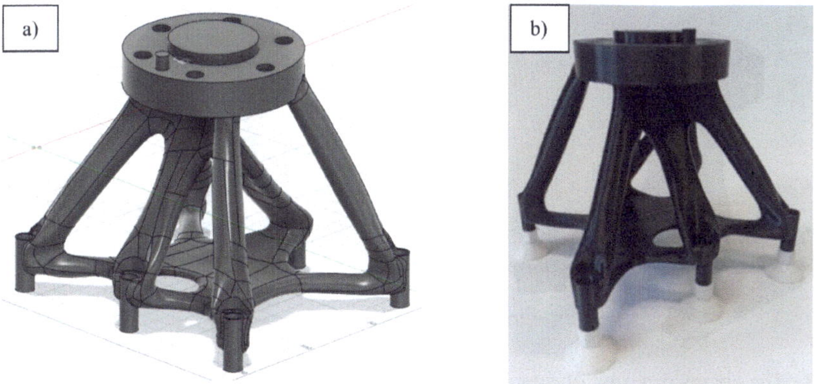

Fig. 6. Component generated using the Generative Design module: a) "Result 3" model; b) prototype manufactured using 3D printing technology (source: own work)

5 Mechanical Properties

To determine the mechanical properties of the generated model and compare them with those of the baseline design manufactured from an aluminium alloy using conventional production technologies, basic strength analyses were carried out. Using the simulation module, calculations were performed for two gripper variants. The first was the baseline version, constructed from aluminium alloy components, while the second was the version generated using the Generative Design module and produced from PA12 (Nylon PA 603-CF). The same loading and support conditions were applied in both cases. The resulting safety factors, von Mises equivalent stresses and displacements are summarised in the figures below (Figs. 7, 8 and 9).

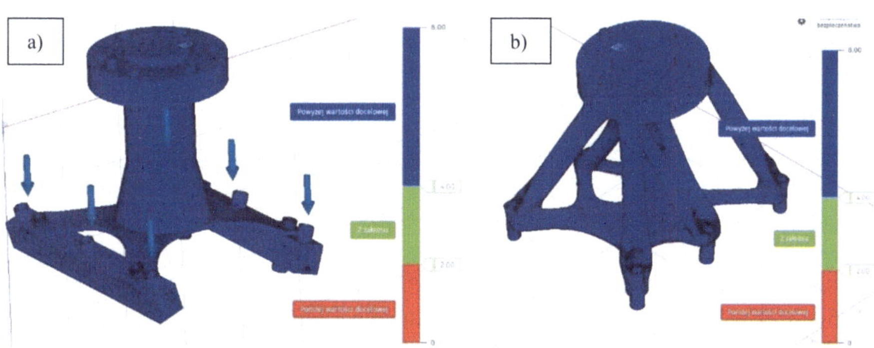

Fig. 7. The obtained results illustrate: (a) safety factor-baseline model; (b) safety factor-generative design model; (source: own work).

The comparison presented in Table 1 contains the values obtained from Autodesk Fusion.

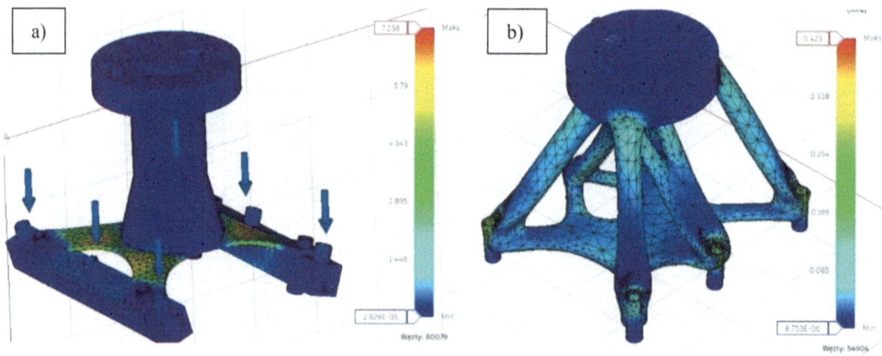

Fig. 8. The obtained results illustrate: (a) von Mises equivalent stress-baseline model; (b) von Mises equivalent stress-generative design model (source: own work).

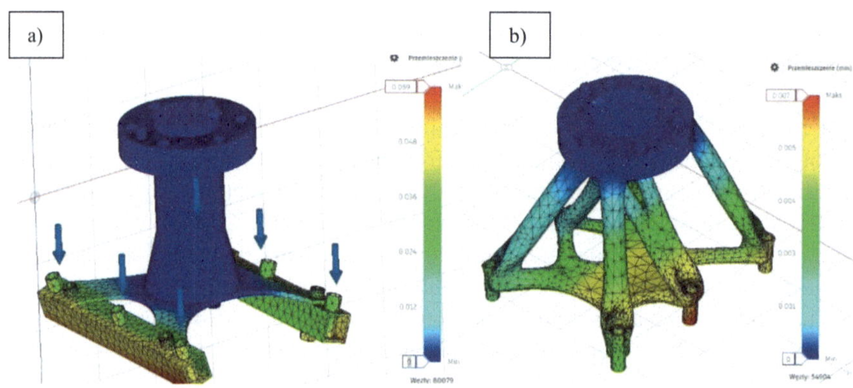

Fig. 9. The obtained results illustrate: (a) displacement-baseline model; (*b*) displacement-generative design model (source: own work).

The analyses indicate a clear advantage of the component designed using Generative Design. Both the von Mises equivalent stresses and the maximum displacements are lower in this case compared with the baseline variant. Particular attention should be paid to the mass of the component-reduced by approximately 70% relative to the baseline design. This reduction results primarily from the use of a material with lower density, which means that the comparison is not fully equivalent; however, producing the proposed geometry in metallic materials would be challenging. Current additive manufacturing technologies capable of processing metallic powders remain very costly and are still rarely used.

In the cost estimation, it was assumed that the design time-and thus the design cost-would be similar in both cases. The unit price of the filament used for 3D printing is approximately three times higher than the price of the selected aluminum alloy. However, considering the nearly threefold greater mass of the baseline variant, the actual material costs are also comparable. Significant differences occur in manufacturing costs,

Table 1. Comparison of the analysed properties of both solutions (source: own work)

	Material		Mass, kg	Safety factor	Maximum von Mises stress, MPa	Maximum displacement, mm
Baseline	AW-2017A	Tensile modulus: 72500, MPa Tensile strength: 225, MPa	1,37	>8	7,238	0,059
Generative Design	PA12 (Nylon PA 603-CF)	Tensile modulus: 7900, MPa Tensile strength: 85, MPa	0,42	>8	0,423	0,007

as the baseline variant, which requires multiple technological operations such as cutting, turning, grinding, welding, and assembly, is several times more expensive than manufacturing using 3D printing technology.

6 Challenges and Limitations

Despite the significant benefits associated with the use of generative design, this technology still faces a number of challenges. One of the key limitations is the high computational cost linked to generating and analysing a large number of design variants. This process requires substantial hardware resources, including access to cloud-based processing power or high-performance computing (HPC) clusters. Moreover, the results produced through generative analysis are often complex and non-intuitive, necessitating informed interpretation by the engineer, both in terms of structural suitability and manufacturability. Another challenge concerns the validation and certification of generatively produced structures. In many industries, such as aerospace and automotive, stringent quality and safety standards apply, and these do not always align with the irregular, organic geometries typical of generative design. Consequently, new verification methods and testing procedures must be developed for full validation and certification. According to unique geometry, verification is frequently limited to simulation-based assessment in the design stages, while experimental testing and long-term durability evaluation remain necessary. These aspects were considered beyond the scope of the present work and are identified as important directions for further investigation.

An equally important aspect is the role of the engineer in the design process. Despite the high degree of automation, AI-based systems do not replace the specialist; rather, they serve as tools that support decision-making. It is the engineer who defines the objectives, constraints, material parameters and boundary conditions, and who ultimately selects the most rational variant from the solutions generated by the algorithm. Maintaining this human–machine interaction is crucial for ensuring quality, safety and accountability throughout the design process [4].

7 Conclusions

Generative design represents a significant stage in the evolution of contemporary engineering methods, combining the computational capabilities of artificial intelligence with human creativity. Unlike traditional CAD approaches or parametric 3D models, this technology enables automatic exploration of the design space through algorithmic generation of multiple structural variants, which are subsequently evaluated against predefined objectives and constraints. The outcome is the development of lighter and more robust solutions, aligning with current trends in sustainable engineering. The integration of generative design with modern simulation tools (FEA, CFD) and additive manufacturing technologies (3D printing) allows for the creation of a complete "design–analysis–production" cycle within a single digital environment. This enables a reduction in design time, material and manufacturing costs, as well as a decrease in the number of prototype iterations. When combined with augmented reality (AR), virtual reality (VR) and digital twins, generative design becomes an interactive and adaptive process capable of responding to real operating data in real time.

A key aspect of the development of this technology is the role of the engineer, not as an operator, but as a creator and analyst of the design process. Artificial intelligence assumes the exploratory and computational functions, while the human designer defines the direction of the search, quality criteria and the context of application. Such synergy between engineering expertise and computational power facilitates the creation of solutions that meet technical requirements while also addressing ecological, ergonomic and economic considerations.

In the context of Industry 4.0 and the emerging Industry 5.0 paradigm, generative design is expected to become a key component of intelligent, integrated production ecosystems. By combining automation, adaptive algorithms and human expertise, this technology contributes to the development of a new design model-one that is more sustainable, flexible and grounded in human–machine collaboration. It represents a step towards the engineering of the future, in which creativity and analytical capability coexist, and the design process becomes not only more efficient but also more environmentally and socially responsible. The completed case study has demonstrated that, in single-piece or prototype manufacturing, this approach enables the achievement of substantial benefits. The workload, cost and, above all, the time required from the emergence of a design challenge to its material realisation are significantly reduced when generative design is combined with additive manufacturing technologies, compared with traditional production methods.

The conducted analyses demonstrate that the application of generative design algorithms in the development of robotic grippers enables the creation of structures that satisfy both strength and functional requirements while achieving a substantial reduction in manufacturing costs. The integration of generative design with a design-for-manufacturing approach tailored to additive manufacturing technologies proves particularly effective for single-unit and low-volume production, offering a promising alternative to conventional design and manufacturing methods.

References

1. Airbus. Pioneering bionic 3D printing—Bionic partition. Airbus News (2016). https://www.airbus.com/en/newsroom/news/2016-03-pioneering-bionic-3d-printing
2. Aisha, A.Y., Rozprawa, D.: New algorithms for multi-objective optimization. promotor: Marcin Studniarski, Wydział Matematyki i Informatyki, Uniwersytet Łódzki (2025)
3. Altair. Altair Inspire—generative design and topology optimization tools [Product documentation and case studies] (n.d.). https://www.altair.com/inspire/
4. Asapu, S., Ravi Kumar, Y.: Design for Additive Manufacturing (DfAM): a comprehensive review with case study insights. JOM **77**, 3931–3951 (2025). https://doi.org/10.1007/s11837-025-07164-x
5. Autodesk. Autodesk and Airbus demonstrate the impact of generative design in aerospace. Autodesk News (2019). https://adsknews.autodesk.com/en-gb/news/autodesk-airbus-generative-design-aerospace-factory/
6. Autodesk. Fusion 360—Introduction to Generative Design (Autodesk University article). Autodesk (n.d.-b). https://www.autodesk.com/autodesk-university/article/Fusion-360-Introduction-Generative-Design
7. Autodesk. General Motors—Generative design in car manufacturing (case study). Autodesk Customer Stories (n.d.-c). https://www.autodesk.com/customer-stories/general-motors-generative-design
8. Autodesk Research/The Living: Bionic Partition—Research project page. Autodesk Research (n.d.). https://www.research.autodesk.com/projects/bionic-partition/
9. Bendsøe, M.P., Ole Sigmund, O.: Topology optimization. Theory, Methods, and Applications, Springer Berlin, Heidelberg (2004). https://doi.org/10.1007/978-3-662-05086-6
10. Cancemi, S.A., et al.: AI-Powered convolutional neural network surrogate modeling for high-speed finite element analysis in the NPPs fuel performance framework. Energies **18**(10), 2557 (2025). https://doi.org/10.3390/en18102557
11. Deb, K., Pratap, A., Agrawal, S., Meyarivan, T.: A fast and elitist multiobjective genetic algorithm: NSGA-II. IEEE Trans. Evol. Comput. **6**(2), 182–197 (2002). https://doi.org/10.1109/4235.996017
12. Etesam, Y., Cheong, H., Ataei, M., Jayaraman, P.K.: Deep generative model for mechanical system configuration design. Proc. AAAI Conf. Artif. Intell. **39**(16), 16496–16504 (2025). https://doi.org/10.1609/aaai.v39i16.33812
13. Islam, M.T., Abdullah, A., Hussin, A., Jamaludin, M.: Weight reduction strategies in passenger car manufacturing: current trends and prospects. In: Automotive Engineering Advances (pp. 1–19) (2022). https://doi.org/10.1201/9781003164241-1
14. Lohan, D.J., Dede, E., Allison, J.T.: Topology optimization for heat conduction using generative design algorithms. 1st World Congress on Structural and Multidisciplinary Optimization (2015)
15. Marković, E., et al.: A Surrogate artificial neural network model for estimating the fatigue life of steel components based on finite elements simulations. Materials **18**(12), 2756 (2025). https://doi.org/10.3390/ma18122756

16. NTOP (blog). What is design for additive manufacturing (DfAM)?—three levels of DfAM. nTopology (2022). https://www.ntop.com/resources/blog/what-is-design-for-additive-manufacturing/
17. Semjén, Á., Szep, J.: Integrating generative and parametric design with BIM: a literature review of challenges and research gaps in construction design. Appl. Eng. Sci. **23**, 100253 (2025). https://doi.org/10.1016/j.apples.2025.100253
18. hu, D., et al.: 3D design using generative adversarial networks and physics-based validation. ASME. J. Mech. Des. July 2020; **142**(7), 071701 (2020). https://doi.org/10.1115/1.4045419
19. Yoo, S., Lee, S., Kim, S., Hwang, K.H., Park, J.H., et al. Integrating deep learning into CAD/CAE system: generative design and evaluation of 3D conceptual wheel. Structural and Multidisciplinary Optimization (2020)

Additive Manufacturing Technologies

Improving Mechanical Performance of FDM-Printed Carbon Fiber-Reinforced PLA via a Novel Infill Pattern

Lam Le[1], Vi Nguyen[1], Chuong Chau[1], and Thanh Tran[2](✉)

[1] Vietnamese-German University, Thoi Hoa, Ho Chi Minh, Vietnam
[2] School of Science, Engineering, Technology, RMIT University Vietnam, Ho Chi Minh, Vietnam
thanh.tran37@rmit.edu.vn

Abstract. This study proposes a combined infill pattern design that is not currently available in standard slicing software. The novel design is described in detail to enhance mechanical performance compared with conventional infill strategies, i.e., rectilinear and concentrated patterns. The paper evaluates the effectiveness of combining different infill patterns by tensile testing. In the paper, a carbon fiber-reinforced PLA, which remains underexplored as compared to standard PLA materials, will be used for studies. The impact of key process parameters, including infill pattern, infill density, and printing speed, on tensile strength is systematically investigated by using the Taguchi method. Results indicate the proposed pattern provides improved strength under the same printing parameters and conditions. A combined infill pattern design makes the printed parts stronger and less brittle across all infill densities. For tensile-strength influence analysis, infill density has the greatest effect. The infill pattern is the next most important factor, while printing speed has the least impact. Besides contributing a novel infill pattern to improve mechanical performance, the paper provides a validated methodology for parameter optimization in fused deposition modeling using carbon fiber-reinforced PLA and offers valuable analysis to enhance structural integrity in additive manufacturing applications.

Keywords: PLA · Infill Pattern; FDM · 3D Printing · DOE

1 Introduction

Additive manufacturing (AM) was first developed approximately 30 years ago by Chuck Hull through the invention of Stereolithography (SLA). Since then, a wide variety of AM technologies have emerged, each based on different principles for layer-by-layer fabrication. While traditional subtractive manufacturing methods, such as machining or milling, remove material from a workpiece to achieve the desired shape, AM technologies can directly form three-dimensional objects from digital models. One of the key advantages of AM is its material efficiency, significantly reducing waste compared to subtractive methods. In addition, AM enables the fabrication of highly complex geometries. Its material includes a wide range of options, from common polymers and resins to

J. Trojanowska et al. (Eds.): MANUFACTURING 2026, LNME, pp. 183–197, 2026.
https://doi.org/10.1007/978-3-032-21654-0_14

advanced composites, metals, and even bio-based or edible materials. In response to user demands for aesthetic and functional customization, novel composite filaments such as PLA blended with wood, metallic, or colored powders have also become increasingly available. AM has revolutionized product development by introducing rapid prototyping, which reduces both development time and cost. Compared to traditional methods like casting or injection molding, which often require costly tooling and long lead times, 3D printing offers a faster, more flexible solution, particularly for early-stage design iterations. Over the past three decades, AM has progressed from printing only with photopolymer resins to now encompassing a diverse array of processes and materials. Technologies such as SLA, Selective Laser Sintering (SLS), Fused Deposition Modeling (FDM), Multi Jet Fusion (MJF), PolyJet, Direct Metal Laser Sintering (DMLS), Digital Light Processing (DLP), and Electron Beam Melting (EBM) have broadened the scope of applications—from micro-scale biomedical devices to large-scale aerospace components. Among these, Fused Deposition Modeling (FDM) and Fused Filament Fabrication (FFF) remain two of the most accessible and widely used AM technologies. Two different technologies with similar material-adding mechanisms still hold substantial innovation potential, particularly in optimizing internal structures such as infill patterns, which reinforce printed objects without requiring solid fills. These internal geometries directly influence a part's strength, weight, and material usage and remain a promising area for further research and enhancement. FDM/FFF materials are typically supplied in filament form.

Polylactic Acid (PLA) is the most used polymer due to its ease of use, low melting point, and availability in a diverse range of colors. While standard PLA is well-studied, newer composite materials such as carbon fiber have gained attention for their enhanced mechanical properties. Carbon fiber is a lightweight material with high mechanical characteristics and chemical resistance. It consists of carbon atoms organized in a specific orientation and is widely used in the aviation and aerospace fields. Mixing chopped carbon fibers into polylactic acid, known as carbon fiber-reinforced PLA (PLA-CF), can moderately increase the strength of the composite material [1]. While continuous carbon fiber filaments require specialized equipment, PLA-CF can be used with standard FDM printers. Therefore, it is more accessible for research and industry applications. Despite the advantages of PLA-CF, there is not much deep research on it. For most engineering applications, optimizing the tensile strength of 3D-printed components is a key focus, as tensile load is one of the most tested and critical mechanical properties. Among the many factors influencing tensile strength, infill pattern plays a significant role. A comparative study by Pandžić, Hodžić, and Milovanović [2] evaluated thirteen common infill patterns for PLA-printed specimens and identified the concentric infill as having the highest tensile strength. The gyroid, octet, and grid patterns followed in descending order of performance. Several other studies have confirmed the superior tensile resistance of the concentric pattern. For example, Rismalia et al. [3] found that concentric infill significantly outperformed both grid and tri-hexagonal patterns. Similarly, research by Son, Minh, and Thanh [4] concluded that concentric infill yielded the highest tensile stress among the evaluated software-generated patterns. Other commonly used 2D patterns, such as rectilinear, grid, and honeycomb, have also been investigated.

While grid is widely used, some studies indicate that rectilinear infill may provide superior tensile performance. One such study demonstrated that rectilinear outperformed the grid in tensile stress capacity [5]. Dave et al. [6] further validated this by comparing rectilinear, concentric, and Hilbert curve patterns; their results showed that rectilinear achieved tensile stress levels comparable to those of concentric and significantly higher than those of the more complex Hilbert pattern. M. Naik, D. G. Thakur, and S. Chandel did a study about the flexural properties of specimens printed with an FDM printer; however, they stacked multiple types of infill patterns vertically in one single sample and collected a positive result in the flexural strength/mass ratio compared to solid samples [7]-a promising signal for multi-infill approaching. Similarly, another study stacked the rectilinear infill and the grid infill vertically in several orders and then investigated the influence of the multi-infill on the part's tensile strength. The final result indicated that the combined infill enhanced the part's tensile strength compared to a single infill structure for a specific raster orientation [8].

In this study, instead of applying different infill types along the vertical axis, we propose a new method of organizing 3D printing infill patterns based on how tensile stress is distributed across the printed part. The core idea is to strengthen the areas under the highest tensile load by using a rectilinear pattern at a higher-than-usual density. This increases the number of continuous printed lines in those critical zones, improving overall strength compared to a uniformly filled part. To maintain a balanced overall density, the areas less affected by tensile stress are filled with a different infill pattern at a lower density. This approach ensures that strength is concentrated where it's needed most, without unnecessarily increasing the part's weight. The study investigates the mechanical performance of chopped carbon fiber-reinforced PLA under tensile stress, with a focus on the design and evaluation of a new infill pattern. The proposed pattern, which is not available in standard slicing software, aims to enhance tensile strength compared to traditional patterns such as rectilinear and concentric. Furthermore, the study evaluates the impact of key FDM parameters-such as infill density, printing speed, and extrusion speed-using the Taguchi method to determine optimal settings for strength maximization. The combination of material innovation and infill design contributes both practically and theoretically to the ongoing development of high-performance 3D-printed components.

2 Proposal of a New Infill Pattern Design

This research utilizes commercially available carbon fiber-reinforced PLA (CF-PLA) containing 15% carbon fiber by weight. The specimen design follows the ASTM D638 Type I standard [9], which is widely recognized for tensile testing of polymer materials (see Fig. 1).

To identify areas of high stress, Finite Element Method (FEM) simulations in Abaqus are performed on a solid ASTM D638 Type I specimen. As shown in Fig. 2, the results clearly show that tensile stress is concentrated in the narrow middle section. This provides a reasonable estimation of the tensile behavior of the sample under axial loadings. It can also support predicting the behavior in the experimental study in Sect. 3.

However, it's important to note that the simulation assumes uniform material properties and infill density. In real parts, the larger cross-sections are naturally stronger, so

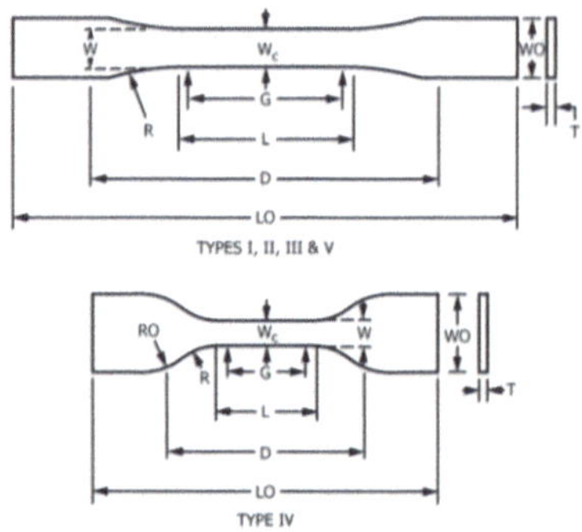

Specimen Dimensions for Thickness, *T*, mm (in.)[A]

Dimensions (see drawings)	7 (0.28) or under		Over 7 to 14 (0.28 to 0.55), incl	4 (0.16) or under		Tolerances
	Type I	Type II	Type III	Type IV[B]	Type V[C,D]	
W—Width of narrow section[E,F]	13 (0.50)	6 (0.25)	19 (0.75)	6 (0.25)	3.18 (0.125)	±0.5 (±0.02)[B,C]
L—Length of narrow section	57 (2.25)	57 (2.25)	57 (2.25)	33 (1.30)	9.53 (0.375)	±0.5 (±0.02)[C]
WO—Width overall, min[G]	19 (0.75)	19 (0.75)	29 (1.13)	19 (0.75)	...	+ 6.4 (+ 0.25)
WO—Width overall, min[G]	...	...	...	...	9.53 (0.375)	+ 3.18 (+ 0.125)
LO—Length overall, min[H]	165 (6.5)	183 (7.2)	246 (9.7)	115 (4.5)	63.5 (2.5)	no max (no max)
G—Gage length[I]	50 (2.00)	50 (2.00)	50 (2.00)	...	7.62 (0.300)	±0.25 (±0.010)[C]
G—Gage length[I]	...	...	...	25 (1.00)	...	±0.13 (±0.005)
D—Distance between grips	115 (4.5)	135 (5.3)	115 (4.5)	65 (2.5)[J]	25.4 (1.0)	±5 (±0.2)
R—Radius of fillet	76 (3.00)	76 (3.00)	76 (3.00)	14 (0.56)	12.7 (0.5)	±1 (±0.04)[C]
RO—Outer radius (Type IV)	...	...	...	25 (1.00)	...	±1 (±0.04)

Fig. 1. Shapes and dimensions of the ASTM D638 standard [9]

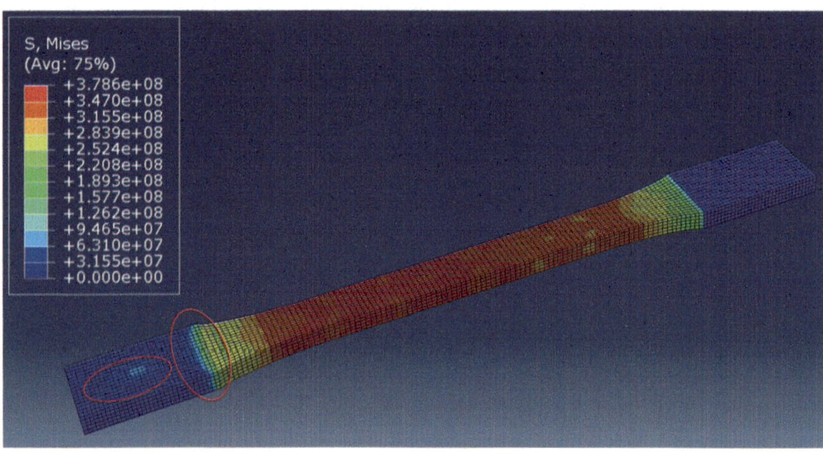

Fig. 2. FEM simulation result of the solid ASTM D638 Type I shape

failure typically occurs at the narrow section. When using the proposed new pattern, which combines denser and lighter regions, the uniform strength assumption no longer holds. If the rectilinear infill is applied only in the most highly stressed (red and orange) areas in Fig. 2, the wider ends of the part may become weaker than intended. Obviously, in Fig. 2, some circled zones outside the narrow section still experience significant tension. If those are filled with lower-density infill, they could fail prematurely. To avoid this, the denser rectilinear infill should be extended slightly beyond the narrow section to cover all areas experiencing significant stress (Fig. 3). Observing, the rectilinear pattern consists of parallel lines that do not bond well because they only connect through the top and bottom layers. To avoid similar weak bonding elsewhere, patterns with lines printed diagonally to the stress direction are preferred for the orange zones in Fig. 3. The Grid pattern (Fig. 4) is a strong candidate, as it offers good tensile strength and has a more interconnected structure, making it suitable for those areas. Therefore, it is used for the new combined infill pattern.

Fig. 3. Infill patterns distribution with density differences consideration

Fig. 4. Grid pattern

3 Experimental Study

3.1 Procedures and Experimental Setup

All specimens are fabricated using the Creality Sermoon D1 3D printer (Fig. 5), a fused deposition modeling (FDM) printer selected for its stability in axial motion and consistent extrusion performance.

Fig. 5. The Creality Sermoon D1 3D printer

The range of adjustable printing parameters is determined by the slicing software. Currently, the market offers a variety of free and paid slicers developed by different providers to support various additive manufacturing technologies. Several studies have evaluated the performance of these slicers, particularly for FDM and FFF technologies. Research findings suggest that Ultimaker Cura delivers superior print quality compared to other free slicing software[10, 11]. Bryła and Martowicz [12] further noted that Cura offers more comprehensive control over the printing process. Based on these findings, this study uses Ultimaker Cura for all slicing operations.

In this study, tensile testing is conducted using the Shimadzu 50 kN Autograph AGS-X series universal testing machine, operated through the TrapeziumX control software. The testing setup includes a fixed clamp oriented upward at the bottom of the testing chamber, a downward-facing movable clamp capable of vertical displacement, and a vertical guide rail system enclosed within the machine's structure (see Fig. 6).

Fig. 6. Shimadzu 50kN Autograph AGS-X series universal testing machine

The experimental procedure begins with the design of 3D specimen models using Autodesk Inventor, categorized into three types based on infill patterns (Figs. 7, 8 and 9).

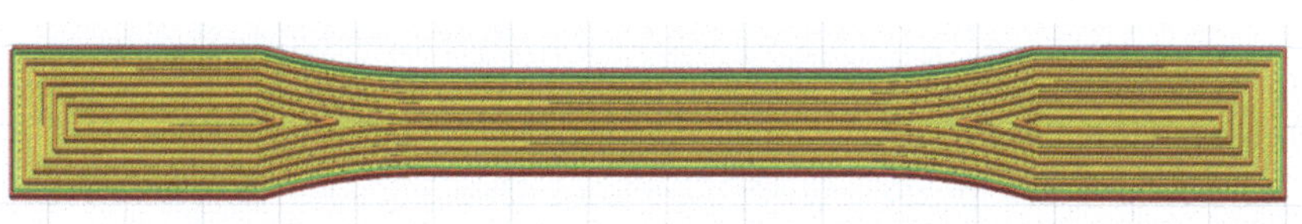

Fig. 7. The Concentric infill pattern inside an ASTM D638 Type I reference shape

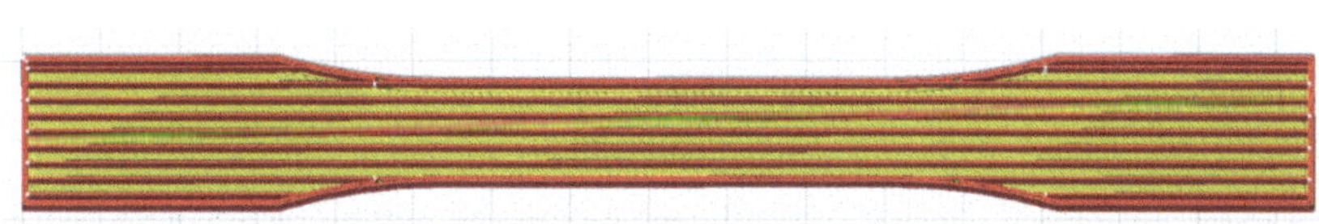

Fig. 8. The Rectilinear infill pattern inside an ASTM D638 Type I reference shape

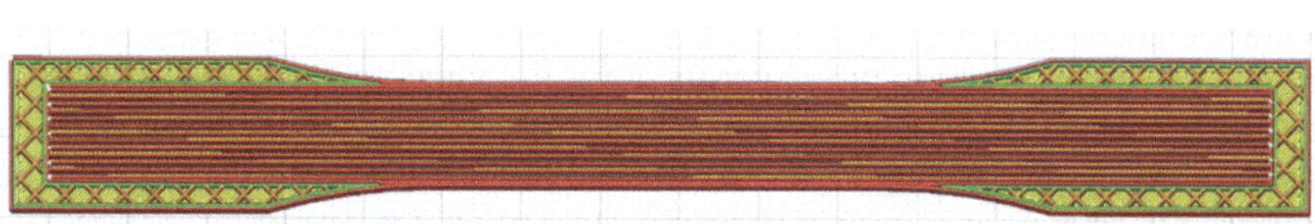

Fig. 9. The Combined infill pattern inside an ASTM D638 Type I reference shape

Normally, the printing lines generated automatically by the slicing software have a width of 0.4 mm, the same as the nozzle diameter, so the real printed lines will be extruded separately from each other without any close-by parallel lines. Instead of printing with the default method, in this research, all the infill patterns will be constructed in the 3D environment with the width of infill lines at 0.8mm each. A sample with the infill section that has two adjacent parallel printed lines per wall will be obtained after slicing the 3D model with a 0.4 mm printing line width setting as the default. This modification enhances tensile strength by improving inter-line bonding, increasing internal material density, and reducing weak points in the printed structure (Figs. 10 and 12).

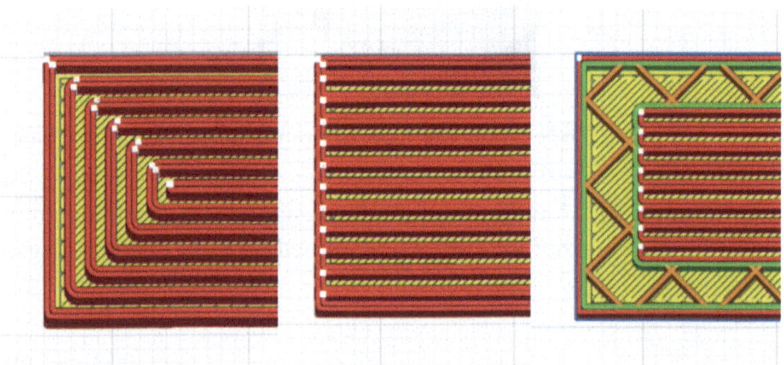

Fig. 10. Infill patterns (from left to right: Concentric, Rectilinear, and Combined infill pattern) with double printing lines

In the following section, the performance of the three infill patterns (concentric, rectilinear, and combined) will be evaluated alongside infill density and printing speed to assess their impact on tensile strength. These infill patterns will then be used with the predicted optimal density and printing speed to fabricate specimens for tensile strength comparison.

3.2 Design of Experiment

The set to be considered consists of infill density, top and bottom raster angle, extruder temperature, layer height, and printing speed. Based on published studies, researchers usually select the infill density up to only 80% but not closer to 100% because they found that the infill density changing from 40% to 80% leads to a significant increase in the tensile strength of the part, which will not remain that effective in higher infill density, and the best tensile strength was obtained at 80% dense obviously [13, 14]. The raster angle of ±45° will be used in this experiment for the same reason as the grid pattern–to enhance the bonding of parallel lines. Carbon fiber-reinforced PLA can be printed ideally at temperatures ranging from 200 °C to 220 °C. However, if the temperature is too low, the nozzle may clog due to chopped carbon fiber; if the temperature is too high, polymer-related defects such as shrinking and over-melted filament may occur. Hence, 210 °C is a safe option. 0.1 mm–0.2 mm is the common range of layer height chosen in various research topics. Many results on the optimal layer height were gathered;

however, these results were not consistent due to the influence of other parameters in each study. However, investigating the effect of layer height is not the purpose of this study; thus, the value 0.15 mm will be chosen as the mean value of layer height. In Cura software for 3D printing, the Flow Rate setting-expressed as a percentage (e.g., 100%)-acts as a multiplier that controls the amount of filament extruded relative to the default value calculated by the slicer. Adjusting this parameter allows users to fine-tune extrusion to ensure proper layer bonding or compensate for material inconsistencies. Decreasing the flow rate makes the bonding weaker and creates porosity inside the part. In contrast, increasing the flow rate too much can cause overflow defects, which can result in dimensions that deviate from the original design. Hence, the flow rate cannot be too high or too low. By default, the flow rate is always 100% (Table 1).

Table 1. Parameter setup for the printer

Parameters	Value
Nozzle temperature	210 °C
Bed temperature	60 °C
Raster angle	±45°
Layer height	0.15 mm
Flow rate	100%

The Printing Speed, measured in millimeters per second (mm/s), defines how quickly the print head moves while extruding filament, directly affecting the balance between detail, strength, and print time. Typically, printing speeds ranging from 20 mm/s to 60 mm/s are selected for stability during printing for experiments. The specimens will be printed with the Creality Sermoon D1–an FDM printer that can print at speeds up to 200 mm/s, so the technical limitation of the machine allows the specimens to be printed in the range from 60 mm/s to 100 mm/s in Table 2. Infill Density refers to the percentage of internal structure filled with material, while infill Pattern determines the geometric layout of that structure. These settings collectively influence the part's strength, weight, and material usage. The following section will explore how these parameters impact the tensile strength. The parameters and their level are designed as in the Tables 2.

Table 2. Variational factors and levels

Factors	Level 1	Level 2	Level 3
Infill pattern	Concentric (A)	Rectilinear (B)	Combined (C)
Infill density	30%	55%	80%
Printing speed	80 mm/s	90 mm/s	100 mm/s

The Taguchi design of experiments is a popular fractional factorial design proposed by Dr. Genichi Taguchi to investigate how different combinations of selected factors, within the ranges of their assigned levels, would influence the experiment's output.

Taguchi orthogonal arrays are a collection of logically structured test cases that enable the efficient and precise determination of the most critical variables. Orthogonal arrays ensure that all elements are examined at various levels while lessening the number of tests performed. It is a reliable experimental design strategy used to optimize processes and enhance product or process quality. In this research, an L9 Taguchi experimental design is employed. Each experiment in L9 was repeated once, and the response value (tensile load) was calculated as the average of the results from the first and second repetitions. L9 Taguchi experimental design and the recorded response results are shown in Table 3.

Table 3. The experiment design Table

Sample Name	Infill pattern	Infill density (%)	Printing speed (mm/s)	Tensile Load (N)
A-30.8.8	A (concentric)	30	80	1165.743
B-55.9.9	B (rectilinear)	55	90	1763.837
C-80.1.1	C (combined)	80	100	2258.635
A-55.1.1	A(concentric)	55	100	1385.681
B-80.8.8	B (rectilinear	80	80	2093.000
C-30.9.9	C (combined)	30	90	1253.184
A-80.9.9	A(concentric)	80	90	1943.278
B-30.1.1	B (rectilinear	30	100	1129.670
C-55.8.8	C (combined)	55	80	1684.245

3.3 Experimental Results and Response Analysis

Figure 11 shows the printed samples. By observation, samples printed with the concentric pattern exhibit a brittle characteristic. The specimens with lower density appear to exhibit higher elongation compared to those with higher density. This brittle property may increase in conjunction with the part's infill density. The specimens gathered from the rectilinear infill pattern group also exhibit a slight brittleness, as evidenced by the broken lines from samples with rectilinear infill, which are not as clean as those from samples with the concentric infill pattern. Besides that, the slightly brittle characteristic only appears on specimens with low and high densities; rectilinear samples with middle-level density have better durability, as their displacement is longer and their inner polymer fibers are torn apart one by one, rather than breaking simultaneously. A curvature trend along the axis of density in the response surface graph of the sample printed with the rectilinear pattern can be predicted based on the phenomenon that brittleness appears at a low level of density, then disappears at a middle level of density, and reappears at a high level of density. The combined infill pattern provides samples with greater strength, as the brittle property appears to be nonexistent at all levels of density. A trend that can easily be observed is that the higher the density, the larger the samples elongate.

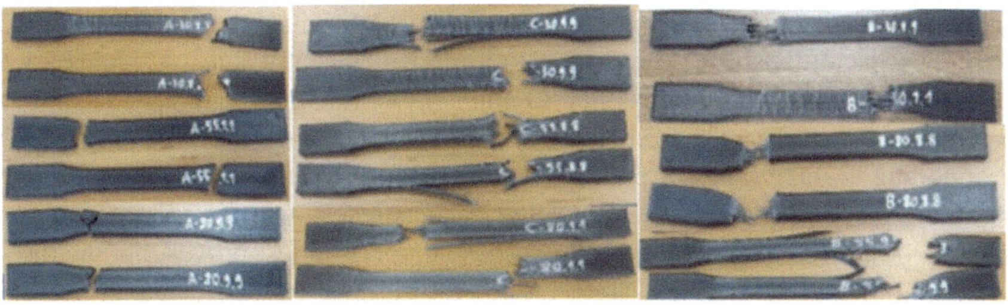

Fig. 11. Tested Samples

Next, we develop the Response Table for Means to help analyze how different levels of each factor influence the performance characteristic. For each level of each factor, the average response is calculated by the formula in Eq. (1):

$$MeanResponse_{FactorA,levelj} = \frac{1}{n}\sum\nolimits_{i=1}^{n} Response_i \tag{1}$$

where,

Factor A = the factor being analyzed;

Level j = the specific level of the factor;

$Response_i$ = the measured response for the ith experiment where Factor A is at Level j;

n = number of experiments with Factor A at Level j;

Delta (Δ) value shows how much influence each factor has and is determined as Δ = Max(mean) - Min(mean). These factors have the greatest effect from highest to lowest Δ value.

Table 4. Response Table for Means

Level	Infill pattern	Infill density	Printing speed
1	1498	1183	1648
2	1662	1611	1653
3	1732	2098	1591
Delta	234	915	62
Rank	2	1	3

Using the Plot 12 tool visualizes how each factor influences the response based on the response Table for the mean value. From the response results in Table 4 and Plot 12, it is evident that infill density has the most significant impact on tensile strength, as indicated by its steep slope. A higher infill density results in greater tensile strength. Regarding the infill pattern, the combined pattern provides the highest tensile strength, followed by the rectilinear pattern, with the concentric pattern yielding the lowest strength. For printing speed, both 80 mm/s and 90 mm/s show similar effects on tensile strength. However, increasing the speed to 100 mm/s leads to a noticeable decline in tensile strength. The predicted optimal parameter combination for maximizing tensile strength is as follows: the infill pattern should be set to combined, the infill density should be 80%, and the printing speed should be either 80 mm/s or 90 mm/s, as both provide similarly strong results. To validate the predicted optimal parameter setup, the three printed parts, corresponding to three types of patterns with the printing setting at infill density 80% and printing speed 80%, are subjected to physical tensile testing using a Shimadzu AGS-X machine. The result of the pattern properties comparison test proves that the design theory was in the right direction since the lowest ultimate strength value collected in the test belongs to the concentric pattern, the middle position belongs to the rectilinear pattern, and the combined pattern performed the strongest maximum strength; the values of them are 2213.637N, 2338.679N, and 2650.722N, respectively. The behavior of each sample used in this experiment can be observed on the force and displacement graphs generated below.

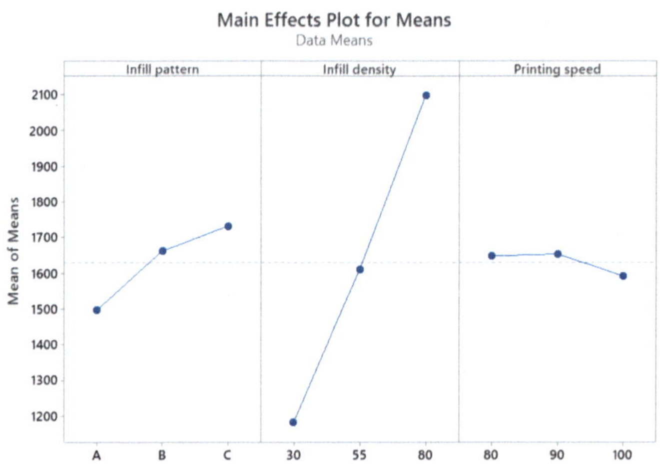

Fig. 12. Main effects plot for Means

The concentric and rectilinear pattern with the higher tension zone concentrated at the middle section of the sample seems to be easier to break at around 3 mm of displacement (Figs. 13 and 14). In contrast, the combined pattern with better load-distributed characteristics can reach an elongation of over 40 mm, although some inner polymer fibers have been damaged or torn apart since the displacement reached around 18 mm (Fig. 15). These results show that the combined pattern performs much better compared with the concentric and rectilinear patterns.

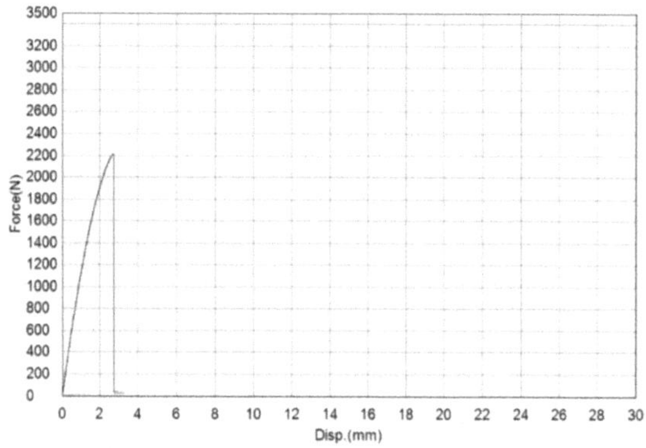

Fig. 13. Force/Displacement graph of the concentric pattern

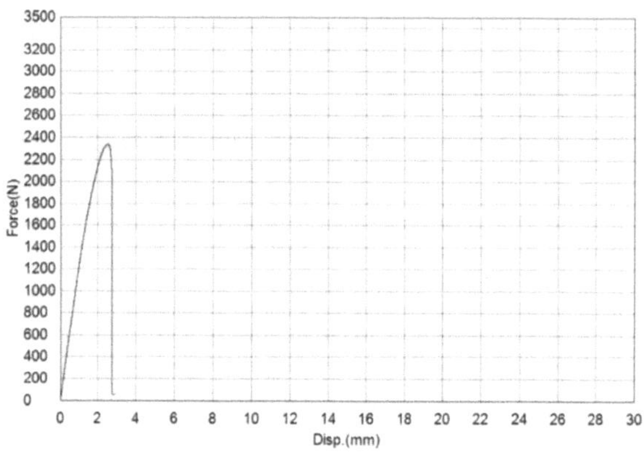

Fig. 14. Force/Displacement graph of the rectilinear pattern

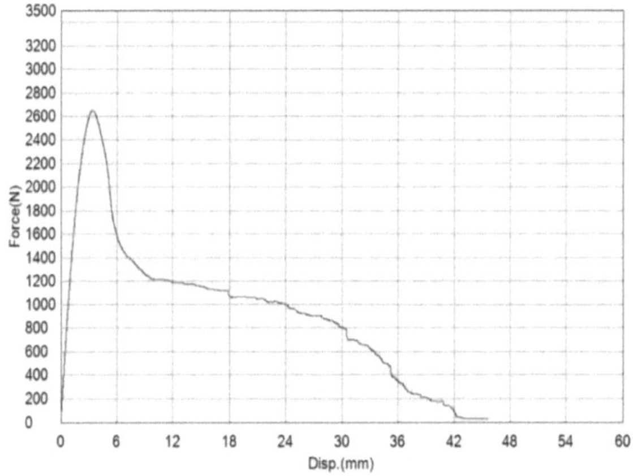

Fig. 15. Force/Displacement graph of the combined pattern

4 Conclusion and Future Works

This study introduces a new infill pattern, named the combined pattern, which is not currently available in popular slicing software. Conducted using carbon fiber-reinforced PLA (PLA-CF), a material that has not yet been widely explored in existing literature, the research offers valuable insights into the mechanical behavior of this composite regarding infill design and process parameters. Observing specimens after the tensile tests, it is evident that the printed samples with concentric and rectilinear patterns are more brittle than those with combined patterns. The combined pattern provides the sample with the strongest tension resistance compared to the two-best tension-loading software-built-in patterns considered in this study (the concentric and the rectilinear patterns). The geometry of the combined pattern also has the best tensile-load-distribution property, while the worst belongs to the rectilinear pattern. The tension loading ability of the rectilinear pattern is better than the concentric pattern if both of the samples are placed parallel to the loading direction. The research also studies the influence of full pattern, full density, and printing speed on tensile strength. The results show that among the three process parameters investigated-infill pattern, infill density, and printing speed-infill density had the most significant positive impact on tensile strength, followed by the infill pattern, and then printing speed. These results demonstrate the mechanical advantages of the combined pattern with PLA-CF material in high-performance applications. This study presents a novel infill strategy and expands the existing body of research on carbon fiber-reinforced PLA in additive manufacturing.

References

1. Ferreira, R.T.L., Amatte, I.C., Dutra, T.A., Bürger, D.: Experimental characterization and micrography of 3D printed PLA and PLA reinforced with short carbon fibers. Compos. Part B: Eng. **124**, 88–100 (2017)

2. Pandžić, A., Hodžić, D., Milovanović, A.: Effect of infill type and density on tensile properties of PLA material for FDM process. In: Proceedings of the 30th DAAAM International Symposium, DAAAM International, pp. 545–554 (2019)
3. Rismalia, M., Hidajat, S., Permana, I., Hadisujoto, B., Muslimin, M., Triawan, F.: Infill pattern and density effects on the tensile properties of 3D printed PLA material. In: Journal of Physics: Conference Series, vol. 1402, no. 4: IOP Publishing, p. 044041 (2019)
4. Minh, P.S.: Optimizing I-Beam by using shape generator tool. Key Eng. Mater. **929**, 117–122 (2022)
5. Cabreira V., Santana, R.M.C.: Effect of infill pattern in Fused Filament Fabrication (FFF) 3D Printing on materials performance. Matéria (Rio de Janeiro) **25**(03), e-12826 (2020)
6. Dave, H.K., Patadiya, N.H., Prajapati, A.R., Rajpurohit, S.R.: Effect of infill pattern and infill density at varying part orientation on tensile properties of fused deposition modeling-printed poly-lactic acid part. Proc. Inst. Mech. Eng. C J. Mech. Eng. Sci. **235**(10), 1811–1827 (2021)
7. Naik, M., Thakur, D., Chandel, S.: Experimental investigation on flexural properties of additively manufactured multi-infill pattern parts fabricated by FDM. In: AIP Conference Proceedings, vol. 2648, no. 1: AIP Publishing LLC, p. 030022 (2022)
8. Dave, H.K., Patel, B.H., Rajpurohit, S.R., Prajapati, A.R., Nedelcu, D.: Effect of multi-infill patterns on tensile behavior of FDM printed parts. J. Braz. Soc. Mech. Sci. Eng. **43**(1), 23 (2021)
9. A. S. D. o. M. Properties: Standard test method for tensile properties of plastics. American Society for Testing and Materials (1998)
10. Šljivic, M., Pavlovic, A., Kraišnik, M., Ilić, J.: Comparing the accuracy of 3D slicer software in printed enduse parts. In: IOP conference series: materials science and engineering, vol. 659, no. 1: IOP Publishing, p. 012082 (2019)
11. Gomez-Alonso, J.-L., Allue, A., De-Marco, I., Retolaza, J., Diez, G.: Influence of slicer software used with 3D printing filament extrusion technology on properties of printed parts with short fiber reinforced thermoplastic composite. DYNA **97**(3), 295–300 (2022)
12. Bryła, J., Martowicz, A.: Study on the importance of a slicer selection for the 3d printing process parameters via the investigation of g-code readings. Machines **9**(8), 163 (2021)
13. Heidari-Rarani, M., Ezati, N., Sadeghi, P., Badrossamay, M.: Optimization of FDM process parameters for tensile properties of polylactic acid specimens using Taguchi design of experiment method. J. Thermoplast. Compos. Mater. **35**(12), 2435–2452 (2022)
14. Kumar, M.A., Khan, M., Mishra, S.: Effect of fused deposition machine parameters on tensile strength of printed carbon fiber reinforced PLA thermoplastics. Mater. Today: Proc. **27**, 1505–1510 (2020)

Study of the Application of Non-Contact Profilometric Methods for Analyzing the Lateral Layers of FDM 3D Prints

Olha Kalman(✉), Martin Kondrát, Jozef Husár, and Patrik Fejko

Faculty of Manufacturing Technologies with a Seat in Prešov, Technical University of Kosice, Presov, Slovakia
olha.kalman@tuke.sk

Abstract. This study compares two non-contact profilometric systems, the LPM-4 and the Keyence LJ-X8020, to evaluate the lateral layer structures of FDM 3D-printed materials. Four specimens were measured—black PLA, white PLA PRO, transparent violet PLA/PHA and orange PET-G—each printed with identical settings to show how material and color affect measurement accuracy. The LPM-4 profilometer yielded high-density numerical datasets that were conducive to the quantitative surface evaluation (Ra, Rz) of the material. The Keyence system, meanwhile, provided high-resolution 2D and 3D visualizations that were instrumental in conducting a qualitative layer diagnostic analysis of the material in question. Both systems showed that optical and material properties affect measurement accuracy. The results show that integrating different profilometric methods improves understanding of 3D-printed surfaces.

Keywords: Additive manufacturing · Topography analysis · Vision systems · Quality control · Surface metrology

1 Introduction

Non-contact profilometric techniques play an increasingly important role in the evaluation of layer structures in FDM 3D printed parts, as they allow precise assessment of dimensional accuracy and surface quality without damaging the specimen. Optical approaches, particularly fringe projection profilometry (FPP), have been successfully applied to non-invasive characterization of layer micro-topography. Recent studies have shown that the integration of machine learning into FPP significantly improves measurement accuracy, reducing the root-mean-square error from approximately 10.57 μm to 4.35 μm [1, 2]. Furthermore, structured-light and fringe-projection systems are widely used for in-situ monitoring of layer formation, enabling real-time surface mapping and detection of process anomalies during FDM printing [3–5].

Customized surface-specific test artefacts have been introduced to support surface texture evaluation and layer-step analysis in additive manufacturing, thereby improving process monitoring and control [6, 7]. In addition, both filament material selection and

J. Trojanowska et al. (Eds.): MANUFACTURING 2026, LNME, pp. 198–212, 2026.
https://doi.org/10.1007/978-3-032-21654-0_15

layer height have a significant impact on dimensional and surface quality in FDM printing. Previous studies report that PET-G exhibits higher accuracy in length and height, whereas PLA performs better in width-related dimensions, highlighting the importance of considering material–geometry interactions when interpreting profilometric results [8, 9]. Furthermore, the profile of support bases has been found to influence Z-axis accuracy, indicating that support strategy and surface conditions of the build plate may impact the final quality of printed parts [10, 11].

Non-contact measurement methods are clearly advantageous in terms of efficiency and data acquisition but have limitations. Contact profilometry, e.g. stylus tracing, is crucial in ensuring dimensional accuracy and providing benchmark measurements, especially in industrial standard metrology. A review shows that non-contact methods are fast and comprehensive, but sensitive to material reflectivity, color, and geometry [12, 13]. A plethora of studies have previously been conducted that have employed non-contact measurement techniques for the purpose of evaluating both the surface quality and the layer structures of FDM printed components. For instance, He et al. developed a machine-vision-based system for real-time geometric quality monitoring during FDM printing, utilizing image processing of layer-by-layer profiles and statistical process control charts [6, 14]. A further study was conducted in which the FDM and DLP technologies were compared. This was achieved by means of optical microscopy, which revealed that FDM parts have rougher and anisotropic surface characteristics in comparison to the smoother, homogeneous structures of DLP parts [15, 16].

Moreover, Medina Sánchez et al. conducted a study that validated laser confocal microscopy against contact profilometry for the purpose of measuring roughness. The study found an average difference of approximately 5% between the two methods. Furthermore, it demonstrated that build orientation and layer height have a substantial influence on surface quality [17, 18]. Finally, Bazan et al. conducted a comparative analysis of contact profilometry with focus-variation microscopy across different geometries and printing techniques. This study concluded that optical methods yielded superior results for the surface topography evaluation of additively manufactured parts [19, 20].

Despite knowledge on this subject being limited, there are few studies focusing on the side-layer (lateral surface) profiles of FDM parts, and even fewer on the comparison of two different non-contact profilometers on the same sample set. This article addresses this gap through a comparative analysis of two non-contact laser profilometers applied to the side-layer structure. The objective is to evaluate measurement workflows, and to assess how material and color influence measurement fidelity in the systems.

2 Research Problem and Objectives

In this research, two non-contacts optical profilometers were used to analyze the side-layer structures of FDM 3D printed samples made from different materials and colors. The first device, LPM-4 (Kvant Ltd., Bratislava, Slovakia), is a laser surface profilometer developed as a modular measurement system intended for optical, contactless evaluation of the 3D surface profile along a defined section. The system consists of a stable aluminum supporting frame that allows manual adjustment of the working distance, a profilometric head mounted on an XZ-axis trolley, and an external control unit. Such configuration enables flexible scanning of surfaces with variable geometry and roughness.

The LPM-4 device has been previously used for surface-quality evaluation in woodworking and manufacturing applications, particularly for analyzing the surface topography of milled medium-density fiberboards and heat-treated wood samples. Its construction allows for manual positioning and a stable scanning path, which makes it suitable for laboratory studies requiring adaptable scanning geometry and sample orientation.

The second system, Keyence LJ-X8020 (Keyence International NV/SA, Mechelen, Belgium), represents an industrial-grade laser profilometer based on the LJ-X8000 series. It utilizes a blue semiconductor laser with a wavelength of 405 nm and achieves a Z-axis repeatability of 0.3 μm, offering high-precision areal and line-profile measurements. The LJ-X8020 provides a measurement range of ±2.2 mm in the Z direction and a data-point density of 3200 points per profile, with an X-axis interval of approximately 2.5 μm.

The Keyence system, unlike the LPM-4, is fully automated, with built-in calibration and focus adjustment, and direct data output for analysis. These characteristics make it good for measuring polymer surfaces, thin layers, and 3D-printed parts. Both systems use lasers and calculators, but the design and purpose differ. The LPM-4 is for research and adaptable to different samples, and the LJ-X8020 is for industry, analyzing surfaces quickly and precisely. Table 1 subsequently provides a concise synopsis of the two non-contact profilometers' key measurement parameters.

Table 1. Comparison of Key Measurement Parameters

Parameter	LPM-4 (Kvant Ltd., Slovakia)	Keyence LJ-X8020 (Keyence, Belgium)
Laser type	Red laser diode (visible spectrum, ≈650 nm)	Blue semiconductor laser (405 nm)
Measurement principle	Laser triangulation (manual scanning system)	Laser triangulation (high-speed line sensor)
Z-axis range	Approx. ±5 mm	±2.2 mm
Z-axis repeatability	~5–10 μm	0.3 μm
X-axis data interval	Variable (manual feed)	2.5 μm (fixed)
Profile resolution	Limited by manual feed accuracy	3200 points per profile
Data acquisition	Manual export, Excel-based	Automated digital export (software interface)
Use case	Laboratory, experimental setup	Industrial metrology, precision analysis

The aim of the experiment was to compare the measurement capabilities of two non-contact laser profilometers–the LPM-4 and the Keyence LJ-X8020–in analyzing the side layers of FDM 3D-printed samples. The experiment used four samples printed with identical processing parameters (layer height, printing speed, nozzle diameter, and infill) but manufactured from different filament materials: PLA (black sample), PLA

PRO (white sample), PLA/PHA (transparent sample), and PET-G (orange sample). The PLA, PLA PRO, and PET-G filaments were supplied by Fillamentum Manufacturing Czech s.r.o., while the PLA/PHA filament was supplied by Devil Design Sp. z o.o. The mechanical, thermal, and optical material properties used in this study were taken from the corresponding Technical Data Sheets (TDS) provided by the manufacturers.

Despite their absence of a standardized geometric form, the samples had been evaluated for geometric accuracy and multispectral imaging response, confirming sufficient dimensional stability for subsequent surface analyses. The following Fig. 1 presents a technical drawing of a specimen utilized in the experimental phase of the research endeavor.

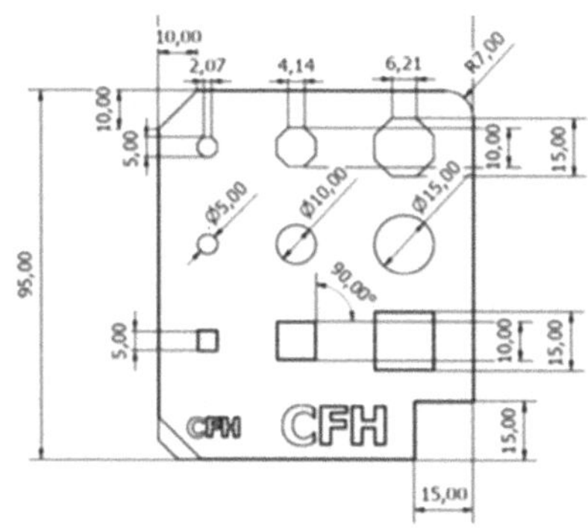

Fig. 1. Technical drawing of the test sample for experimental analysis of layer structures in printed parts

The reflectivity and transmittance of the laser beam vary between these materials, affecting the quality and readability of the profilometric data. The LPM-4 facilitates manual layer acquisition and data export to Excel for numerical analysis of height profiles and roughness. Concurrently, the Keyence LJ-X8020 provides automated digital outputs with 2D and 3D profile display via a software interface. The experiment aimed to compare these two approaches to identify numerical and visual data processing differences and evaluate material and color influences on measurement stability. The final aim was to make recommendations for selecting a profilometer for layer analysis of FDM parts.

3 Materials and Methods

3.1 Tested Samples

Four samples made with three different thermoplastics using the fused deposition modelling (FDM) 3D printing process were investigated. The aim was to examine how filament composition and optical characteristics affect non-contact profilometric measurements. The materials selected for this study were PLA PRO (white), PLA (black), PLA/PHA (transparent violet), and PET-G (orange). All samples were fabricated using a Bambu Lab X1 Carbon FDM 3D printer manufactured by Bambu Lab (Shenzhen,

China). Each was chosen based on its unique optical behavior, ranging from opaque to semi-transparent. It was hypothesized that this would affect laser reflection and data acquisition quality. All specimens were printed identically (layer height, nozzle diameter, printing speed, and extrusion temperature) to ensure comparability and eliminate the influence of the print parameters on the surface structure (see Table 2).

Table 2. The manufacturing process parameters

Fan speed	Build platform temperature	Infill	Layer [mm]	Extrusion temperature
200	55	15%	0.20	220

Despite their absence of a standardized geometric form, the samples had been evaluated for geometric accuracy and multispectral imaging response, confirming sufficient dimensional stability for subsequent surface analyses [21, 22].

For the study, the side perimeter was selected as the region of interest as it represents the structure most susceptible to optical scattering. The methodology concentrated on analyzing the uniformity of deposited layers and the precision of the surface profile. The material was reduced to four filaments for observation under laser illumination and comparison of surface-profile responses from two non-contact profilometers.

3.2 Measurement Procedure

Each of the four FDM samples was analyzed using two non-contacts optical profilometers—the Keyence LJ-X8020 and the LPM-4. The same lateral surface areas of the printed parts were measured to ensure direct comparability of the data. Only corresponding profile segments from the LPM-4 dataset were selected to match the measurement locations captured by the Keyence LJ-X8020. In the setup with the Keyence LJ-X8020 (Fig. 2a), the laser line scanned the surface perpendicularly, capturing continuous 2D profiles. The collected data were automatically processed in the Keyence Analysis Software, generating 3D surface reconstructions, cross-sectional profiles, and visual maps of layer uniformity. For the LPM-4 system (Fig. 2b), the sample was positioned along the X-axis. The profilometric head scanned the same lateral section of the printed wall, acquiring height data. The data was exported as numerical values into Microsoft Excel. The analysis included extraction of the average surface roughness and maximum profile height, along with visual inspection of the scanned profile curve.

The comparison of both measurement approaches revealed distinct advantages: while the Keyence LJ-X8020 provided high-resolution and fully automated visualization suitable for identifying micro-defects and layer irregularities, the LPM-4 allowed flexible manual control and direct quantitative evaluation of the layer profile. Together, these complementary datasets enabled a comprehensive assessment of the side-layer quality of the FDM-printed samples.

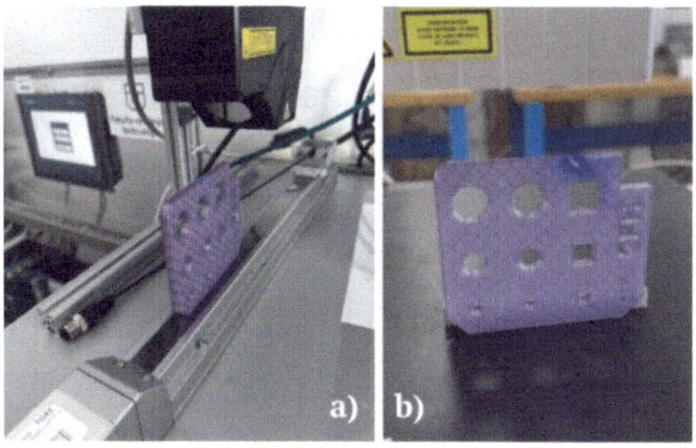

Fig. 2. Demonstration of the lateral perimeter measurement process utilizing (a) the Keyence LJ-X8020 and (b) the LPM-4

4 Results

4.1 Results from Profilometer a LPM-4 (Excel Analysis)

The LPM-4 system scanned the lateral surfaces of the FDM samples and exported the data to Microsoft Excel for quantitative analysis. Only profile segments corresponding to the measurement locations of the Keyence LJ-X8020 were selected for comparison, each containing 2037 height values. Repeated measurements were performed to assess stability, revealing low variability for opaque materials and increased deviations for the transparent PLA/PHA due to optical effects. Height-profile graphs were generated for each sample, illustrating layer uniformity and local surface fluctuations. The calculated roughness parameters (Ra, Rz) confirmed that surface deviations and peak distributions vary with material composition and optical reflectivity, influencing the reliability of non-contact profilometric measurements.

The initial specimen analyzed was fabricated from black PLA filament, a material that is optically opaque, exhibits high laser reflectivity, and possesses stable surface properties. This particular sample was utilized as the standard against which the measurement precision and response of the two profilometers were evaluated. Following the extraction of data from the LPM-4 device, the measured height points were plotted to generate the surface height-profile curve (see Fig. 3).

Figure 3 illustrates the repeating pattern of the printed wall's deposition of successive FDM layers along the perimeter. The curves demonstrate height variations, reflecting the extrusion lines produced during layer-by-layer fabrication, with a narrow range indicating uniformity.

The quantitative analysis yielded a maximum profile height Rz = 0.052 mm and an arithmetic mean roughness Ra = 0.012 mm, indicating a smooth and uniform surface. These values are consistent with typical PLA layer formation under optimized FDM conditions and reflect stable filament deposition with minimal surface irregularities. The low Ra represents the average surface deviation, while Rz captures peak-to-valley variations characteristic of FDM layer geometry, confirming reliable optical performance of the black PLA sample during non-contact laser scanning.

The PET-G (orange) sample displayed a regular periodic height profile associated with FDM layer deposition. Compared to PLA, smoother transitions between layer

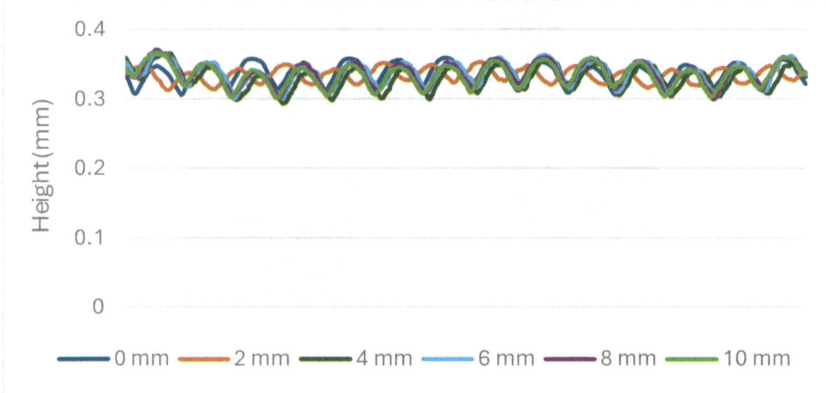

Fig. 3. Graph of measured outer-perimeter values for samples printed from black PLA filament

peaks were observed, indicating uniform material flow and good interlayer adhesion. Minor local irregularities were attributed to the material's thermal behavior and optical interaction with the laser beam. The reduced roughness values (Rz = 0.047 mm, Ra = 0.009 mm) confirm improved surface homogeneity and lower height deviations relative to PLA (Figs. 4 and 8).

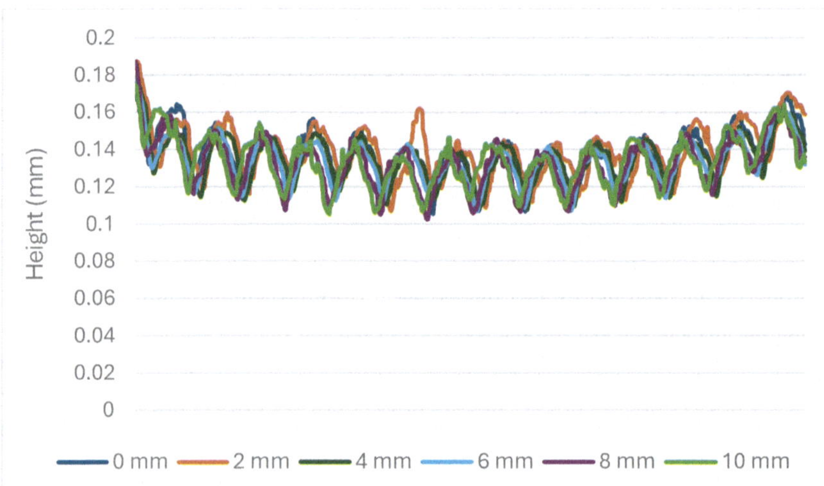

Fig. 4. Graph of measured outer-perimeter values for samples printed from orange PET-G filament

The third sample, made by PLA PRO (white), had a different surface behavior compared with the other materials. The height-profile graph (Fig. 5) shows a less uniform signal with increased local fluctuations, but the overall amplitude is low. Unlike the black PLA sample, which had a smooth and continuous pattern, the PLA PRO profile has small irregular oscillations caused by optical scattering effects from the light-colored, semi-glossy surface.

The calculated parameters—Rz = 0.009 mm and Ra = 0.001 mm—show the lowest average roughness values of all the samples. This result is because of the fine extrusion and dense layer bonding of the PLA PRO filament. However, the low roughness metrics and the less stable graph show that surface reflectivity of white material affects the reliability of laser signal capture, producing locally fluctuating readings even on geometrically smooth surfaces.

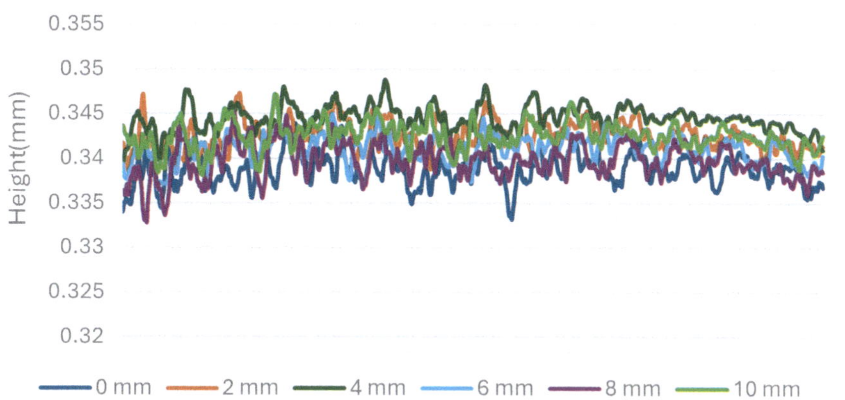

Fig. 5. Graph of measured outer-perimeter values for samples printed from white PLA PRO filament

The PLA/PHA sample exhibited an irregular profilometric response characterized by increased noise and unstable height fluctuations (Fig. 6). Unlike the smooth and uniform profiles observed for PLA and PET-G, the transparent PLA/PHA material caused partial laser transmission and internal scattering, leading to inconsistent readings and peak overestimation. This behavior is reflected in the higher roughness values (Rz = 0.069 mm, Ra = 0.009 mm), despite visually smooth layer deposition. These results highlight the limitations of laser-based optical profilometry when applied to semi-transparent polymer materials.

The height profile graphs showed differences between the materials: the black PLA was stable with low noise and consistent layer spacing, while the transparent PLA/PHA was irregular and unstable due to laser transmission. These findings corresponded to the 3D visualizations—where the PLA surface displayed uniform extrusion ripples (Fig. 7b) and the transparent PLA/PHA surface revealed distortions resulting from signal refraction and scattering inside the material (Fig. 7a).

The LPM-4 furnished detailed numerical data at specific points, which was suitable for quantitative surface evaluation. Furthermore, the Keyence LJ-X8020 will complement this analysis through high-resolution visual mapping of the same regions. This will facilitate a qualitative assessment of layer boundaries, continuity, and potential micro-defects.

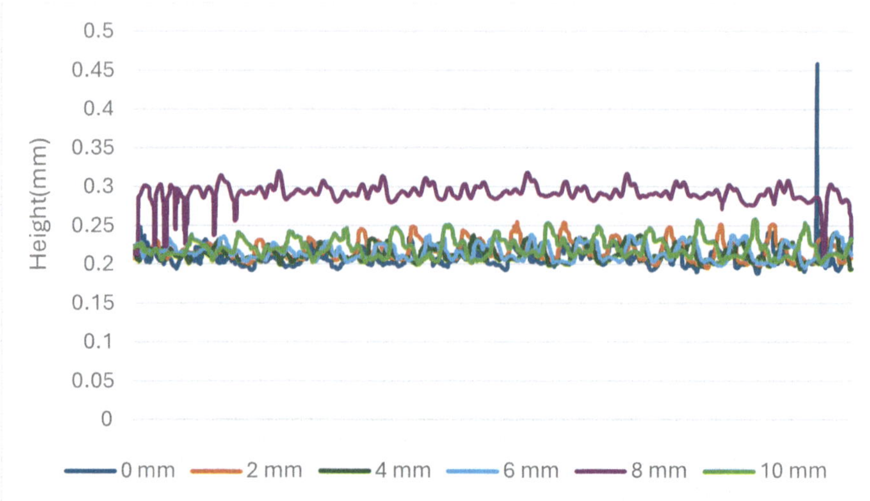

Fig. 6. Graph of measured outer-perimeter values for samples printed from transparent PLA/PHA filament

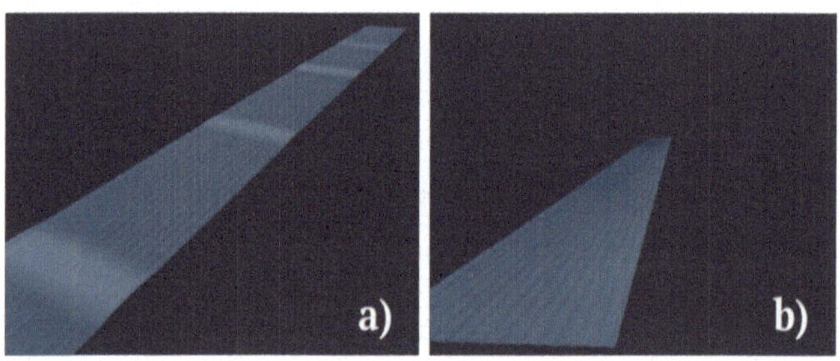

Fig. 7. 3D visualizations of lateral layer structures: (a) transparent PLA/PHA sample; (b) black PLA sample

4.2 Results from Profilometer B the Keyence LJ-X8020 (Interface Image Analysis)

The surface scan obtained using the Keyence LJ-X8020 profilometer for the PLA PRO (white) sample provided a detailed representation of the lateral layer structure. The measurement line, delineated along the lateral periphery of the specimen, exhibited a conspicuously regular layer pattern, with minimal deviations in height. The recorded maximum height value was approximately 1.568 mm, while the minimum was 1.452 mm, resulting in a total vertical variation of about 0.116 mm within the scanned section.

In the lower panel, the waveform in the software interface shows a consistent, periodic signal for uniform layer deposition. The amplitude is small and evenly distributed, confirming smooth extrusion and stable bonding between layers. The 2D top view shows a continuous measurement line without interruptions, confirming a good optical response of the white PLA PRO surface under the blue laser of the Keyence system.

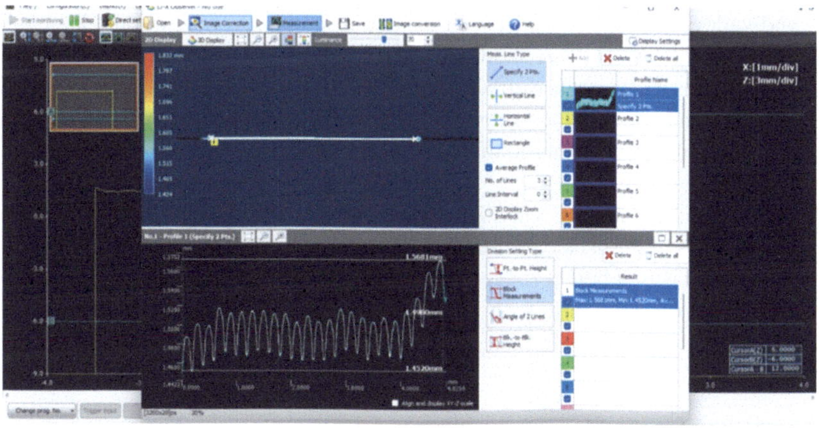

Fig. 8. Visualization of the PLA PRO (white) sample showing periodic layering with clearly defined extrusion lines and optical noise

The Keyence measurement confirms the LPM-4 results. The PLA PRO surface is low-noise and consistent, despite its light color, validating non-contact profilometry for FDM materials. The measurement of the PLA (black) sample using the Keyence LJ-X8020 profilometer produced one of the most stable and well-defined surface profiles among all the materials that were tested. The captured 2D and 3D visualizations (see Fig. 9) reveal a clear, continuous measurement line, indicating excellent laser reflection on the matte black surface, with no signal distortion or optical interference.

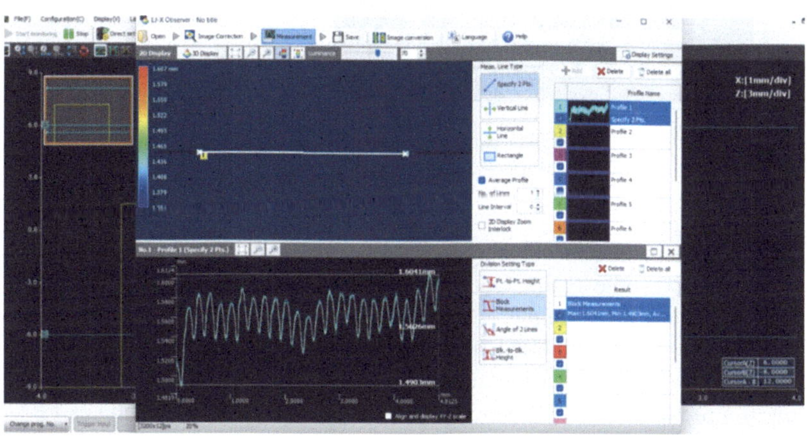

Fig. 9. Visualization of the PLA (black) sample showing periodic layering with clearly defined extrusion lines and optical noise

The profile data of the black PLA sample exhibits a regular waveform corresponding to uniform FDM layer deposition (Fig. 9), with a maximum height of approximately 1.604 mm, a minimum of 1.493 mm, and a total variation of 0.111 mm. The evenly spaced peaks and valleys indicate consistent extrusion and stable layer adhesion. Compared to white PLA PRO, the black PLA shows a smoother and more stable signal with minimal optical noise, as its matte surface effectively absorbs the blue laser light, ensuring precise surface capture and complete scan coverage. Consequently, black PLA provides optimal optical contrast and serves as a reliable reference material for non-contact profilometric evaluation.

Similarly, the Keyence LJ-X8020 analysis of the PET-G (orange) sample revealed a well-defined and stable surface profile characteristic of consistent FDM layer formation (Fig. 10). The measured height ranged from 1.293 mm to 1.390 mm, resulting in a total vertical variation of 0.097 mm, with minimal optical artefacts and a clearly discernible layer structure.

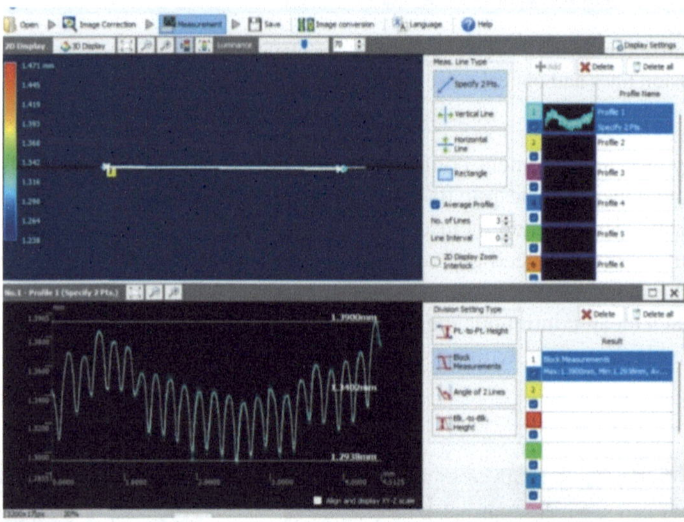

Fig. 10. Visualization of the PET-G (orange) sample showing periodic layering with clearly defined extrusion lines and optical noise

The waveform in the panel displays regular peaks and valleys, representing extruded layers. The PET-G profile exhibits reduced amplitude and smoother transitions between layers than the black PLA sample.

The PET-G sample exhibited high measurement stability and uniform optical response, with balanced surface texture and low deviation in height values. The combination of a regular sinusoidal pattern and minimal optical noise validates PET-G as a material with reliable profilometric characteristics suitable for optical evaluation of FDM-printed components.

Keyence LJ-X8020 profilometry revealed substantial signal instability in the PLA/PHA (transparent violet) sample due to its optical transparency. The 2D profile view shows irregular measurement lines, scattered intensity zones, and fluctuating reflection points. The height profile displays noisy waveform variations, reaching a maximum of 1.595 mm and a minimum of 1.420 mm, indicating a total height variation of approximately 0.175 mm, the largest among the analyzed materials.

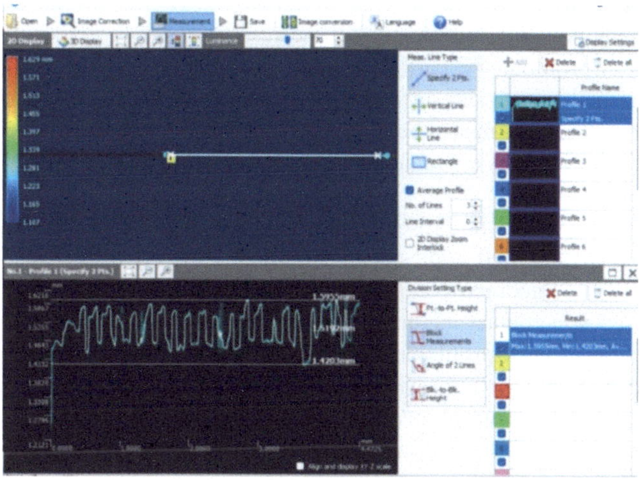

Fig. 11. Visualization of the PLA/PHA (transparent violet) sample showing periodic layering with clearly defined extrusion lines and optical noise

In contrast to the opaque PLA and PET-G samples, the transparent PLA/PHA material exhibited partial laser transmission through its volume, thereby reducing the accuracy of reflected data. This phenomenon is distinctly observable in the LED observation mode (see Fig. 11), wherein the laser light penetrates the surface, yielding diffused internal reflections and prominent scattering points along the upper layer line. Consequently, the profilometric curve manifests random intensity spikes and discontinuities, rather than the smooth sinusoidal structure observed in previous samples.

The results obtained demonstrate that semi-transparent materials, such as PLA/PHA, induce substantial distortions in laser-based optical profilometry, owing to internal refraction and scattering effects. Despite the Keyence system's capacity to capture the general layer pattern, the instability of the reflected signal resulted in overestimated peak values and reduced precision in surface reconstruction.

5 Discussion

A comparison of the LPM-4 and Keyence LJ-X8020 revealed clear differences in measurement stability and precision. While the LPM-4 provided detailed numerical datasets for surface evaluation, its manual operation resulted in higher variability for low-reflectivity materials. In contrast, the Keyence LJ-X8020 delivered stable, high-density 2D and 3D data supported by automated calibration and real-time imaging, making it suitable for qualitative layer diagnostics. Material color and composition strongly influenced laser signal accuracy: black PLA exhibited the most stable profile due to effective light absorption, whereas white PLA PRO showed minor fluctuations caused by surface scattering. The transparent PLA/PHA sample produced the least stable results as laser penetration and internal refraction generated peaks and distortions, confirming the limitations of optical profilometry for semi-transparent polymers in agreement with Jin et al. and He et al. [6, 23].

From a methodological perspective, these results highlight certain limitations of non-contact laser profilometry, particularly its sensitivity to optical material properties such as color, translucency, and surface reflectivity. While laser-based systems provide rapid and non-destructive measurements, their accuracy may be compromised when applied to optically challenging materials, which must be considered when interpreting profilometric data.

A comparison with existing research indicates that the obtained results are consistent with previously reported studies demonstrating the effectiveness of machine vision systems for real-time geometric control of FDM layers [24]. These findings confirm that visual-based non-contact methods enable continuous monitoring of print quality. Similarly, the use of laser profilometry has been shown to be effective in identifying dimensional deviations and improving process stability [25]. Moreover, the observed variations in geometric accuracy among different materials are in agreement with prior research reporting that extrusion flow rate and layer thickness significantly influence the geometric consistency of surface profiles [26].

From an industrial perspective, the findings provide practical guidance for profilometer selection. The LPM-4 is better suited for laboratory-based studies focused on quantitative roughness evaluation and detailed numerical analysis, whereas the Keyence LJ-X8020 is more appropriate for industrial or semi-industrial applications requiring rapid inspection, automated operation, and intuitive 2D/3D visualization. The complementary use of both systems can therefore support robust quality control and process optimization in FDM manufacturing.

While non-contact profilometry remains an indispensable tool for layer analysis and process optimization in FDM manufacturing, this study confirms that material optical properties must be considered to ensure accurate interpretation of profilometric data, as also highlighted by DeFisher & Fess [27].

6 Conclusions

The present study demonstrated the effectiveness of non-contact profilometric analysis for evaluating the lateral layer structures of FDM 3D printed parts. A comparison of the LPM-4 and Keyence LJ-X8020 profilometers was undertaken, with the objective of determining the relative merits of the two systems. It was established that both systems offer complementary insights, with the LPM-4 enabling quantitative evaluation of height parameters (Ra, Rz), while the Keyence system offers high-resolution qualitative visualization of surface uniformity and defects. It was determined that material color and optical properties were critical factors influencing measurement accuracy. The black PLA sample yielded the most stable results, whereas the transparent PLA/PHA exhibited significant signal distortion due to laser transmission through the material. The findings demonstrate that the precision of optical profilometry is contingent not only on the accuracy of the device, but also on the reflectivity and translucency of the material being measured.

It is evident that the combined utilization of both profilometers provides a comprehensive framework for the assessment of FDM surface quality. This, in turn, supports improved process control, surface modelling and material evaluation in the domain of additive manufacturing.

Acknowledgments. This publication was created within the project *Research of Advanced Technologies for Increasing the Efficiency of Multivalent Energy Systems for Sustainable Industrial Development* (ITMS21+ code: 401101C504), co-funded by the European Union under the Programme Slovakia. The work was also supported by the Slovak Research and Development Agency under Contract No. APVV-23-0591, and by the projects VEGA 1/0317/25 and KEGA 028TUKE-4/2025 funded by the Ministry of Education, Science, Research, and Sport of the Slovak Republic.

References

1. Zhang, H., Vallabh, C.K.P., Zhao, X.: Machine learning enhanced high dynamic range fringe projection profilometry for in-situ layer-wise surface topography measurement during LPBF additive manufacturing. Precis. Eng. **84**, 1–14 (2023). https://doi.org/10.1016/j.precisioneng.2023.06.015
2. Zhang, H., Vallabh, C.P., Xiong, Y., Zhao, X.: A systematic study and framework of fringe projection profilometry with improved measurement performance for in-situ LPBF process monitoring. Measurement **191** (2022). https://doi.org/10.1016/j.measurement.2022.110796
3. Zhang, H., Caputo, A.N., Zhang, H., Hasan, M., Vallabh, C., Zhao, X.: In-situ fringe projection profilometry and spatter monitoring data fusion to predict mechanical properties in laser powder bed fusion additive manufacturing
4. O'Dowd, N.M., Wachtor, A.J., Todd, M.D.: Effects of digital fringe projection operational parameters on detecting powder bed defects in additive manufacturing. Addit. Manuf. **48**, 102454 (2021). https://doi.org/10.1016/j.addma.2021.102454
5. Lishchenko, N., Piteľ, J., Larshin, V.: Online monitoring of surface quality for diagnostic features in 3D printing. Machines **10**, 541 (2022). https://doi.org/10.3390/machines10070541
6. He, K., Zhang, Q., Hong, Y.: Profile monitoring based quality control method for fused deposition modeling process. J. Intell. Manuf. **30**, 947–958 (2019). https://doi.org/10.1007/s10845-018-1424-9
7. Oleff, A., Küster, B., Stonis, M., Overmeyer, L.: Process monitoring for material extrusion additive manufacturing: a state-of-the-art review. Prog Addit Manuf **6**, 705–730 (2021). https://doi.org/10.1007/s40964-021-00192-4
8. Bolat, Ç., Ergene, B., Ispartalı, H.: A comparative analysis of the effect of post production treatments and layer thickness on tensile and impact properties of additively manufactured polymers. Int. Polym. Process. **38** (2023) https://doi.org/10.1515/ipp-2022-4267
9. Zidek, K., Pitel, J., Hosovsky, A.: Machine learning algorithms implementation into embedded systems with web application user interface. In: 2017 IEEE 21st International Conference on Intelligent Engineering Systems (INES), pp. 000077–000082. IEEE, Larnaca (2017)
10. Surface Analysis in Additive Manufacturing Using Image Processing–IJSREM. https://ijsrem.com/download/surface-analysis-in-additive-manufacturing-using-image-processing/. Accessed 3 Nov 2025

11. Tauberova, R., Lazorik, P., Nazarejova, J., Knapcikova, L.: Monitoring of printing quality in additive manufacturing. In: 2023 14th International Conference on Measurement. pp. 1–4 (2023)
12. Buchenau, T., Mertens, T., Lohner, H., Bruening, H., Amkreutz, M.: Comparison of optical and stylus methods for surface texture characterisation in industrial quality assurance of post-processed laser metal additive Ti-6Al-4V. Materials (Basel) **16**, 4815 (2023). https://doi.org/10.3390/ma16134815
13. Baumann, S.: Profilometry (Contact and Non-Contact). In: Materials Characterization Services (2023). https://mat-cs.com/xxr-x-ray-reflectivity/. Accessed 29 Oct 2025
14. Petsiuk, A.L., Pearce, J.M.: Open source computer vision-based layer-wise 3D printing analysis. Addit. Manuf. **36**, 101473 (2020). https://doi.org/10.1016/j.addma.2020.101473
15. Hanon, M.M., Zsidai, L.: Sliding surface structure comparison of 3D printed polymers using FDM and DLP technologies. IOP Conf. Ser.: Mater. Sci. Eng. **749**, 012015 (2020). https://doi.org/10.1088/1757-899X/749/1/012015
16. Fidlerová, H., Makyšová, H., Sklenárová, L., Bajdor, P.: Streamlining packaging as part of sustainable reverse logistics processes. AL **8**, 423–433 (2021). https://doi.org/10.22306/al.v8i4.249
17. Medina Sánchez, G., Ordóñez Partal, J.A., Vicente, R.D., Guerrero-Villar, M.F.: Surface quality optical measurement of part inclined planes manufactured by FDM technology. KEM **959**, 45–55 (2023). https://doi.org/10.4028/p-agoJ9v
18. Jurko, J., Miškiv-Pavlík, M., Husár, J., Michalik, P.: Turned surface monitoring using a confocal sensor and the tool wear process optimization. Processes **10**, 2599 (2022). https://doi.org/10.3390/pr10122599
19. Bazan, A., Turek, P., Przeszłowski, Ł: Comparison of the contact and focus variation measurement methods in the process of surface topography evaluation of additively manufactured models with different geometry complexity. Surf. Topogr. Metrol. Prop. **10**, 035021 (2022). https://doi.org/10.1088/2051-672X/ac85cf
20. Trojanowska, J., Żywicki, K., Pająk, E.: Influence of selected methods of production flow control on environment. In: Golinska, P., Fertsch, M., Marx-Gómez, J. (eds.) Information Technologies in Environmental Engineering, pp. 695–705. Springer, Berlin Heidelberg, Berlin, Heidelberg (2011)
21. Kalman, O., Husár, J., Lazorik, P., Fejko, P.: Surface quality assessment using a multispectral camera focusing on accuracy and color variations of printed parts made with FDM technology. MM Sci. J. (2025) https://doi.org/10.17973/MMSJ.2025_06_2025054
22. Antosz, K., Knapčíková, L., Husár, J.: Evaluation and application of machine learning techniques for quality improvement in metal product manufacturing. Appl. Sci. **14**, 10450 (2024). https://doi.org/10.3390/app142210450
23. Jin, Y., Li, H., He, Y., Fu, J.: Quantitative analysis of surface profile in fused deposition modelling. Addit. Manuf. **8**, 142–148 (2015). https://doi.org/10.1016/j.addma.2015.10.001
24. Singh, M., Sharma, P., Sharma, S.K., Singh, J.: A novel real-time quality control system for 3D printing: a deep learning approach using data efficient image transformers. Expert Syst. Appl. **273**, 126863 (2025). https://doi.org/10.1016/j.eswa.2025.126863
25. Kalman, O., Husár, J., Lazorík, P.: Using laser profilometry to investigation FDM printing parameters for outer-perimeter analysis and surface quality improvement. Appl. Sci. **14**, 9405 (2024). https://doi.org/10.3390/app14209405
26. Boschetto, A., Bottini, L.: Accuracy prediction in fused deposition modeling. Int. J. Adv. Manuf. Technol. **73**, 913–928 (2014). https://doi.org/10.1007/s00170-014-5886-4
27. DeFisher, S., Fess, E.M.: Comparison of contact and non-contact asphere surface metrology devices. In: Optifab 2013. SPIE, pp. 160–169 (2013)

Analysis of Additive Manufacturing Process Errors Using Intelligent Monitoring Techniques

Gregor Sopko, Martin Pollák(✉), Peter Gabstur, and Jakub Kaščak

Faculty of Manufacturing Technologies With a Seat in Presov, Technical University of Kosice, Presov, Slovakia
martin.pollak@tuke.sk

Abstract. Additive manufacturing represents one of the key technological pillars of industrial transformation in the context of Industry 4.0, where production quality, precision and repeatability depend not only on material selection and design, but primarily on the optimisation of process parameters and the implementation of advanced monitoring systems. FDM/FFF technologies, among the most widely used additive manufacturing methods, are highly sensitive to variations in printing conditions, which often result in different types of process defects. Continuous monitoring and early detection of such errors are therefore essential for improving production reliability and stability. This paper focuses on the analysis of typical process defects in FDM/FFF technology, with emphasis on their identification and prediction using intelligent monitoring tools based on artificial intelligence and IoT elements. An experimental monitoring system was developed using a Raspberry Pi 3 microcomputer, a Raspberry camera and the OctoPrint software platform extended with AI-based modules, particularly the Detector 2 plugin. The system was experimentally validated through a series of printing tests performed on a Creality CR-30 PrintMill printer, targeting defects such as stringing, layer deformation and material flow instabilities.

Keywords: 3D printing · Creality CR-30 PrintMill · process errors · stringing · artificial intelligence · monitoring

1 Introduction

In recent years, additive manufacturing has become one of the pillars of Industry 4.0 [1], which combines digitalisation, automation and intelligent control systems. Since its inception, 3D printing has evolved from a tool for rapid prototyping to a sophisticated manufacturing technology used in the aerospace, automotive, medical and consumer industries. Its greatest benefits include production flexibility, waste reduction, product customisation and the ability to produce complex shapes without the need for special tools.

According to ISO/ASTM 52900, additive manufacturing is defined as the process of joining material to create objects from 3D model data, usually layer by layer, as opposed to subtractive technologies. The standard also divides AM processes into seven

J. Trojanowska et al. (Eds.): MANUFACTURING 2026, LNME, pp. 213–224, 2026.
https://doi.org/10.1007/978-3-032-21654-0_16

main categories: material extrusion, vat photopolymerisation, powder bed fusion, binder jetting, material jetting, directed energy deposition and sheet lamination. Among these, material extrusion technology, implemented through FDM/FFF printers, has become the most widespread in the field of polymer and composite materials [2].

Despite the significant growth of FDM/FFF, one of the main challenges remains ensuring the quality and accuracy of prints. The printing process is sensitive to a number of variables, such as extrusion temperature, feed rate, retraction parameters, geometric accuracy of the axis, and filament quality. Deviations in these parameters often lead to process errors such as deformation, layer adhesion, layer shifting, blobs (material accumulation) or stringing (formation of plastic threads). These errors have a direct impact on the surface quality, mechanical properties and service life of printed components [3].

Current trends in research and development are therefore focused on automated quality control systems and intelligent process monitoring. These systems use a combination of camera sensors, accelerometers, temperature sensors and neural networks that can evaluate the printing process in real time and identify anomalies. The aim of these solutions is not only to detect errors during production, but also to predict their occurrence before the process fails, thereby minimising waste and downtime [4].

An interesting direction in the development of additive systems is belt printing, represented by the Creality CR-30 model [12]. This concept uses an endless moving belt as a printing surface, with the print head positioned at a 45° angle. This principle enables continuous production without the need for manual removal of prints, dramatically increasing efficiency and suitability for small-batch or automated production. On the other hand, however, this system brings new technical complications: altered gravitational effects, non-standard layer geometry, and a higher risk of stringing and deformation.

It is therefore essential that these processes are actively monitored and optimised using intelligent algorithms. The integration of artificial intelligence and IoT into belt 3D printing provides the possibility of linking data from sensors with adaptive parameter control, which creates the conditions for the emergence of autonomous production cells capable of learning and responding to process changes in real time.

This article aims to:

- analyse process errors in FDM/FFF technology,
- to present the design and testing of a new monitoring system for the Creality CR-30 PrintMill 3D printer,
- and evaluate the effectiveness of AI-based intelligent monitoring in real printing conditions.

2 Process Anomalies and the Occurrence of Mechanical Errors in FFF/FDM Manufacturing Technology

FDM/FFF technology, known as Fused Deposition Modelling or Fused Filament Fabrication, is one of the most widely used methods of additive manufacturing of polymer components [5]. Its principle is based on the gradual application of molten filament in layers, which are deposited on top of each other according to data created in a CAD system and subsequently exported to STL format. The process is based on the precisely controlled movement of the extruder in three axes, with the material changing from a solid to a viscous state in the heating chamber's and being applied in a defined trajectory. The advantages of this technology are flexibility, low production costs and the possibility of producing complex shapes, but its accuracy and reliability are significantly affected by process errors that occur during printing.

Process errors in FDM/FFF are the result of the interaction of several factors, such as temperature, speed, mechanical stability [11] of the structure, material quality, and correct parameter settings. The most common problems include nozzle clogging, insufficient adhesion of the first layer, incorrect cooling, filament damage, layer shifting, the formation of thin plastic threads known as stringing, and the separation or deformation of individual layers. Each of these errors has a specific cause and requires a different approach to optimisation. Nozzle clogging often occurs as a result of material residue after changing filaments, which is particularly common when switching from flexible to stiffer materials such as TPU and PLA. Prevention involves regular cleaning of the nozzle, maintaining a stable extrusion temperature, and using filament with a uniform diameter.

Insufficient adhesion of the print (Fig. 1) to the print bed is one of the most common errors leading to print failure. This phenomenon can be caused by dirt, unevenness of the bed, or inappropriate heating temperature. The solution is precise calibration, correct Z-axis height adjustment, and the use of adhesive auxiliary layers such as brim or raft, which increase the contact between the model and the bed. In addition, the speed of printing the first layer also has a significant impact and should be reduced by about half compared to the standard value so that the plastic has time to adhere to the surface [7].

Fig. 1. Error manifested by the edges of the object lifting off the bed

Temperature stability is a key factor for proper bonding between layers. If the extruder temperature is too low, the layers will not bond properly, while if the temperature is too high, the material may flow excessively and deform the details. Maintaining a constant bed temperature prevents rapid cooling of the material, which is particularly important for materials such as ABS in order to prevent deformation and layer separation (Fig. 2). At the same time, cooling fans must be monitored, as they can cause the plastic to solidify too quickly, thereby weakening the bonds [8].

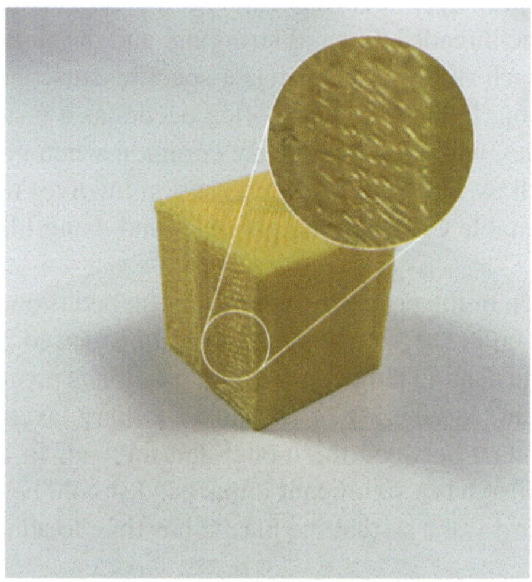

Fig. 2. Incorrect material dosing

Damaged or worn filament (Fig. 3) is a common source of inaccuracies, as changes in its diameter, moisture content or structure can cause irregular material flow through the nozzle. It is recommended to store filament in a dry place and regularly check its surface to prevent breakage or wear. Layer shifting in the X or Y axis is usually the result of mechanical vibrations, loose belts or excessive acceleration during extruder movements. This problem can be eliminated by optimising speed parameters, reducing acceleration and checking the strength of the printer frame [9].

Another common visual defect is the formation of thin plastic threads (Fig. 4), known as stringing. This occurs when molten material escapes from the nozzle during extruder movement without extrusion. This phenomenon causes unwanted filaments to form on the surface of the model, which compromises both the aesthetics and accuracy of the model. The solution is to correctly set the retraction parameters, i.e. the withdrawal of material from the nozzle before movement. Retraction optimisation depends on the type of extruder and the viscosity of the material, and it is advisable to experimentally adjust the speed and length of the retraction to minimise plastic leakage. In addition to retraction, it is important to maintain the extrusion temperature at the lower end of the recommended range to reduce the fluidity of the material [6].

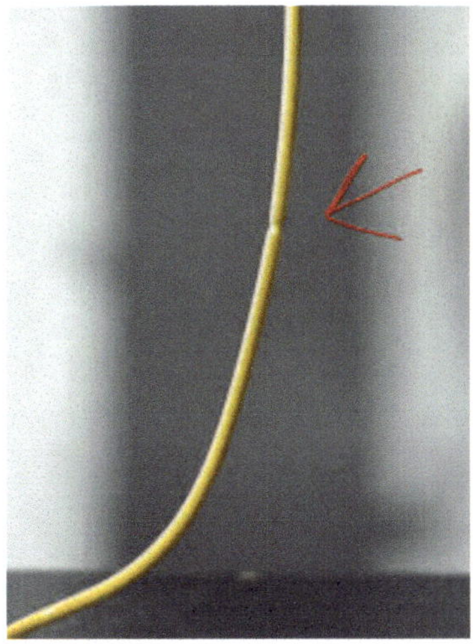

Fig. 3. Degraded material

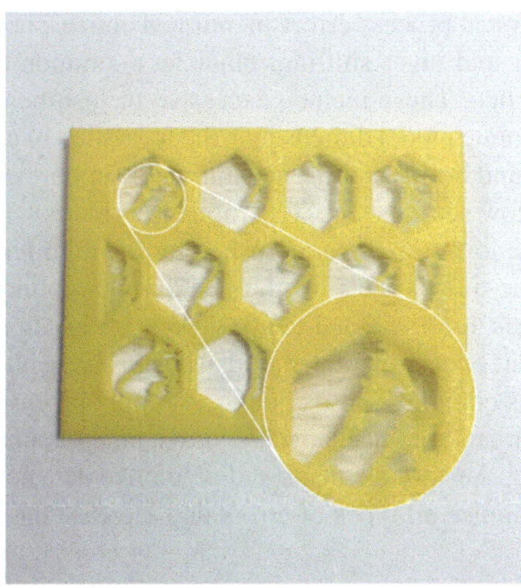

Fig.4. Stringing as a surface defect in FDM 3D printing.

Another phenomenon that affects surface quality is vibration and rippling. These defects most often occur in sharp changes in the direction of extruder movement, where inertia and frame resonance cause irregular layers and visible rippling on the surface (Fig. 5). An effective solution is to implement "input shaping" algorithms that use accelerometers to measure vibrations and adjust motion profiles to compensate for them [6]. This approach is typical of modern control systems, including open architecture models such as the Creality Ender 3 V2 or CR-30.

Fig. 5. Layers separated from each other

In addition to the typical process errors mentioned above, such as clogged nozzles, deformations, stringing, and layer shifting, other less common but significant deviations also occur in practice. These include excessive or insufficient extrusion causing uneven layers and inaccurate model dimensions, the formation of air cavities in the infill, poor quality overhangs and bridges, or the separation of support structures due to incorrect temperature or airflow settings (Fig. 6). Another problem is the so-called elephant foot effect, in which the lower layers are crushed by the high temperature of the bed, or warping, in which the model peels off the bed during cooling and deforms. In the case of multiple extruders, errors related to nozzle synchronisation, inaccurate material exchange timing or colour contamination may also occur. These phenomena confirm that the FDM/FFF process is complex and its quality can only be ensured by simultaneous real-time control of temperature, mechanical and dynamic parameters. The integration of intelligent monitoring, sensory elements and predictive data analysis thus represents a promising way to minimise all types of errors and increase the reliability of additive manufacturing.

The analysis and optimisation of process errors are key prerequisites for improving the quality of additive manufacturing. Combining a physical understanding of phenomena with intelligent monitoring paves the way for adaptive control systems that can respond dynamically to deviations. In the case of belt printing, such as Creality CR-30 PrintMill, this adaptability is even more important given the specific kinematics of movement and the need for continuous evaluation of layer stability. The goal of modern research in this area is to achieve high repeatability, minimise deviations, and create a process-stable environment that enables fully automated and reliable production within the concept of intelligent additive manufacturing.

Fig. 6. Temperature-damaged areas on the model

3 Methodology and Experimental Procedure

A system integrated into the Creality CR-30 PrintMill printer shown in Fig. 7 was used for experimental verification of the possibility of intelligent monitoring.

The Creality CR-30 PrintMill printer works on the principle of FDM/FFF 3D printing technology, using a belt printing system with continuous 45-degree pad movement. This design principle allows for continuous production and printing of objects with theoretically unlimited length, but at the same time brings new challenges in terms of accuracy and process stability. The tilted Z-axis changes the direction of gravitational and thermomechanical forces, which can lead to layer deformation and reduced adhesion of the model to the belt. In addition, the constant movement of the belt changes the cooling conditions and can affect the uniformity of the layers. Stringing is more pronounced in this type of printing because the nozzle moves at an angle and the retraction distances vary depending on the direction of movement. Therefore, it is necessary to individually adjust the retraction and cooling speed parameters for each model geometry.

Belt printing also increases the system's susceptibility to vibrations and deviations in long models, so it is essential to pay attention to the strength of the structure, belt tension and regular calibration. In addition, it is recommended to use temperature, vibration and material flow sensors for continuous process monitoring. This data can then be processed using artificial intelligence algorithms that can identify emerging errors in real time and adjust the print parameters.

Fig. 7. Creality CR-30 PrintMill printer

In order to ensure real-time visual control, a custom camera mount was designed, which was produced using 3D printing and attached directly to the frame of the device. A Raspberry Pi camera connected to a Raspberry Pi 3 microcomputer was placed in this mount (Fig. 8).

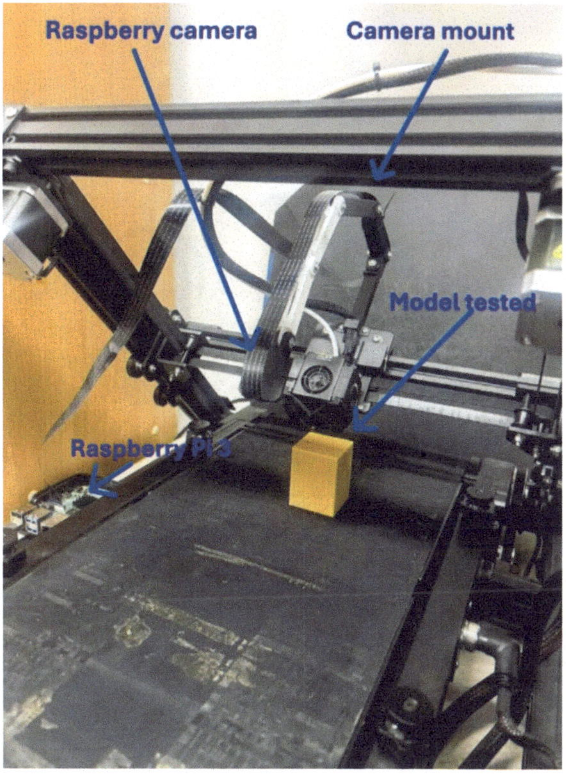

Fig. 8. Printer assembly and camera mounting

Recording and monitoring were controlled using OctoPrint software (OctoPrint is software that allows you to control a 3D printer via a web interface. Thanks to this software, we can monitor the status of the 3D printer and, if necessary, control it remotely from any device with internet access.) which enabled automatic image capture throughout the printing process. The system was set up to take a picture every five seconds and then send it to a machine vision algorithm that evaluated the quality of the ongoing print. The output of the classification model was a percentage indicator of the success of the process, based on which it was possible to assess whether the print was progressing correctly or whether errors were occurring.

Testing was performed on a series of print jobs simulating various types of faults typical of FFF/FDM technology. As part of the experiments, printing was first performed without deliberately disrupting the conditions (Fig. 9), which proceeded correctly and the system repeatedly evaluated the quality above 90%.

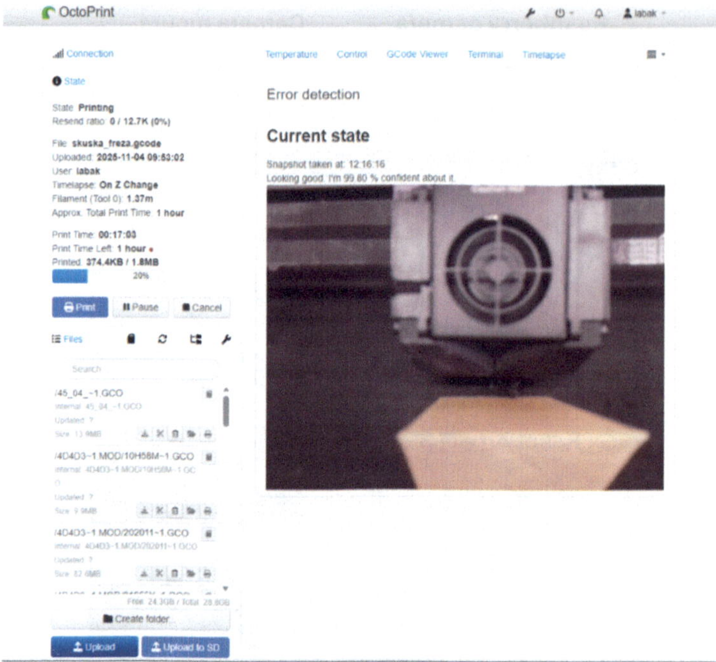

Fig. 9. Stable printing process

Subsequently, a failure was induced in the initial phase of applying the first layer, for example by slightly restricting the filament flow, which caused the formation of thin stringing deposits shown in Fig. 10. Artificial intelligence was able to detect this error within the first few seconds of printing [10], with the quality rating suddenly dropping by about half.

The last test scenario focused on the impact of poor-quality visual imaging. The deterioration of the lighting in the workspace reduced the contrast of the images, which had a negative impact on the accuracy of the prediction. In this case, the system responded late or incorrectly evaluated the print as problem-free, even though obvious defects were forming. This result points to the high sensitivity of the model to image quality, which was one of the biggest limitations in real production conditions.

Fig. 10. Occurrence of stringing defects

The experiments confirmed that automated monitoring using cameras and machine learning can effectively identify process errors and significantly reduce material and time losses. At the same time, however, they pointed out that the system needs to be further developed, particularly in terms of visual robustness, for example by improving lighting, using multiple cameras or adding additional sensory methods. The combination of these elements with adaptive control represents a promising path towards fully autonomous 3D printing quality control in the spirit of Industry 4.0.

4 Conclusion

In recent years, additive manufacturing has moved from experimental and prototype applications to industrial practice, where the emphasis is on reliability, automation, and real-time quality control. Despite technological advances, the 3D printing process remains susceptible to errors, with defects such as stringing, deformation and material flow failure still a significant problem in FFF/FDM technology. Modern manufacturing environments therefore require systems that can autonomously monitor the printing process and respond to deviations in a timely manner.

Research and testing of the proposed solution for online print monitoring has shown that the combination of camera sensors and machine learning algorithms is an effective way to increase production success and eliminate unnecessary losses. An intelligent system must be adaptive, self-learning and capable of providing feedback to the process. This trend is already being confirmed by the latest solutions from 3D printer manufacturers, which integrate AI modules directly into their software and hardware.

Advances in the Industrial Internet of Things (IIoT), edge computing, digital twins and multimodal monitoring will fundamentally influence how 3D printing is managed in the coming years. Systems equipped with thermal cameras, vibration sensors and advanced image processing are already capable of predicting defects before they are visually identifiable. Fully autonomous closed-loop quality control, where the printer can correct parameters itself, is expected to become standard soon, even outside of highly industrialised facilities.

Acknowledgement. This research was funded by project VEGA 1/0121/23 and KEGA 002TUKE-4/2023 granted by the Ministry of Education, Research, Development and Youth of the Slovak Republic and by the EU NextGenerationEU through the Recovery and Resilience Plan for Slovakia under projects No. 09I05-03-V02-00042 and No. 09I03-03-V05-00015. This work was supported by the Slovak Research and Development Agency under contract no. APVV-24-0470.

References

1. Dilberoglu, U.M., Gharehpapagh, B., Yaman, U., Dolen, M.: The role of additive manufacturing in the era of Industry 4.0. In: Procedia Manufacturing (2017). https://doi.org/10.1016/j.promfg.2017.07.148
2. ISO/TC 261: ASTM Committee F42: ISO/ASTM 52900:2021 (2021)
3. Koltsaki, M., Mavri, M.: A comprehensive overview of additive manufacturing processes through a time-based classification model (2024). https://doi.org/10.1089/3dp.2022.0167
4. Fu, T.H., Li, D.R.: Real-time process monitoring and error correction in material extrusion-based additive manufacturing via multi-output machine learning. J. Manuf. Process. **152**, 638–656 (2025). https://doi.org/10.1016/J.JMAPRO.2025.08.025
5. Jadhav, A., Jadhav, V.S.: A review on 3D printing: an additive manufacturing technology. Mater Today Proc. **62**, 2094–2099 (2022). https://doi.org/10.1016/J.MATPR.2022.02.558
6. Paraskevoudis, K., Karayannis, P., Koumoulos, E.P.: Real-time 3D printing remote defect detection (Stringing) with computer vision and artificial intelligence. Processes **8**, 1464 (2020). https://doi.org/10.3390/pr8111464
7. Schröder, S., et al.: Tuning the selectivity of metal oxide gas sensors with vapor phase deposited ultrathin polymer thin films. Polymers (Basel) **15**, 524 (2023). https://doi.org/10.3390/polym15030524
8. Caltanissetta, F., Dreifus, G., Hart, A.J., Colosimo, B.M.: In-situ monitoring of Material Extrusion processes via thermal videoimaging with application to Big Area Additive Manufacturing (BAAM). Addit. Manuf. **58**, 102995 (2022). https://doi.org/10.1016/j.addma.2022.102995
9. Lendvai, L., Fekete, I., Jakab, S.K., Szarka, G., Verebélyi, K., Iván, B.: Influence of environmental humidity during filament storage on the structural and mechanical properties of material extrusion 3D-printed poly (lactic acid) parts. Results Eng. **24**, 103013 (2024). https://doi.org/10.1016/J.RINENG.2024.103013
10. Charia, O., Rajani, H., Ferrer Real, I., Domingo-Espin, M., Gracias, N.: Real-Time stringing detection for additive manufacturing. J. Manuf. Mater. Process. **9**, 74 (2025). https://doi.org/10.3390/jmmp9030074
11. REALvision Online–powered by Create it REAL A/S. https://realvisiononline.com/blog/the-12-most-common-problems-in-3d-printing-and-how-to-fix-them. Last accessed 4 Nov 2025
12. https://3dprintpgh.com/blog/post/creality-cr-30-belt-adhesion-issues/. Last accessed 04 Nov 2025

Training Application for the Operation and Selection of Parameters of the Additive Manufacturing Process Using the FDM Method

Krzysztof Łukaszewski(✉), Anna Karwasz, and Tymoteusz Puchalski

Faculty of Mechanical Engineering, Poznan University of Technology, Poznan, Poland
krzysztof.lukaszewski@put.poznan.pl

Abstract. This article presents research results on the use of a 3D training application for operating an FDM printer and selecting process parameters. A training application was developed in the Unity environment, featuring a realistic model of the FlashForge Adventurer 4 printer, parameter selection modules, and instructional videos. Comparative studies were conducted with 10 participants divided into an experimental group (training via the application) and a control group (training using the standard method). The results showed that the experimental group was, on average, 74% shorter training time, 47% more accurate in performing tasks, and achieved over three times higher knowledge gains than the control group. The practical findings confirmed that the use of interactive applications enhances the educational process in the field of additive manufacturing technology.

Keywords: Additive manufacturing · Fused Deposition Modeling (FDM) · Training applications · Unity 3D

1 Introduction

Additive manufacturing technology, specifically the FDM (Fused Deposition Modeling) method, is becoming increasingly important in industry and education. As 3D printing systems become more efficient and accessible, there is an increasing need for user training in printer operation and process parameter selection. Traditional training methods, such as printed manuals or video tutorials, often lack interactivity and active user engagement.

Technological advances have made it possible to create interactive applications combining three-dimensional models with intuitive user interfaces. Such applications can improve the learning process by enabling hands-on experience and practical understanding without the need for direct access to real equipment. Articles increasingly discuss the impact of virtual tools on process learning, particularly in the context of technological applications.

The aim of this study was to develop and evaluate a training application for learning how to operate a 3D printer and select FDM process parameters, and to compare its effectiveness with a traditional paper-based training method.

J. Trojanowska et al. (Eds.): MANUFACTURING 2026, LNME, pp. 225–239, 2026.
https://doi.org/10.1007/978-3-032-21654-0_17

2 Literature Review

2.1 IT Tools in Training

IT tools used in education have highlighted the importance of alternative teaching methods during the COVID-19 pandemic [1]. Platforms such as Zoom and Microsoft Teams enabled the continuation of the educational process; however, research has shown limitations related to the lack of direct physical interaction [2].

Maor et al. conducted a study on teachers' learning outcomes from online classes delivered via the Zoom application. They observed that online classes tended to be more diverse and creative in structure, yet less consistent in teaching methodology. The authors attributed this to the sudden transition to online education and the marginalization of teachers in the process of adapting to new digital tools [3].

Ramadani and Xhaferi carried out a study among secondary school teachers regarding their experiences with online teaching. The teachers reported that online instruction cannot fully replicate face-to-face learning, as body language, facial expressions, and gestures play a crucial role in students' understanding. However, they also emphasized that remote learning offers greater comfort and easier access to online resources [4].

A common solution for distance education has been the use of Learning Management Systems (LMS) such as Moodle, Blackboard, and Google Classroom. These systems support online education by allowing teachers to create and manage courses, facilitate communication, and monitor student progress. Moreover, they can be adapted to meet individual educational needs [5, 6].

To further enhance classroom engagement, tools such as Kahoot!, a game-based learning platform that enables the creation of interactive quizzes, are widely used [7]. Wang and Tahir presented a literature review showing that the use of Kahoot! improves learning outcomes by promoting active participation and providing immediate feedback. The platform also helps reduce stress and anxiety commonly associated with assessments [8].

Koh and Daniel demonstrated in their work that the shift to remote learning tools ensures the continuity of information delivery during global crises. However, they also emphasized the need to address issues such as cheating during online assessments and unequal access to technology among students [9].

2.2 Virtual Reality Applications

Virtual reality (VR) applications are becoming increasingly significant in online learning environments. Due to their interactivity and immersive potential, such applications provide an engaging and innovative approach to acquiring knowledge [10].

Predescu et al. examined the impact of virtual reality on the educational process. Their study involved three groups of instructors: the first group received traditional video-based instruction, the second group also received video-based training with interactive elements, and the third group was trained using VR headsets. Participants trained with VR reported higher satisfaction and demonstrated improved learning performance compared to those who followed traditional instruction methods [11].

In the context of VR applications, the concept of the "I^3 triangle" is often applied. It encompasses three fundamental characteristics of virtual reality: immersion, interaction, and imagination [12]. Immersion refers to the user's sense of being present within the virtual environment. Interaction describes how users engage and manipulate objects or elements within the application. Imagination, in turn, reflects the user's ability to connect virtual experiences with realistic interpretations and creative problem-solving [13].

Beyond the educational context, IT and VR tools are increasingly used in professional environments to train employees. Many organizations now offer e-learning and VR-based training programs, which are valued for their flexibility and the possibility of remote access to learning materials. These systems enable employees to acquire new skills without leaving the workplace, enhancing both accessibility and efficiency [14].

Unity 3D has become one of the leading platforms for creating both gaming and educational applications [15–17]. The Unity environment features an intuitive interface that allows developers to design 3D applications without extensive programming knowledge. Key interface components include the Hierarchy, Project, Scene, Game, and Inspector windows. Application development in Unity relies on the C# programming language, which provides flexibility and control over interactive elements [18].

VR applications are now widely used across various industries and throughout different stages of product development—from concept design to implementation. Virtual prototypes allow for simulation, testing, and modification before physical production, significantly reducing time and costs [19, 20].

One of the major advantages of VR-based training is the ability to simulate real-world scenarios without exposing participants to potential hazards. Such applications are used to train personnel in high-risk professions, including emergency responders such as police officers and firefighters. VR training has also been successfully applied in the military, mining, industrial safety, medical education, and other professional fields [21–23].

Buń, Górski, and colleagues developed a VR-based training application for teaching human anatomy at medical universities. The system allows users to explore 3D anatomical models, view animations of organs and tissues, and collaborate in shared virtual sessions [24]. Similarly, Górski et al. created a virtual city bus configurator that enables users to modify and visualize various bus component configurations while interacting with both internal and external elements of the virtual model [25].

2.3 Testing Training Applications

Testing training applications is essential for obtaining feedback from users and evaluating the overall training effectiveness. Such testing provides insight into whether the author's objectives have been achieved and whether the proposed solution is beneficial in practice. Evaluations typically include participant surveys and knowledge tests conducted before and after the training [26, 27].

The testing process should be carried out in two distinct phases. The first phase involves developer testing, aimed at verifying the technical functionality, usability, and reliability of the application. The second phase engages research participants, including both users involved in the application's development and new users unfamiliar with it.

This approach ensures a comprehensive assessment of the application's educational and practical performance [13].

Agbo et al. developed a scientific application designed to support computational thinking. To evaluate its effectiveness, two participant groups were formed: an experimental group, which used the application, and a control group, which followed conventional teaching methods. Both groups completed a pre-test and post-test to measure changes in knowledge and understanding. Additionally, participants were surveyed about their learning experience. The study revealed that learners who used the application demonstrated significantly higher engagement and motivation compared to the control group. The authors also included questions assessing participants' comprehension of the underlying principles of the application [28].

2.4 FDM (Fused Deposition Modeling) Technology

3D printing is an additive manufacturing process in which material is deposited layer by layer to create a three-dimensional object. FDM/FFF (Fused Deposition Modeling / Fused Filament Fabrication) is one of the most commonly used. The process involves extruding successive layers of melted thermoplastic material, known as filament. The filament is fed into an extruder, where it is heated until it melts, and then deposited onto the build platform through a nozzle to form the desired geometry [29, 30].

The FDM 3D printing process requires the careful selection of multiple parameters. These include the extruder temperature, bed temperature, printing speed, layer height, first-layer height, number of shells, top and bottom infill density, and the type and density of supports. Incorrect parameter selection can lead to print defects such as model detachment from the build plate, uneven layer distribution, or surface irregularities [3].

Materials commonly used in FDM technology include thermoplastics such as ABS, PLA, PC, PETG, and carbon fiber–reinforced composites such as PLA-CF and PETG-CF. ABS is valued for its strength and durability but requires precise temperature control and strong bed adhesion due to its high shrinkage rate. PLA offers excellent dimensional stability and good adhesion to the build platform, making it suitable for beginners. PC (polycarbonate) is one of the strongest materials available for FDM printing, providing excellent mechanical properties. PETG combines high strength with good impact resistance and chemical stability [32, 33].

Printing defects may also arise from improper temperature control or poor-quality materials. It is crucial to maintain uniform heat distribution between the extruder and the print bed, as this directly affects print quality. Different filaments require specific operating temperatures: for ABS, the recommended extruder temperature is 220–250 °C and bed temperature 100–110 °C, whereas for PLA, the extruder temperature should range between 180–220 °C and the bed temperature between 20–60 °C [34].

2.5 Applications of Additive Manufacturing Technology

Rapid prototyping is one of the main advantages of 3D printing technology, as it enables the visualization of product concepts and the implementation of design modifications before final production. The high speed and low cost of prototyping make this technology particularly valuable in product development and industrial design [35, 36].

3D printing–based prototyping also has significant applications in tooling and manufacturing process development [37].

3D printing is medicine, the technology is used to produce implants, protective devices, and wearable components, many of which are customized to meet individual patient needs. It has also found significant applications in tissue engineering, where 3D-printed structures serve as biocompatible scaffolds for cell growth. Moreover, 3D printing is increasingly employed in the production of medical aids, such as prosthetic and orthotic devices, as well as disposable medical equipment [38, 39].

3D printing technology is now applied across a wide range of industries, including manufacturing, logistics, art, automotive, aerospace, and electronics. Owing to the flexibility of CAD software, designers can freely create and modify 3D models according to specific functional or aesthetic requirements, making additive manufacturing a versatile tool for innovation [40, 41].

3 Research Methodology

3.1 Research Object

The subject of the study was a FlashForge Adventurer 4 Series 3D printer. The device features a build volume of 220 × 200 × 250 mm, a fully enclosed build chamber, and a heated build plate capable of reaching temperatures up to 110 °C. The print head can be fitted with interchangeable nozzles of 0.3 mm, 0.4 mm, and 0.6 mm diameters, with maximum operating temperatures of 240 °C or 265 °C, depending on the nozzle type.

3.2 Building the Training Application

The development of the training application was divided into the following stages:

Step 1: 3D Modeling. Detailed measurements of the actual printer were taken, and a complete 3D model of all its components was created in Autodesk Inventor. The model included the housing, print head, build plate, extruder, filament feeding system, and all mechanical parts. In total, over 30 components were modeled and assembled into a single, user-friendly 3D assembly.
Step 2: Import to Unity 3D. The model was exported in OBJ format and imported into Unity 3D (version 2021.3). Material properties were added to individual parts of the printer, including metallic surfaces, switches, and structural elements, to enhance visual realism.

Figure 1 illustrates the user interface of the application, showing the parameter selection and support configuration panel.

Stage 3: Functionality Implementation. The following interactive functions were implemented:
A navigation system allowing free movement in the virtual environment using the W, A, S, D keys and camera control with the mouse.
Animated printer doors (front and side) controlled by the T and Y keys.

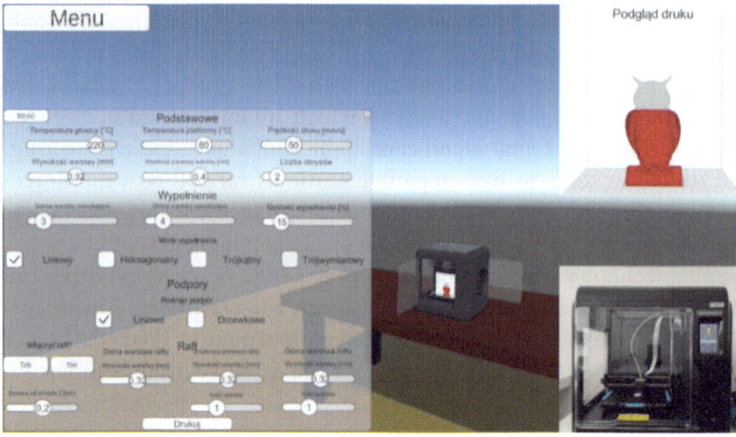

Fig. 1. User interface of the training application with a panel for selecting parameters, filling and supports for the Filament application [43]

An interactive nozzle selection module with real-time visualization of changes on the 3D model.
A filament selection system enabling users to choose filament type (ABS, PLA, PC, PETG, PLA-CF, PETG-CF) and color (nine options).
An advanced control panel for setting print parameters such as nozzle and bed temperature, print speed, layer height, infill percentage, and support/raft options.
A notification system displaying recommended parameter values based on the selected filament type.
Video modules demonstrating key operations such as loading filament, calibrating the printer, and initiating the printing process.
A real-time simulation of the printing process displayed within the virtual printer model.
Step 4: Script Implementation. C# scripts were written to implement interactive functionality, including:
Displaying real-time values on parameter sliders.
Showing tooltips with parameter descriptions when hovering over interface elements.
Managing animations and logical dependencies between components.

3.3 Research Scenario

The study was conducted with 10 male participants aged 20–30, all of whom had little or no prior experience with FDM technology. Participants were randomly assigned to two groups of five:

- Experimental group: trained using the developed interactive training application.
- Control group: trained using the standard FlashForge Adventurer 4 operation manual.

Research procedure:

1. **Pre-training knowledge test**, 25 multiple-choice questions covering FDM technology, parameter selection, and printer operation.

2. **Training phase**, participants completed the training using either the application or the printed manual, depending on group assignment.
3. **Practical task,** participants were instructed to prepare and print a 3D model of an owl under the following conditions:
   - nozzle diameter: 0.4 mm,
   - filament type: PLA,
   - maximum printing time: 30 min.
4. **Post-training knowledge test**, identical to the initial test to measure knowledge gain.
5. **User feedback survey**, participants completed a questionnaire evaluating the training method.

The owl model (file name: ofw_face_front.STL) was downloaded from the Thangs platform (thangs.com) [42]. Figure 2 shows the 3D model view and a sample printout of the owl.

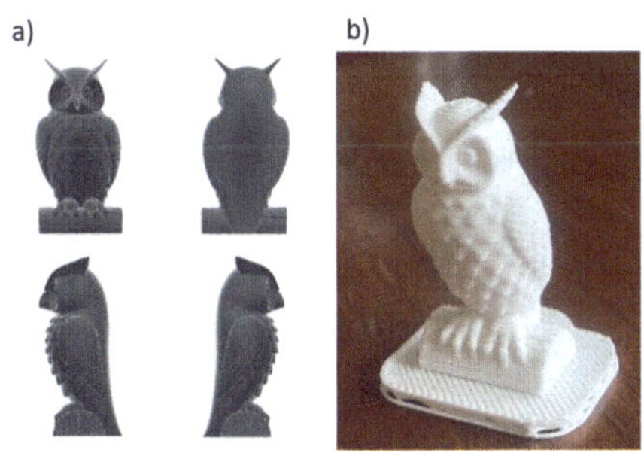

Fig. 2. Model: a – 3D model, view in the Inventor environment, b – sample owl printout [43]

During the practical phase, the following parameters were measured:

- Time spent using the application or manual.
- Time required to prepare the model in **FlashPrint 5** and start printing.
- Number of attempts required to obtain a correct print.
- Number of times training materials were reviewed (as a measure of independence).
- Quality of the printed model, evaluated on a 0–3 point scale:
- 1 point - the print was successfully completed,
- 1 point - the model geometry was preserved,
- 1 point - the surface was smooth and free from visible defects.

If an incorrect print occurred, participants were allowed to repeat the task once, and the total number of attempts was recorded.

4 Research Results

4.1 Training Time

Table 1 presents a summary of the training and task execution times achieved by both groups.

Table 1. Comparison of training and task execution times [43]

Parameter	Experimental group (application)	Control group (manual)
Average time using the application/manual	12:36 min	23:55 min
Average time spent in slicing software and printer setup	5:07 min	7:02 min
Total average time	17:43 min	30:57 min

Participants trained with the interactive application spent, on average, 13 min and 14 s less time completing the training and printing tasks compared to those using the printed manual - representing a 43% reduction in total time.

Moreover, participants in the experimental group spent 47% less time preparing the model in the slicing software and initiating the print job, confirming the higher efficiency of the interactive method.

4.2 Number of Attempts to Achieve a Correct Print

Table 2 presents the number of attempts required to achieve a successful print.

Table 2. Number of attempts to complete the required printout [43]

Participant	Testimonial group	Control group
Person 1	2	3
Person 2	2	3
Person 3	2	4
Person 4	1	2
Person 5	1	3
Mean	**1.6**	**3.0**

The average number of attempts in the experimental group was 1.6, whereas the control group required an average of 3.0 attempts. This means that participants trained with the application made 47% fewer attempts to obtain a correct print.

Notably, two participants (40%) in the experimental group achieved a successful print on their first attempt, while none in the control group did so.

The most common causes of print failure in the control group were:

- excessive print time due to incorrect parameter settings (5 cases),
- model detachment from the build platform caused by insufficient or excessive bed temperature (2 cases),
- surface defects resulting from improper print speed or parameter configuration (3 cases).

4.3 Independence in Task Performance

Participant independence was assessed based on the number of times they referred to training materials during task execution (Table 3).

Table 3. Independence of participants (number of material references) [43]

Participant	Testimonial group	Control group
Person 1	1	6
Person 2	3	4
Person 3	2	7
Person 4	1	3
Person 5	3	4
Mean	**2.0**	**4.8**

Participants in the experimental group referred back to the application an average of 2 times, whereas those in the control group consulted the manual 4.8 times on average.

This indicates that participants trained with the interactive application demonstrated 58% greater independence when performing the practical tasks.

4.4 Knowledge Gain

Table 4 summarizes the results of the knowledge test administered before and after the training (maximum score: 25 points).

Table 4. Results tests knowledge [43]

Group	Average score before training	Average score after training	Knowledge increase
Experimental	15.0	21.8	6.8
Comparative	17.6	19.6	2.0

Although the control group achieved slightly higher pre-training scores (by 2.6 points), the experimental group demonstrated a significantly larger knowledge gain.

The improvement in the experimental group (+6.8 points) was over three times greater than that of the control group (+2.0 points), representing an overall increase of approximately 240% in learning efficiency.

4.5 Survey Results

All participants completed a post-training survey evaluating the clarity, usefulness, and satisfaction with their respective training methods. The experimental group also responded to additional questions about the functionality of the interactive application.

Comprehensibility of graphical materials (scale 1–5, where 5 = very clear):

- Experimental group: mean = 4.4
- Control group: mean = 1.8

Completeness of information regarding task execution:

- Experimental group: 100% indicated that the application contained all necessary information.
- Control group: 60% reported that the printed manual did not provide sufficient information.

Recommendation for use in other training contexts:

- Experimental group: 100% would recommend using the application (60% "definitely yes", 40% "probably yes").
- Control group: 80% would not recommend the manual-based method (20% "definitely not", 60% "probably not").

In the detailed application survey, all participants (100%) agreed that:

- the application accurately reflects the FDM printing process,
- the content scope was comprehensive and relevant,
- the application met their expectations (40% fully, 60% to a large extent),
- the interface usability was rated as high or very high,
- the interactive elements significantly enhanced learning effectiveness.

The most appreciated features were:

- contextual information about print parameters with suggested values,
- instructional videos,
- interactive nozzle and filament selection modules.

Suggestions for future improvement included:

- expanding the user interface functionality,
- adding more detailed information about the slicing software,
- enabling video playback controls (pause and rewind options).

5 Discussion

The results of the study indicate that the use of the interactive training application significantly improves the educational process related to FDM printer operation and selection of printing parameters.

The experimental group outperformed the control group across all measured aspects, training time, number of attempts to achieve a correct print, level of independence, and knowledge gain.

Although participants trained with the application spent 47% less total learning time, they engaged with the educational content for over three times longer than those using the printed manual. This finding suggests that interactive presentation of material, combined with direct visualization of parameters on a 3D model, enhances both understanding and retention of practical skills. These observations align with the results of Predescu and Rej [11], who demonstrated that interactive and visual learning tools strengthen both virtual and traditional training outcomes.

The 47% reduction in failed print attempts achieved by the experimental group has considerable practical implications. In industrial settings, print failures often result in wasted material, extended production time, and increased energy costs. Improved operator training through interactive applications directly contributes to more cost-effective and reliable manufacturing processes.

Furthermore, the 58% higher level of independence among participants trained with the application underscores its effectiveness in developing procedural knowledge. Interactive training requires users to follow sequential steps and make parameter selections autonomously, promoting the formation of stronger cognitive patterns and practical habits. Similar conclusions were reported by Agbo et al. [28], who found that users of interactive educational tools demonstrated greater engagement and performance improvements compared to those using static materials.

Survey feedback confirmed the positive reception of the application. Participants particularly appreciated the suggested parameter values for different filament types, which were missing from the printed manual, identified as one of the most frequent challenges by the control group. Many users reported that the manual-based approach required trial-and-error parameter adjustments, often leading to print failures, especially for longer prints exceeding 30 min.

When analyzing the causes of print failures, errors in the control group were primarily due to incorrect parameter settings (five cases of excessive printing time), while in the experimental group, the most common issue was model detachment from the build platform (two cases). This indicates that participants trained with the application developed a better conceptual understanding of the relationships between print parameters and print quality.

In the control group, surface defects and irregular layer distribution were frequently observed, often caused by a lack of understanding of how top and bottom layer configurations affect surface finish. In contrast, participants using the application were guided by automated notifications and visual cues, which helped them prevent such defects during printing.

Despite the overall positive results, participants suggested several improvements for future versions of the application, including:

- enhanced guidance on using the user interface,
- additional information on the slicing software,
- and the option to pause and rewind instructional videos to better control the learning pace.

When compared with other studies on interactive educational tools, the results confirm a broader trend toward improved learning outcomes through interactivity and visualization. Górski et al. [25] observed similar engagement effects in their study on a virtual bus configurator, while research on the Kahoot! platform demonstrated that game-based learning tools enhance motivation and retention of knowledge [8].

The present study was limited by its small and homogeneous sample size (10 participants, all male, aged 20–30), as well as by the participants' basic prior knowledge of FDM technology. Future research should include a larger and more diverse participant group, representing different age ranges, educational backgrounds, and experience levels. Longitudinal studies, collecting data several weeks or months after training, could also provide valuable insight into knowledge retention over time.

Looking ahead, the application could be further enhanced through integration with Virtual Reality (VR) technology, offering a higher level of immersion and interactivity. As noted by Predescu et al. [11], full VR immersion provides an even more effective learning environment, enabling users to experience realistic operational conditions within a safe virtual space.

6 Conclusions

Based on the research results obtained from the conducted study, the following conclusions can be drawn:

1. The training application designed for learning how to operate an FDM printer and select printing parameters significantly streamlines the educational process, reducing the average training time by 43% compared to traditional paper-based instruction.
2. Participants trained with the application demonstrated 47% fewer printing errors and achieved correct print results more efficiently, which translates into measurable time and cost savings in practical applications.
3. The knowledge gain in the application-trained group was over three times higher than in the control group, despite the shorter overall training time. This confirms the effectiveness of interactive and visual learning methods in technical education.
4. The independence level of participants trained using the application was 58% higher, indicating that interactive training supports the acquisition of procedural knowledge and the formation of more durable cognitive patterns.
5. A key feature contributing to the effectiveness of the application was the suggested parameter values for different filament types. The absence of such guidance in the printed manual was directly associated with a higher rate of print failures in the control group.
6. All participants in the experimental group (100%) reported positive feedback and stated that they would recommend the application to other users, whereas 80% of participants in the control group said they would not recommend the traditional paper-based method.

7. The study confirms that 3D training applications can serve as effective alternatives to conventional training methods in additive manufacturing education, particularly in contexts requiring remote learning or limited equipment access.
8. The Unity 3D environment, in combination with the C# programming language, proved to be an effective platform for creating advanced, interactive 3D training applications featuring realistic models and user-friendly interfaces.

Notes, Additional Information

Survey participants consented to the publication of survey results.

The presented results are derived from a scientific statutory research project number 0613/SBAD/4940 conducted by Faculty of Mechanical Engineering, Poznan University of Technology, Poland, supported by the Polish Ministry of Science and Higher Education.

References

1. Jaganbabu, J., Majula, M.: Impact of e-Learning and its aplication and challenges. Int. J. Apl. Challenges **6**(1), 50–55 (2024)
2. George, M.L.: Effective teaching and examination strategies for undergraduate learning during COVID-19 school restrictions. J. Educ. Technol. Syst. **49**(1), 23–48 (2020)
3. Maor, R., et al.: Difference between zoom-based online versus classroom lesson plan performances in creativity and metacognition during COVID-19 pandemic. Learn. Environ. Res. **26**, 803–822 (2023)
4. Ramadani, A., Xhaferi, E.: Teachers' experiences with online teaching using the zoom platform witf EFL teachers in high schools in Kumanova. SEEU Rev. **15**(1), 142–155 (2020)
5. Gamage, S.H., Ayres, J.R., Behrend, M.B.: A systematic review on trends in using Moodle for teaching and learning. Int. J. STEM Educ. **9**(1) (2022)
6. Dias, S.B., Hadjileontiadou, S.J., Diniz, J., Hadjileontiadis, L.J.: DeepLMS: a deep learning predictive model for supporting online learning in the Covid-19 era. Sci. Rep. **10**(1) (2020)
7. Salas-Rueda, R.A., Eslava-Cervantes, A.L., Alvarado-Zamorano, C.: Use of Kahoot! and Jamboard for the realization and organization of creative school activities during the COVID-19 pandemic. Creativity Stud. **16**(2), 784–804 (2023)
8. Wang, A.I., Tahir, R.: The effect of using Kahoot! for learning-A literature review. Comput. Educ. **149** (2020)
9. Koh, J.H.L., Daniel, B.K.: Shifting online during COVID-19: A systematic review of teaching and learning strategies and their outcomes. Int. J. Educ. Technol. High. Educ. **19**(56) (2022)
10. Mikołajczyk, K.: VR w edukacji-subiektywny przegląd możliwości/VR in education-a subjective review of the possibilities. e-Mentor **2**(79), 33–40 (2019)
11. Predescu, S.-L., Caramihai, S.I., Moisescu, M.-A.: Impact of VR application in an academic context. Appl. Sci. **13**(8) (2023)
12. Burdea, G.C., Langrana, N.A.: Virtual force feedback: lessons, challenges, future applications. J. Robot. Mechatron. **5**(2) (1993)
13. Górski, F.: Metodyka budowy otwartych systemów rzeczywistości wirtualnej / Methodology for building open virtual reality systems. Wydawnictwo Politechniki Poznańskiej, Poznań (2019)
14. Giannakos, M., Mikalef, P., Pappas, I.O.: Systematic literature review of E-Learning capabilities to enhance organizational. Learning **24**, 619–635 (2022)

15. Eslami, M., Pirmoradian, M., Mokhtarian, A., Baghaei, S.: Design and manifacture of a soft robot with dual-interaction in virtual reality. Heliyon **9** (2023)
16. Rossler, K.L., Sankaranarayanan, G., Ryason, A., Czoski, A.: Developing and exploring the usability of an immersive virtual reality simulator on medication administration safety. Clin. Simul. Nurs. **94** (2024)
17. Koch, J., Gomse, M., Schüppstuhl, T.: Digital game-based examination for sensor placement in context of an Industry 4.0 lecture using the Unity 3D Engine-a case study. Procedia Manuf. **55**, 563–570 (2021)
18. Hardman, C.: Programowanie gier przy użyciu Unity i C#. Podręcznik dla całkiem początkujących/Game Programming with Unity and C#: A Guide for Absolute Beginners. APress Media, LLC, Warszawa (2020)
19. Jimeno, A., Puerta, A.: State of the art of the virtual reality applied design and manufacturing processes. Int. J. Adv. Manuf. Technol. **33**(9–10) (2007)
20. Cecil, J., Kanchanapiboon, A.: Virtual engineering approaches in product and process design. Int. J. Adv. Manuf. Technol. **31**(9), 846–856 (2007)
21. Grabowski, A.: Wykorzystanie współczesnych technik rzeczywistości wirtualnej i rozszerzonej do szkolenia pracowników / Using modern virtual and augmented reality techniques for employee training. Bezpieczeństwo Pracy **4**, 18–21 (2012)
22. Grajewski, D., Górski, F., Hamrol, A., Zawadzki, P.: Immersive and haptic educational simulations of assembly workplace conditions. Procedia Comput. Sci. **75**, 359–368 (2015)
23. Xie, B., et al.: A review on virtual reality skill training applications. Front. Virtual Reality **2** (2021)
24. Buń, P., Górski, F., Wichniarek, R., Kuczko, W., Hamrol, A., Zawadzki, P.: Application of professional and low-cost head mounted devices in immersive educational application. Procedia Comput. Sci. **75**, 173–181 (2015)
25. Górski, F., Buń, P., Wichniarek, R., Zawadzki, P., Hamrol, A.: Immersive city bus configuration system for marketing and sales education. Procedia Comput. Sci. **75**, 137–146 (2015)
26. SafetyCulture: Training evaluation (2024). https://safetyculture.com/topics/training-evaluation/. Accessed 07 Aug 2024
27. Tign, L., She, J.: How does virtual reality (VR) facilitate design? A review of VR usage in early-stage engineering design. Proc. Des. Soc. **3**, 2115–2124 (2023)
28. Agbo, F.J., Oyelere, S.S., Suhonen, J., Tuklainen, M.: Design, development, and evaluation of a virtual reality game-based application to support computational thinking. Educ. Technol. Res. Dev. **71**(5) (2022)
29. PKN: Drukowanie 3D–wytwarzanie przyrostowe/3D printing–additive manufacturing (2019). https://www.pkn.pl/informacje/2019/06/drukowanie-3d-wytwarzanie-przyrostowe. Accessed 07 Aug 2024
30. B3D: Technologie przyrostowe-podstawowe informacje/Additive Manufacturing Technologies-Basic Information (2024). https://b3d.com.pl/technologie-przyrostowe-podstawowe-informacje/. Accessed 07 Aug 2024
31. Gokhare, V.G., Raut, D.N., Shinde, D.K.: A review paper on 3D-Printing aspects and various processes used in the 3D-Printing. Int. J. Eng. Res. Technol. **6** (2017)
32. ATMAT: Kompendium wiedzy-rodzaje filamentów/
33. Compendium of knowledge-types of filaments (2024). https://atmat.pl/uploads/Kompendium/Kompendium%20wiedzy_Filamenty_PL.pdf. Accessed 08 Aug 2024
34. 3D Insider: PLA filament (2024). https://3dinsider.com/pla-filament/. Accessed 08 Aug 2024
35. FlashForge Polska: FlashForge adventurer 4 (2024). https://flashforge-polska.pl/drukarki-3d/flashforge-adventurer-4/. Accessed 14 Aug 2024

36. Górski, F., Wichniarek, R.: Symulacja procesów wytwarzania przyrostowego z wykorzystaniem środowiska wirtualnego/Simulation of additive manufacturing processes using a virtual environment. Postępy Nauki i Techniki **7**, 176–185 (2011)
37. Dodziuk, H.: Druk 3D/AM. Zastosowania oraz skutki społeczne i gospodarcze/3D/AM Printing: Applications and Social and Economic Impacts. PWN, Warszawa (2019)
38. Cader, M.: Druk 3D w przemyśle. Najnowsze trendy i sposoby wykorzystania/3D printing in industry. The latest trends and applications. Automatyka **11**, 26–33 (2018)
39. Zhang, Q., et al.: 3D Printing method for bone tissue enginnering scaffold. Med. Novel Technol. Dev. **17** (2023)
40. Mendaza-DeCal, R., Peso-Fernandez, S., Rodriguez-Quiros, J.: Orthotics and prosthetics by 3D-printing: accelerating its fabrication flow. Res. Vet. Sci. **162** (2023)
41. Alzoubi, L., Aljabali, A.A.A., Tambuwala, M.M.: Empowering precision medicine: the impact of 3D printing on personalized therapeutic. AAPS PharmSciTech **24**(8) (2023)
42. CADXpert: Zastosowania druku 3D (2024). https://cadxpert.pl/drukowanie-3d/zastosowania-druku-3d/. Accessed 08 Aug 2024
43. joshuatadesso, „ofw_face_front.STL. Available: https://thangs.com/designer/joshuatadesso/3d-model/ofw_face_front.STL-300974, Date accessed: 18 Aug 2024
44. Puchalski, T.: Aplikacja szkoleniowa obsługi oraz doboru parametrów procesu wytwarzania przyrostowego metodą FDM, praca magisterska/Training application for operation and selection of parameters of the FDM removal process, master's thesis, Wydział Inżynierii Mechanicznej, Politechnika Poznańska, 17 Sep 2024

Monitoring and Control Systems

Digitalization of Manufacturing Process Monitoring and Control Using Digit.io Software

Krzysztof Żywicki[1(✉)], Agnieszka Kujawińska[1], Jędrzej Iglewski[2], Magdalena Hryb[1], Paulina Rewers[1], and Przemysław Łuczak[2]

[1] Faculty of Mechanical Engineering, Poznan University of Technology, Poznan, Poland
krzysztof.zywicki@put.poznan.pl
[2] HIT-Kody Kreskowe Iglewscy, Suchy Las, Poland

Abstract. The article presents a case study of the implementation of an IT system for supervising the production process. As part of the work, digitalizing solutions enabling automatic data acquisition were also implemented. The example concerns the process of printing wallpaper and posters, which is characterized by high product variability. As part of the research, an analysis of the current state was carried out, including the identification of key factors affecting the quality of the process. On this basis, decision rules were adopted to enable effective control of the production process. On this basis, the IT system was configured and implemented. Its validation showed that its implementation contributed to a significant reduction in the fraction of defective products.

Keywords: monitoring production processes · digitalization of production · pro-duction control

1 Introduction

With growing market competition and increasing demands for efficiency, stability and quality in manufacturing processes, companies are increasingly implementing digital solutions to support monitoring and adaptive production control. This transformation, referred to as Industry 4.0, involves the integration of cyber-physical systems, the Internet of Things (IoT), advanced data analytics and automation of operations, enabling the construction of accessible and stable production systems [1, 2]. The literature emphasises that the digitisation of processes is no longer just an element of operational support, but is becoming a key component of companies' development strategies and building their resilience to environmental volatility [3].

At the same time, increased demand volatility, pressure to shorten lead times, the need to maintain high production flexibility and the need to respond quickly to disruptions mean that traditional process monitoring methods–based mainly on periodic checks and post-process analyses–are proving insufficient [4]. In such conditions, systems that enable continuous acquisition of process data, its analysis in real time, and semi-automatic or automatic corrective decision-making are becoming increasingly important [5]. These solutions form the foundation of the Smart Manufacturing concept, which

J. Trojanowska et al. (Eds.): MANUFACTURING 2026, LNME, pp. 243–251, 2026.
https://doi.org/10.1007/978-3-032-21654-0_18

integrates automation, data analytics and flexible process control throughout the entire manufacturing cycle [6–8].

The results of studies presented in the literature confirm that the implementation of smart production systems involves significant technical and organisational challenges. The most frequently mentioned ones include problems with the integration of heterogeneous data sources, ensuring the quality and consistency of process data, the interoperability of IT/OT systems, and communication security [9, 10]. Equally important are organisational barriers, such as limited digital skills of employees, lack of standardisation of decision-making procedures, and difficulties in adapting management structures to work in a data-driven environment [11]. These challenges are particularly evident in the small and medium-sized enterprise sector, where limited financial and IT resources often hinder the implementation of complex MES or SCADA systems [12].

Despite these limitations, the results of numerous studies (especially empirical ones) confirm that the implementation of real-time production monitoring and control systems leads to measurable organisational and quality benefits [13–15]. These include, among others, improved utilisation of production resources, increased productivity, reduced operating costs and reduced material waste [16–19]. In addition, the positive environmental effects resulting from greater process stability and precise control of technological parameters, which contribute to reducing energy and raw material consumption, are increasingly being emphasised.

In recent years, the development of advanced methods for analysing data from multiple sources, such as machine signals, visual, acoustic and environmental data, has become an important direction. Particular attention is paid to machine learning methods, including unsupervised approaches, which enable the identification of process anomalies without the need for complete training data sets [20, 21]. These solutions are considered particularly useful in production environments with high product variability and short series, where classic predictive models based on historical data lose their effectiveness.

However, it should be emphasised that a significant part of the research to date has focused on implementations in large industrial enterprises, especially in industries such as automotive, metallurgy and electronics [22]. A review of the literature indicates that despite the dynamic development of the Industry 4.0 concept and the growing number of studies on digital monitoring and control of production processes, the printing and packaging industry remains relatively underrepresented in empirical studies. Available publications focus mainly on the technological aspects of printing (e.g. materials, imaging techniques, colour quality) or on conceptual approaches to digitisation, while rarely analysing comprehensive implementations of production monitoring systems that integrate process, quality and decision-making data in real production conditions [23, 24]. At the same time, printing and packaging processes are characterised by high product variability, short production runs, frequent changeovers and a significant contribution of subjective visual assessment (print quality assessment), which significantly distinguishes them from the industries dominating the Industry 4.0 literature, such as automotive and electronics [23]. Research on digitisation in this sector indicates a lack of tools that enable the rapid definition of quality rules, the contextualisation of data in relation to a specific order, and effective support for operational decisions without the need to implement costly and complex enterprise-class systems [24, 25]. As a result, there is a clear research gap regarding practical, scalable and replicable solutions for digital process monitoring in the printing and packaging industry, especially in the SME sector. This gap justifies the need to present case studies showing not only the effects of implementation, but also the system architecture, data integration methods and the logic of quality decision-making in a highly diverse and small-scale production environment.

Based on the above conditions, the development and implementation of the Digit.io system is a response to the identified research gap concerning practical, scalable solutions for the digitisation of production supervision in the SME sector, particularly in industries with high product range variability. The system integrates data from sensors and machines, automates the acquisition of process information, and enables the definition of quality parameters and decision rules, supporting continuous production monitoring, deviation identification, and corrective action in near real time.

The aim of this article is to present a case study of the implementation of the Digit.io system in a company producing wallpapers and large-format posters, with particular emphasis on the architecture of data acquisition and integration, the configuration of quality rules, and the identification of critical factors affecting product quality. The paper also presents an assessment of the effects of the implementation, including the impact of the system on reducing the number of non-compliant products and improving process stability.

2 Digit.io IT System-Control Module

The IT system known under the trade name Digit.io enables active production control through dynamic production scheduling and ongoing monitoring of material flow and manufacturing process parameters. The system operates on the basis of analysis and decision-making using simulation and decision support methods.

One of the main modules of the system is the Control module, which enables the analysis of manufacturing processes through the use of analysis and decision support methods. Data analysis may concern:

- product quality,
- process quality,
- machines–machine monitoring,
- tooling–technological tooling monitoring.

For each of these elements, you can define any number of characteristics to be monitored and analysed. Each of them can be assigned a type: nominant, maximant, min-imant. Each feature type has specific limit values that the user defines for a given feature, e.g. for the Quantitative nominant type:

- nominal dimension (WN),
- upper tolerance line GLT)
- lower tolerance line (DLT).

Each feature type has an assigned group of analysis tools within which specific tools are defined, e.g.: descriptive statistics, feature value distribution chart, quality capability indicators.

Each group has tools defined in relation to the type of feature. The data sources can be:

- measurements taken manually by employees,
- signals from measuring sensors,
- external databases–hardware solutions for measuring systems.

An important feature of the module is the notification of deviations from the accepted limit values of the characteristics of the objects being monitored. Alerts are triggered automatically by the system and indicate any problems that arise during the execution of a production order. This feature allows the user to take action to eliminate the deviation or problem.

3 Object and Methodology of the Research

The subject of this article is the implementation of the Digit.io IT system and the introduction of digitization in the supervision of manufacturing processes at Bluebird Design, a company producing wallpapers, large-format posters and other wall decorations (Fig. 1). The aim is to demonstrate that the implementation of such solutions allows for an improvement in the efficiency indicators of the production process.

Wallpaper and other wall markings and decorations are printed using digital roll-to-roll printing machines. The company also uses a wallpaper cutting machine, which al-lows wallpaper to be cut to the size required by the customer. This machine operates on the principle of digital size design, but requires the use of cutting tools, which wear out, causing damage to the material being cut.

The research methodology included:

- analysis of the current production process and definition of critical factors necessary for effective process supervision,
- analysis of the configuration requirements of the Digit.io Control module,
- selection of sensors for automatic identification of process data,
- development of decision rules for supervising and controlling the production process,
- implementation and testing, including determination of the effectiveness of the solutions.

The production process for both types of products is carried out according to the same operations but using different machines:

- printing: the printing machine operator proceeds with the planned order, downloads the files with the print templates previously approved by the customer; selects the appropriate material in accordance with the requirements, loads the machine with a roll of material and performs initial calibration,
- cutting: the printed products are then sent to the cutting machine, the operator sets the cutting tools for the required product dimensions (width and length), then checks the quality and current condition of the cutting tool, replaces the tool in the cutting machine spindle if necessary, and proceeds with cutting,
- quality control: checking the printout for graphic quality (poor image quality visible through blurred pixels), checking the print quality and creases on the wallpaper, checking the joints (width, cutting precision, no burrs),

Fig. 1. View of the production hall and printing machines

- packaging: the operator packs the products and places them in the shipping area.

An analysis of the current situation showed that there were areas in the production process where data was not collected or was recorded on paper by machine operators. This resulted in inaccuracies, errors and omissions, which led to a lack of effective supervision and control of manufacturing processes.

4 Implementation of Digitalisation in the Monitoring of Production Processes

The implementation of the Digit.io system and digitization solutions began with the identification of data generation points in the production process and data flow, along with the information created on their basis.

For selected input and output values, process parameters and process variables, the method and technical solutions enabling their automatic acquisition were determined. Appropriate measuring sensors were selected and a network infrastructure adequate for the process was designed and implemented. Communication interfaces based on the OPC UA protocol were also programmed, allowing data exchange with the Digit.io Control analytical module (Fig. 2).

In the next stage of system implementation, tests were carried out to determine the relationship between process parameters and input values and output values, taking into account process variables. Based on the results obtained, a knowledge base was created, which forms the basis for making decisions about the status of the process and the need to take control measures. The main elements of the knowledge base are:

- cause-and-effect relationships: defined and recorded relationships between various process parameters and product quality,

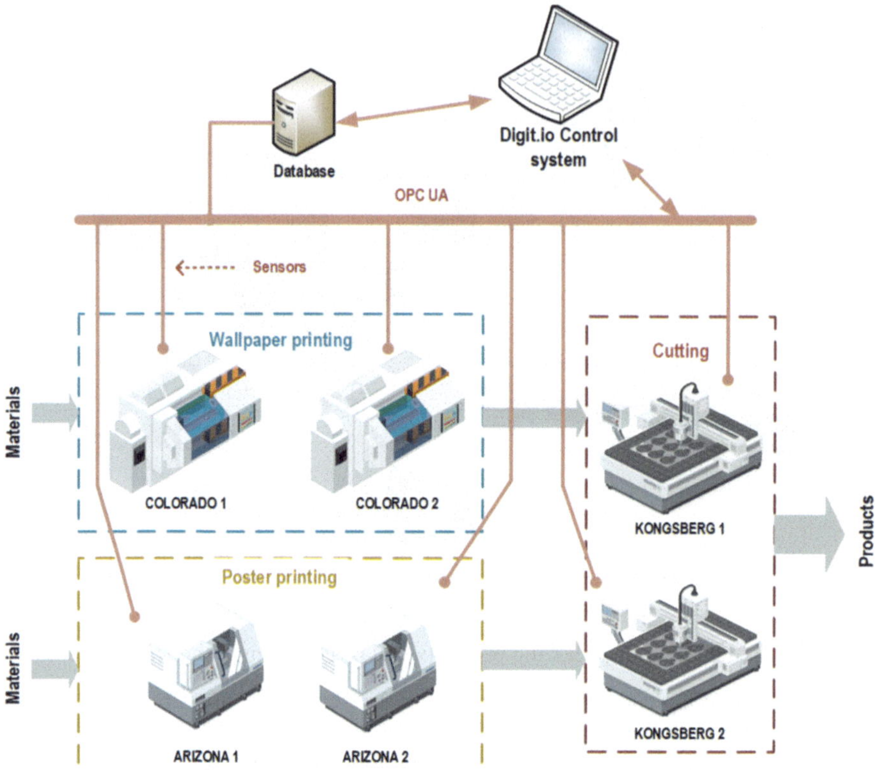

Fig. 2. Information flow diagram in the production process.

- decision rules: specifying what steps should be taken in response to specific production conditions (the rules included adjusting process parameters to current data on output values or process variables),
- limit values that should not be exceeded in order for a part at a given stage of the production process to be considered compliant with the requirements,
- events: information about unforeseen events, e.g. failures, etc., and actions taken to solve the problems that arise.

Based on the research conducted, it was assumed that the decision models would be based on regression and heuristic methods. The results of regression technique analyses (linear and multiple regression) and practical experience (knowledge of operators and technologists) were used to develop control rules.

5 Summary

The implementation of the IT system with automatic data acquisition resulted in a significant improvement in process quality. The results are presented in Fig. 3 for the five weeks prior to implementation and the five weeks following implementation of the new production process monitoring system.

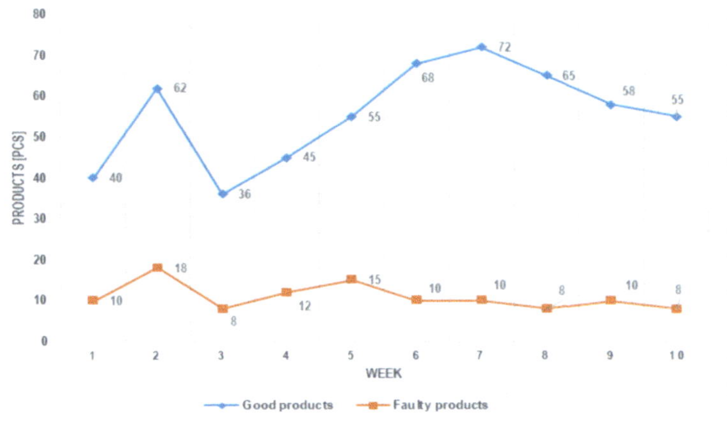

Fig. 3. Number of defective products before and after implementation of the Digit.io system.

The results indicate an average 12% reduction in the number of defective products for production orders (Fig. 4).

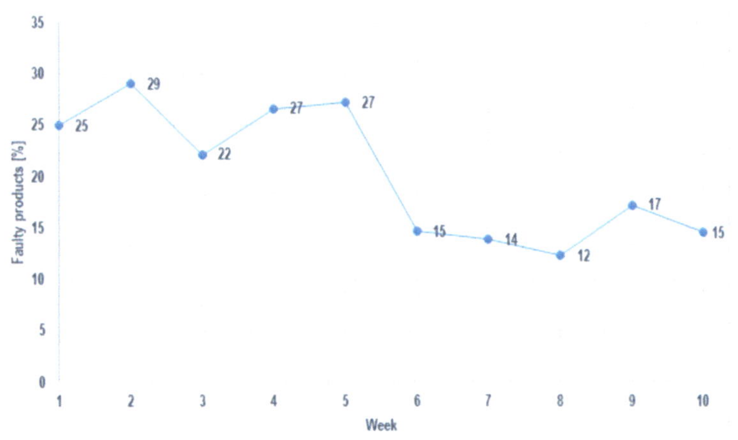

Fig. 4. Percentage of defective products in the fulfilment of production orders.

During implementation, factors influencing the occurrence of quality errors were identified. These areas primarily include: control of colour profiles related to printing on a specific printing machine, using raw materials and other key printing parameters, and control of tools used during post-print cutting, enabling the printout to be adjusted to the customer's requirements. This has significantly reduced the risk of quality errors. The implementation of automatic data acquisition has enabled effective control of production processes using decision rules.

In subsequent stages, the system is planned to be expanded to include scheduling and production flow control.

Acknowledgments. This article was created and financed as part of research and development work carried out by a consortium comprising the Department of Production Engineering at Poznań University of Technology and HIT–Kody Kreskowe Iglewscy sp.j., project: The project is co-financed by the European Union under the European Regional Development Fund program under the Intelligent Development Program. The project is carried out as part of the National Center for Research and Development's Program "Szybka ścieżka" (project no. POIR.01.01.01-00-0352/22).

References

1. Tashkinov, A.G.: The application of industry 4.0 into the company's production activities through effective decision-making. Sci. Rep. **15**, 34202 (2025). https://doi.org/10.1038/s41598-025-15688-0
2. Barari, A., Tsuzuki, M.S.G.: Smart Manufacturing and Industry 4.0. Appl. Sci. **13**, 1545 (2023). https://doi.org/10.3390/app13031545
3. Sufian, A.T., Abdullah, B.M., Miller, O.J.: Smart manufacturing application in precision manufacturing. Appl. Sci. **15**, 915 (2025). https://doi.org/10.3390/app15020915
4. Fekrisari, M., Kantola, J.: Integrating Industry 4.0 in manufacturing: overcoming challenges and optimizing processes (case studies). TQM J. **36**, 347–370 (2024). https://doi.org/10.1108/TQM-12-2023-0411
5. Masood, T., Sonntag, P.: Industry 4.0: adoption challenges and benefits for SMEs. Comput. Ind. **121**, 103261 (2020). https://doi.org/10.1016/j.compind.2020.103261
6. Stentoft, J., Adsbøll Wickstrøm, K., Philipsen, K., Haug, A.: Drivers and barriers for Industry 4.0 readiness and practice: empirical evidence from small and medium-sized manufacturers. Prod. Plan. Control. **32**, 811–828 (2021). https://doi.org/10.1080/09537287.2020.1768318
7. Abdul-Hamid, A.-Q., Ali, M.H., Osman, L.H., Tseng, M.-L., Omar, A.R.C.: An industry 4.0 adoption in the circular economy application hierarchical model: driver, enable and barrier aspects. Ind. Manag. Data Syst. **124**, 386–415 (2024). https://doi.org/10.1108/IMDS-12-2022-0780
8. Ali, I., Phan, H.M.: Industry 4.0 technologies and sustainable warehousing: a systematic literature review and future research agenda. Int. J. Logist. Manag. **33**, 644–662 (2022). https://doi.org/10.1108/IJLM-05-2021-0277
9. Wang, S., Wan, J., Zhang, D., Li, D., Zhang, C.: Towards smart factory for industry 4.0: a self-organized multi-agent system with big data based feedback and coordination. Comput. Netw. **101**, 158–168 (2016). https://doi.org/10.1016/j.comnet.2015.12.017
10. Ferrag, M.A., Shu, L., Djallel, H., Choo, K.-K.R.: Deep learning-based intrusion detection for distributed denial of service attack in Agriculture 4.0. Electronics **10**, 1257 (2021). https://doi.org/10.3390/electronics10111257
11. Arruda, H.M., Bavaresco, R.S., Kunst, R., Bugs, E.F., Pesenti, G.C., Barbosa, J.L.V.: Data science methods and tools for Industry 4.0: a systematic literature review and taxonomy. Sensors **23**, 5010 (2023). https://doi.org/10.3390/s23115010
12. Gao, Y., Chai, C., Li, H., Fu, W.: A deep learning framework for intelligent fault diagnosis using AutoML-CNN and image-like data fusion. Machines **11**, 932 (2023). https://doi.org/10.3390/machines11100932
13. de Paula Ferreira, W., Armellini, F., De Santa-Eulalia, L.A.: Simulation in industry 4.0: a state-of-the-art review. Comput. Ind. Eng. **149**, 106868 (2020). https://doi.org/10.1016/j.cie.2020.106868

14. Awan, U., Gölgeci, I., Makhmadshoev, D., Mishra, N.: Industry 4.0 and circular economy in an era of global value chains: what have we learned and what is still to be explored? J. Clean. Prod. **371**, 133621 (2022). https://doi.org/10.1016/j.jclepro.2022.133621
15. Wu, K., Xu, J., Zheng, M.: Industry 4.0: review and proposal for implementing a smart factory. Int. J. Adv. Manuf. Technol. **133**, 1331–1347 (2024). https://doi.org/10.1007/s00170-024-13839-7
16. Rojek, I., Kujawińska, A., Burduk, R., Mikołajewski, D.: Improving process control through decision tree-based pattern recognition. Electronics **13**, 4823 (2024). https://doi.org/10.3390/electronics13234823
17. Kujawińska, A., Diering, M.: The impact of the organization of the visual inspection process on its effectiveness. Int. J. Adv. Manuf. Technol. **112**, 1295–1306 (2021). https://doi.org/10.1007/s00170-020-06543-9
18. Pavlenko, I., Trojanowska, J., Ivanov, V., Radchenko, S., Husár, J., Mižáková, J.: Signal decomposition for monitoring systems of processes. Processes **12**, 1188 (2024). https://doi.org/10.3390/pr12061188
19. Kaźmierczak, P., Żywicki, K., Rewers, P.: The impact of downtime on the stability of the production schedule. Appl. Sci. **15**, 150 (2025). https://doi.org/10.3390/app15010150
20. Awan, U., Sroufe, R., Shahbaz, M.: Industry 4.0 and the circular economy: a literature review and recommendations for future research. Bus. Strateg. Environ. **30**, 2038–2060 (2021). https://doi.org/10.1002/bse.2731
21. Chalapathy, R., Chawla, S.: Deep learning for anomaly detection: a survey (2019). https://doi.org/10.48550/arXiv.1901.03407
22. Paul, J., et al.: Digital transformation: a multidisciplinary perspective and future research agenda. Int. J. Consum. Stud. **48**(2), e13015 (2024). https://doi.org/10.1111/ijcs.13015
23. Soľavová, M., Bolek, V., Zelina, M.: Digitalization, automation and sustainability of occupations in the printing industry in Slovakia. **11**, 33 (2024)
24. Abdallah, Y., Shehab, E., Al-Ashaab, A.: Digital transformation challenges in the manufacturing industry (2021). https://doi.org/10.3233/ATDE210004
25. Szentgyörgyvölgyi, R.: Effect of the digital technology to the print production processes. Acta Polytechnica Hungarica (2008)

Vision-Based Supervisory Concept for Manual Assembly Operations Using TinyML on an STM32 Microcontroller

Jakub Kaščak(✉), Lucia Knapčíková, Martin Pollák, Zuzana Mitaľová, Gregor Sopko, and Peter Gabštur

Faculty of Manufacturing Technologies, Technical University of Kosice, Presov, Slovakia
jakub.kascak@tuke.sk

Abstract. Manual assembly of high-precision components remains a critical point in otherwise highly automated production systems, with quality and traceability depending strongly on operator performance and ad-hoc visual checks. This paper presents a vision-based supervisory concept for assembly operations implemented with TinyML on an STM32 microcontroller. The proposed approach uses a low-cost STM32H747I-DISCO board with an attached camera module to monitor selected assembly steps and evaluate their conformity with a pre-trained image classification model. A complete end-to-end workflow is described, starting from 4K image acquisition at the assembly workplace, dataset preparation, and model training in Teachable Machine, through export to TensorFlow Lite for Microcontrollers and conversion by STM32Cube.AI, up to deployment and execution on the target hardware. The embedded application performs real-time image capture, classification, confidence estimation and on-device visualization, while supporting optional local time-stamped logging of recognised operations (e.g., to SD card) for later analysis. A proof-of-concept study on a representative reducer/actuator assembly demonstrates that, under stable illumination and a constrained set of classes, the TinyML based supervisor can distinguish between key assembly states and provide interpretable feedback to engineers. The work is positioned as a practical building block towards lightweight, retrofittable digitalization of assembly workplaces and as a basis for further research on robust datasets, integration with plant IT systems, and deployment in safety- and security-constrained industrial environments.

Keywords: TinyML · STM32 microcontroller · Vision-based supervision · Assembly operations · Image classification · Edge computing · Industry 4.0

1 Introduction

Modern manufacturing is undergoing a gradual transition from isolated, manually supervised production cells to cyber-physical systems with tightly integrated sensing, computation, and control functions along the entire value chain. This transformation is often framed under the concept of Industry 4.0, which emphasizes vertical and horizontal

J. Trojanowska et al. (Eds.): MANUFACTURING 2026, LNME, pp. 252–266, 2026.
https://doi.org/10.1007/978-3-032-21654-0_19

integration of production systems, widespread connectivity, and the use of data-driven methods to support decision making at the shop-floor level [1–3]. Within this context, assembly systems are increasingly recognized as critical nodes where product variability, manual operations, and quality requirements intersect, and where systematic structuring and monitoring of operations are required to maintain robustness and traceability [4]. Deep learning based computer vision has become a key enabler for automatic inspection, object recognition, and monitoring of human-machine interaction in such environments [5]. However, most industrial computer vision solutions still rely on PC-based or server-based processing, often combined with high-end industrial cameras and substantial infrastructure. These architectures are powerful but come with non-negligible hardware costs, integration effort, and cybersecurity exposure, especially when cloud connectivity is involved. For many small and medium-sized enterprises, as well as for retrofitting of existing assembly stations, these requirements represent a significant barrier to adoption.

Recent advances in TinyML address this challenge by tailoring machine learning models and runtimes to resource-constrained microcontrollers, enabling on-device inference at milliwatt power levels and under tight memory budgets [6–8]. In this paradigm, trained models are quantized and compiled into static libraries that execute directly on embedded targets, often without a conventional operating system. Such deployments inherently support low-latency inference, eliminate the need to stream image data to external compute nodes, and reduce the attack surface of networked production systems. At the same time, they require careful co-design of network architecture, quantization strategy, and memory layout to satisfy both accuracy and real-time constraints [9].

Edge computing concepts further reinforce this trend by promoting the placement of computation as close as possible to the physical process, thereby reducing bandwidth usage, increasing resilience to connectivity issues, and enabling localized control loops [10]. When applied to assembly operations, edge-based computer vision can provide continuous supervision of workpiece state, tool usage, and sequence compliance, while respecting constraints on data governance and operator privacy. For many use cases, only compact metadata need to be stored or transmitted-such as timestamps, predicted operation labels, and confidence scores-rather than raw image streams. The present contribution focuses on a vision-based supervisory layer for an assembly workstation that is implemented entirely on a low-power microcontroller using TinyML. The target scenario involves the assembly of gear reducers and actuators in a production environment, where selected visual checkpoints along the assembly workflow are monitored by a fixed camera. Instead of replacing human operators or existing PLC logic, the concept adds an additional, non-intrusive layer that verifies whether a given assembly state matches one of several predefined reference configurations and logs this information for later analysis. The overall architecture is depicted in Fig. 1, which illustrates the interaction between the vision sensor, the embedded inference engine, the human machine interface, and the logging mechanisms.

In this architecture, the core of the supervisory function is a compact convolutional neural network trained to classify a limited set of assembly states (e.g., "housing open," "cover mounted," "fasteners installed") from camera frames. The network is trained offline and then deployed to the microcontroller in a quantized form compatible with

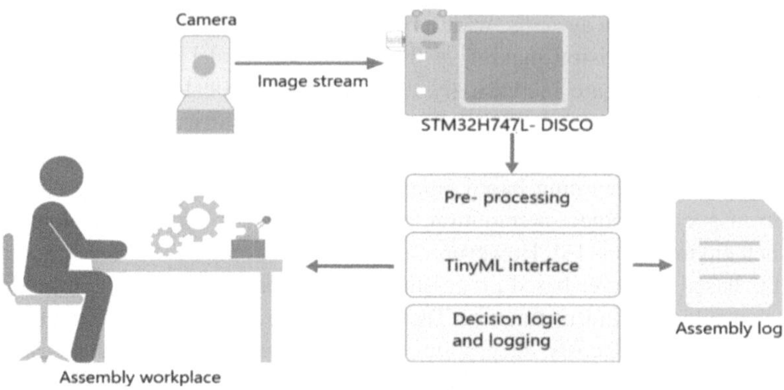

Fig. 1. Overall architecture of the vision-based assembly supervisor.

TinyML runtimes. During operation, the system continuously acquires images from the camera, preprocesses them to match the model input format, and performs on-device inference. The predicted class and associated confidence are then mapped to semantic labels representing assembly operations, which are combined with timestamps to form a lightweight event log. This log can subsequently be used to reconstruct the chronological evolution of the assembly sequence and to identify deviations, delays, or repeated corrective actions at selected checkpoints.

Despite the significant progress in TinyML tooling and embedded AI runtimes, the barrier for introducing such concepts into industrial assembly remains high. Many existing studies on vision-based assembly monitoring consider powerful GPUs or industrial PCs as the primary execution platform and do not explicitly address the constraints and integration challenges of microcontroller based implementations. At the same time, existing TinyML demonstrations often focus on generic benchmarks that are only loosely related to real-world assembly workflows. There is therefore a need for application-oriented case studies that connect concrete industrial assembly processes with a complete, reproducible toolchain-from data collection and model training to deployment on a specific microcontroller platform-under realistic constraints on hardware, lighting, and operator interaction.

The goal of this paper is to address this gap by providing a structured proof-of-concept for a vision-based assembly supervisor implemented on an STM32 microcontroller using TinyML, with a particular emphasis on:

- defining an assembly-oriented image classification task based on discrete visual checkpoints within a real industrial assembly workflow,
- describing a reproducible software toolchain that starts from camera-based data acquisition and leads to a quantized model deployed on a microcontroller,
- demonstrating that such a system can operate in a fully embedded manner, using only on-device computation and lightweight logging of predicted assembly states.

The concept is intentionally scoped as a proof-of-concept rather than a fully validated industrial system. It focuses on a limited number of assembly states, a controlled lighting setup, and a static camera viewpoint, yet it is designed to be extensible toward more complex configurations. The description concentrates on technical transparency and methodological clarity, so that the approach can serve as a template for further experiments within the broader framework of Industry 4.0 oriented assembly system design [1–4], deep learning and TinyML deployment [5–9], and edge computing in manufacturing [10].

2 System Concept and Use Case

In line with Industry 4.0 and cyber-physical production concepts, the device is intended as an additional sensing and decision layer that augments conventional quality control and work instruction systems rather than replacing the human operator [1–4]. The system observes the assembly area from an overhead position and evaluates whether the visible configuration of parts corresponds to a valid assembly state or to a selected set of typical deviations (missing component, wrong orientation, omitted step). The decision is produced directly on the microcontroller to minimise latency and to avoid permanent transmission of image data beyond the workstation boundaries, which is beneficial both for network load and for protection of industrial know-how [5–10].

The software toolchain that enables this concept from image acquisition and dataset preparation through Teachable Machine training to deployment of a TensorFlow Lite for Microcontrollers model on STM32 via STM32Cube.AI-is shown in Fig. 2 [6, 7, 11–13, 16].

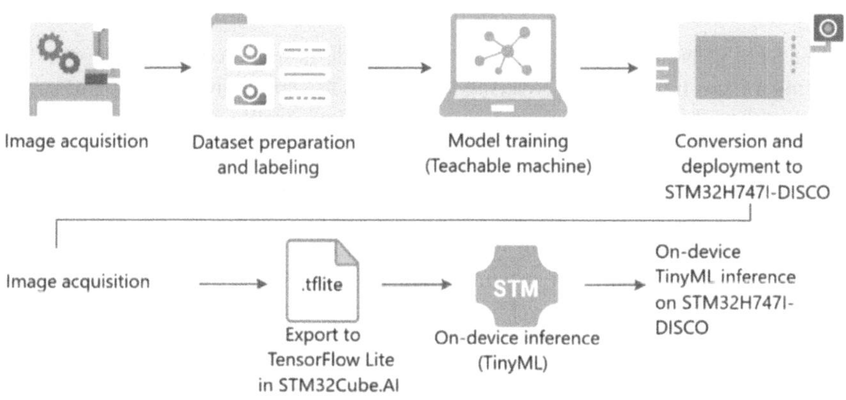

Fig. 2. Software workflow from data collection to TinyML deployment on STM32.

The industrial motivation for the supervisor stems from a gear-reducer assembly line where a dedicated area is reserved for experimental digitalisation activities in cooperation with an external industrial partner. The complete assembly comprises several manual and semi-automated steps, however, the present work deliberately focuses on a reduced subset suitable for a proof-of-concept implementation while respecting constraints on disclosure of company specific process details. A simplified assembly workflow is therefore defined with three main zones: component infeed, assembly area and finished product / outgoing area. Within the assembly area, three visual checkpoints are considered:

(i) placement/orientation of the housing, (ii) mounting/presence of the cover or lid, and (iii) a transient tool-in-use state associated with tightening actions. These checkpoints represent states where a wrong component, incorrect orientation, omitted operation, or tightening activity manifest as observable configuration differences in the camera image and are reflected in the class taxonomy used for training (Table 1). The corresponding simplified material and information flow, together with the locations of the critical checkpoints, is illustrated in Fig. 3, which serves as the basis for dataset design and model training.

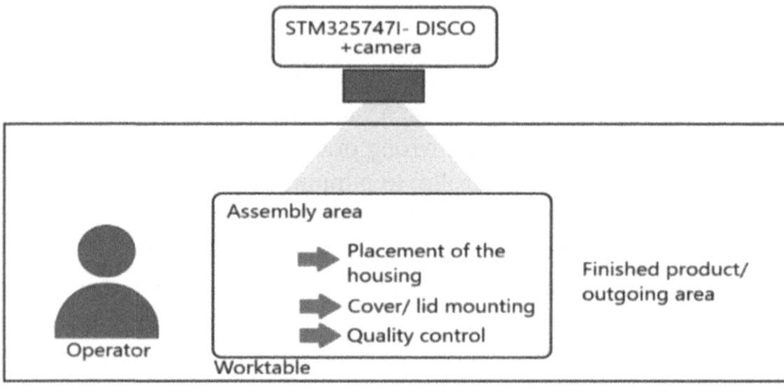

Fig. 3. Simplified assembly workflow with critical visual checkpoints.

The STM32 board with attached camera module is mounted above the worktable, providing a fixed field of view covering the defined assembly area. The operator works in front of the table, while the board operates as a low-profile supervisor that analyses the process without interfering with task execution. The concept assumes a compact protective enclosure for the microcontroller and camera, as well as a simple mounting frame allowing adjustment of the viewing angle and distance to the assembly area. These mechanical aspects are not the focus of the present contribution but influence illumination conditions and image perspective, and are therefore considered during data acquisition.

3 Dataset Creation and TinyML Workflow

3.1 Image Acquisition and Class Definition

The TinyML model is trained on images representing distinct assembly states derived from the simplified workflow in Fig. 3. Images are captured using a 4K camera positioned approximately at the intended location of the STM32 supervisor. High resolution facilitates later down-sampling and cropping while preserving sufficient detail for the relatively small input size of the embedded model. For the proof-of-concept, the dataset is organised into six visually distinctive classes capturing nominal states and typical deviations, as summarised in Table 1: (C1) Housing_correct, (C2) Housing_incorrect, (C3) Cover_mounted, (C4) Cover_missing, (C5) Tool_in_use, and (C6)

Background_or_other. This taxonomy supports checkpoint-oriented supervision while keeping the class set compatible with microcontroller memory and compute constraints.

The dataset is intentionally limited to a small number of visually distinctive classes, reflecting the memory and compute constraints of the microcontroller and focusing on the feasibility of the end-to-end workflow rather than exhaustive coverage of all assembly errors [6, 8]. Images are captured under representative shop floor lighting with moderate variation due to shadows and reflections. Background clutter is reduced, but no artificial markers or fiducials are used, in order to remain close to realistic deployment scenarios.

Teachable Machine is used as the front-end tool for dataset management and model training [11, 16]. Images are uploaded, automatically resized and labelled according to the defined classes, with basic data augmentation (random rotation and horizontal flipping) applied to improve robustness to small pose variations of the components on the worktable.

Table 1. Dataset structure for assembly-state classification.

Class ID	Class label	Description	Imaging conditions/notes
C1	Housing_correct	Correct placement and orientation of housing/base	4K overhead camera, uniform background, normal lighting
C2	Housing_incorrect	Misaligned or rotated housing, visible assembly error	Same setup as C1, intentional misplacement
C3	Cover_mounted	Correctly mounted cover/lid with all screws inserted	Final state before torqueing (if applicable)
C4	Cover_missing	Cover or lid not present although previous step finished	Used to detect skipped operation
C5	Tool_in_use	State where operator is tightening screws / using tool	Blur/motion may be present, used mainly for temporal context
C6	Background_or_other	Frames not belonging to monitored operations (transition, idle)	Helps reduce false positives outside assembly area

3.2 Model Training in Teachable Machine

Teachable Machine provides a browser-based interface for configuring and training a convolutional neural network tailored to image classification tasks. In this article, the default image project configuration is employed, which internally relies on a compact convolutional backbone suitable for subsequent conversion to TensorFlow Lite [11, 12]. Parameters are adjusted only within the limits exposed by the interface (number of epochs, learning rate and batch size), as the focus is on reproducibility of an accessible workflow rather than on extensive architecture search.

Training is performed until convergence of the validation accuracy, typically within a few tens of epochs due to the limited dataset size. The platform provides confusion matrices and per-class accuracy estimates, which are used to verify that all assembly states are distinguishable. Once a satisfactory model is obtained, the project is exported as a TensorFlow Lite model in.tflite format, representing a quantised version suitable for deployment on resource constrained devices [7, 12].

3.3 Conversion to STM32Cube.AI and C Code Generation

The exported.tflite model is next imported into STM32Cube.AI, a toolchain extension that analyses the neural network, performs additional optimisations and generates C code compatible with STM32 microcontrollers [13].

Configuration of STM32Cube.AI is guided by the targeted STM32H747I-DISCO board and by constraints on flash and RAM size. Fixed-point compression is selected to minimise memory footprint while preserving recognition accuracy [7–9]. The generated code is integrated into an STM32CubeIDE project that also contains application-level files responsible for camera interfacing, display output and logging. The key parameters of the software and hardware environment used for the prototype are listed in Table 2, including firmware versions and library components [13–15].

Table 2. Hardware and software configuration.

Category	Item	Specification/version/notes
Embedded platform	Microcontroller board	STM32H747I-DISCO (STM32H747XIH6, dual-core Cortex-M7/M4)
Embedded platform	On-board display	4.3" TFT LCD, 480 × 272 px (used for live preview + class/confidence)
Embedded platform	Camera module	BCM-style camera connected to STM32H747I-DISCO (QVGA/VGA, grayscale or RGB, downsampled to 96 × 96)
Data acquisition	Imaging device	4K camera mounted above worktable (off-board, used only for dataset collection)
Data acquisition	Acquisition setup	Fixed tripod/rig, constant distance, static illumination, neutral background
Software stack	Training tool	Google Teachable Machine - Image project
Software stack	Model export	TensorFlow Lite (*.tflite, quantized, 96 × 96 grayscale input)
Software stack	Middle layer	STMicroelectronics X-CUBE-AI (STM32Cube.AI) for C code generation
Software stack	Project configuration	STM32CubeMX-configuration of clocks, middleware, and AI pack

(continued)

Table 2. (*continued*)

Category	Item	Specification/version/notes
Software stack	Firmware IDE	STM32CubeIDE (C/C++ project for STM32H747I-DISCO)
Deployment	Inference mode	On-device TinyML inference
Deployment	Output	Class label + confidence [%] rendered on LCD; optional logging to UART / SD card

The final TinyML inference engine running on the microcontroller is characterised by the number of layers, activation functions, input resolution and memory requirements. These aspects are documented in Table 3, based on the configuration reported by STM32Cube.AI during code generation.

Table 3. STM32 AI neural network configuration.

Parameter	Value	Notes
Input resolution	96 × 96 pixels	Image resized and converted to grayscale on-device/during preprocessing
Input channels	1 (grayscale)	Reduces memory footprint compared to RGB
Input datatype	int8 (compressed)	Post-training quantization in TensorFlow Lite
Number of classes	6	Matches assembly states defined in Table 1
Base model	Teachable Machine (TFLite)	Exact architecture exported as *.tflite
Inference time (Cortex-M7 @ 480 MHz)	9–12 ms	Range over n = 30 runs, single-frame inference, preprocessing included
Quantization scheme	Symmetric per-tensor (weights + activations)	
According to TFLite export		
Output representation	Softmax probabilities (confidence per class)	TinyML inference returns confidence for each class
Triggering mode	Frame-by-frame classification	Later extendable to temporal smoothing

4 Concept of a Vision-Based Assembly Supervisor

The proposed concept targets a lightweight supervisory layer for manual assembly, implemented directly on an STM32H747I-DISCO microcontroller equipped with a B-CAMS camera module [13–15]. The supervisor observes a dedicated region of the worktable and continuously evaluates the current assembly state using a compact convolutional neural network deployed through STM32Cube.AI and TensorFlow Lite for Microcontrollers [6, 12, 13]. All processing is executed on the microcontroller, which eliminates the need for permanent network connectivity and supports integration into existing industrial environments with strict constraints on infrastructure and data access [3, 10].

The supervisor is designed as an additional, non-intrusive information layer rather than as a hard interlock. The output is stored in the form of a temporal sequence of predicted states, which can later be correlated with production data from the cyber-physical production system [1, 3]. This representation enables subsequent analysis of cycle time, sequence adherence and the occurrence of atypical patterns in the assembly process, while keeping the on-device logic simple enough for real-time execution on a microcontroller class platform [6–9].

4.1 Methodology

Based on discussions with process experts and inspection of the workstation layout, three visual checkpoints were defined and mapped to the class taxonomy in Table 1:

1. housing placement/orientation (C1 vs C2),
2. cover mounting/presence (C3 vs C4),
3. tightening context captured as Tool_in_use (C5), used to indicate progression between static checkpoints. Frames outside monitored operations are grouped as Background_or_other (C6) to reduce false positives.

4.2 Software Workflow from Data Collection to STM32 Deployment

The software toolchain follows a structured workflow from image acquisition to deployment of the TinyML model on the STM32 platform, as summarised in Fig. 2. The workflow combines the Teachable Machine environment for rapid prototyping [11] with TensorFlow Lite Micro and STM32Cube.AI for embedded deployment [12, 13, 16].

1. Image acquisition. Image data are recorded using a 4K camera positioned similarly to the intended STM32-based supervisor.
2. Dataset preparation and labelling. Images are organised into folders representing the six target classes (Table 1) corresponding to the visual checkpoints. Obvious artefacts such as motion blur or occlusions are removed. The resulting dataset structure and class definitions are summarised in Table 1.
3. Model training in Teachable Machine. The labelled dataset is imported into Teachable Machine's image classification template [11]. A MobileNet-derived backbone with transfer learning is used, and the training configuration (number of epochs, learning rate, image resolution) is tuned to achieve a satisfactory trade-off between accuracy and model size suitable for microcontroller deployment [6–9].

4. Export to TensorFlow Lite. The trained model is exported as a TensorFlow Lite.tflite file, compressed to 8-bit integers to reduce memory footprint and inference latency, in line with common TinyML practice [6–8, 12].
5. Conversion with STM32Cube.AI. The.tflite model is imported into STM32Cube.AI, which performs compatibility checks, generates an optimized C representation of the network, and provides estimates of RAM and flash requirements for the target STM32H747I device [13].
6. Integration in STM32CubeIDE. The generated C code is integrated into an STM32CubeIDE project for the STM32H747I-DISCO board [14]. The application code handles camera capture, pre-processing, invocation of the inference function, interpretation of the output vector, and logging of the predicted state to SD card. The high-level structure of this pipeline follows the pre-processing, TinyML inference, and decision logic. This workflow provides a reproducible path from a browser-based prototyping environment to fully embedded execution on an STM32 microcontroller, while keeping each individual step accessible to engineers with limited machine learning background.

4.3 Hardware and Software Configuration

The proof of concept supervisor is implemented on the STM32H747I-DISCO discovery board, which integrates a dual-core STM32H747XI microcontroller (Cortex-M7 and Cortex-M4), on-board display and peripheral interfaces [14]. A B-CAMS-OMV camera module is attached via the dedicated connector and positioned above the worktable in a simple 3D-printed enclosure [15]. The camera is oriented to cover the assembly, with a fixed field of view and working distance chosen to minimise occlusions by the operator's hands.

The STM32 board and camera are mounted outside the operator's primary working envelope in order not to interfere with normal handling of parts. Figure 4 illustrates the experimental hardware setup, including the discovery board, camera module and external power supply.

Fig. 4. Experimental setup without case.

The STM32H747 core operates the main application firmware, while the second core remains unused in the current implementation but offers headroom for future extensions, such as parallel logging or communication tasks. The software stack is based on STM32Cube HAL drivers, the X-CUBE-AI middleware for neural network inference [13], TensorFlow Lite for Microcontrollers conversion tools [12] and the Teachable Machine training front-end [11, 16].

4.4 On-Device Implementation: Pre-Processing, TinyML Inference and Logging

Each frame captured by the B-CAMS camera is first resized and converted to the grayscale resolution expected by the model (96 × 96 pixels in the current configuration). Simple normalisation and, if necessary, contrast adjustment are applied in fixed-point arithmetic to meet the requirements of the int8-quantised network [6–9, 12]. Pre-processing is intentionally kept lightweight in order to guarantee real-time performance on the core without reliance on external accelerators.

The pre-processed image is passed to the auto-generated inference function produced by STM32Cube.AI [13]. This function implements the forward pass of the convolutional neural network in optimised C code and returns a vector of class scores. The inference routine is invoked periodically or upon an event trigger (e.g. change detection in the image stream), depending on the application requirements.

The class scores are converted into a discrete assembly state by selecting the maximum (top-1) score. To suppress uncertain predictions, a confidence threshold can be applied; frames below the threshold are treated as Background_or_other (C6). If the confidence exceeds the threshold, the corresponding state and a timestamp are written to an SD-card log file together with an identifier of the current workpiece. This produces a time-ordered sequence of states that reflects the evolution of the assembly process from housing placement through cover mounting and tightening actions.

Optionally, the predicted state and confidence are rendered on the on-board display of the STM32H747I-DISCO together with a small preview of the camera image. Figure 5 illustrates an example of such on-device feedback, showing the recognised assembly state and its confidence value.

Fig. 5. On-device preview with recognised state and confidence.

4.5 Offline and On-Device Classification Performance

Teachable Machine provides validation feedback during training, including class-wise separability indicators and a confusion-matrix view. In this work, the evaluation is reported primarily through qualitative observations from online operation and through deployment-oriented performance indicators (on-device latency and memory footprint), as the contribution focuses on demonstrating an end-to-end, microcontroller-deployable workflow for assembly-state supervision rather than a comprehensive statistical benchmarking study. During deployment, misclassifications were observed mainly under partial occlusions and atypical hand positions, most notably between Background_or_other (C6) and operational states, and occasionally between borderline cases of Housing_correct (C1) and Housing_incorrect (C2).

After conversion to TensorFlow Lite for Microcontrollers [12] and deployment via STM32Cube.AI [13], the predictions obtained on the STM32H747I-DISCO matched the desktop inference results on the validation set, indicating that quantisation and C code generation did not introduce relevant deviations for the considered model size and input resolution. This observation is in line with results reported in related TinyML studies [6–9]. The end-to-end processing chain (pre-processing, inference and decision logic) operated comfortably within the manual assembly cycle time, so the supervisor acted as a frame-by-frame monitoring aid rather than a process bottleneck. On the STM32H747I-DISCO (Cortex-M7 @ 480 MHz), the single-frame end-to-end latency ranged from 9 to 12 ms ($n = 30$, preprocessing included), which supports frame-by-frame supervision without becoming a cycle-time bottleneck [8, 9].

4.6 Qualitative Observations During Online Operation

During online tests with an operator performing the simplified assembly sequence, the supervisor produced a coherent time series of class labels corresponding to the expected progression of states: housing placement (C1/C2), cover mounting/presence (C3/C4), and tightening context (C5), with non-relevant frames grouped as C6. Occasional low-confidence predictions were observed when the operator's hand fully occluded the assembly area or when components were held above the work surface. In these situations, the output typically shifted to Background_or_other (C6) or remained low-confidence rather than producing clearly incorrect operational labels, which is preferable from a usability and safety perspective.

The on-board display preview (Fig. 5) was useful for verifying the field of view, lighting and focus during setup and for communicating the system behaviour to non-expert stakeholders [1–4, 10].

5 Discussion

The presented concept shows how a vision-based supervisory layer can be added to a conventional manual assembly workstation without invasive changes to tools or workflow. A TinyML model executed directly on a resource-constrained STM32 microcontroller follows current trends in edge and cyber-physical manufacturing systems [3, 6, 10]. The STM32H747I-DISCO platform provides sufficient computational resources for the considered workload while remaining compact and cost effective for industrial pilot installations [14, 15].

The proof-of-concept is intentionally restricted to a small dataset and a narrow set of visual conditions. Variations in lighting, camera pose, background or component variants would require additional data collection and fine tuning. The model currently reasons on individual frames, which increases sensitivity to temporary occlusions and unclear in-between poses. Scaling the approach to more complex assemblies with branches or optional steps would likely require sequence based models or a closer coupling between visual inference and process logic.

At the present stage, logging is performed locally on an SD card, which is sufficient for demonstrating feasibility but does not exploit the full potential of online analytics, traceability and integration with planning or quality systems. In a future implementation, the STM32-based supervisor could act as an edge node streaming condensed event data to a central platform [1–4, 10]. From a security and robustness perspective, on-device inference reduces the need to transmit raw video streams, which is advantageous for bandwidth and privacy. Nevertheless, authenticated firmware updates, secure communication channels and appropriate physical protection of the device remain essential prerequisites for industrial deployment.

5.1 Conclusions and Future Work

This paper presented a vision-based supervisory concept for manual assembly operations deployed on an STM32H747I-DISCO board using a TinyML workflow. A lightweight image classifier trained in Teachable Machine and converted via STM32Cube.AI enables on-device recognition of discrete assembly states from a single overhead camera, together with optional time-stamped event logging. The concept was demonstrated on a simplified actuator assembly task with a small representative dataset and a non-intrusive monitoring setup.

The results show that even modest embedded hardware can provide useful contextual information about assembly progression and record it in a form suitable for later analysis or integration into digital manufacturing frameworks [1–4, 6, 10–13]. The architecture is intentionally generic and can be transferred to other discrete assembly scenarios by retraining the model on application-specific image data and adjusting the decision logic.

Future work will address (i) dataset expansion under real industrial variability to support robustness assessment, (ii) integration with shop-floor digital infrastructure for live data exchange with planning and quality systems, and (iii) extensions toward temporal modelling and multi-view sensing, potentially combined with additional modalities (e.g., torque or position). Overall, the STM32-based vision supervisor is positioned as a low-cost edge component supporting process documentation and continuous improvement in industrial practice.

Acknowledgements. This research was funded by project KEGA 009TUKE-4/2026, VEGA 1/0210/25 and APVV-24-0470 granted by the Ministry of Education, Research, Development and Youth of the Slovak Republic, and by the EU NextGenerationEU through the Recovery and Resilience Plan for Slovakia under the project No. 09I03-03-V05-00015, No. 09I05-03-V02-00042 and project DRP0200194, Moving PLastics and mAchine iNdustry towards Circularity (PLAN-C) under the Interreg Danube Region Program, Co-funded by the European Union.

References

1. Kagermann, H., Wahlster, W., Helbig, J.: Recommendations for implementing the strategic initiative INDUSTRIE 4.0. Final Report of the Industrie 4.0 Working Group. acatech-National Academy of Science and Engineering, Munich (2013)
2. Lasi, H., Fettke, P., Kemper, H.-G., Feld, T., Hoffmann, M.: Industry 4.0. Bus. Inf. Syst. Eng. **6**(4), 239–242 (2014)
3. Lee, J., Bagheri, B., Kao, H.-A.: A cyber-physical systems architecture for Industry 4.0-based manufacturing systems. Manuf. Lett. **3**, 18–23 (2015)
4. Bortolini, M., Faccio, M., Gamberi, M., Pilati, F., Ferrari E.: Assembly system design in the Industry 4.0 era: a general framework. IFAC-PapersOnLine **49**(12), 570–575 (2016)
5. Goodfellow, I., Bengio, Y., Courville, A.: Deep Learning. MIT Press, Cambridge (2016)
6. Warden, P., Situnayake, D.: TinyML: machine learning with tensorflow lite on arduino and ultra-low-power microcontrollers. O'Reilly Media, Sebastopol (2019)
7. David, R., Duke, J., Jain, A., Reddi, V.J., Jeffries, N., Li, J., et al.: TensorFlow Lite Micro: embedded machine learning for TinyML systems. In: Proceedings of the 4th Conference on Machine Learning and Systems (MLSys 2021), pp. 1–15 (2021)
8. Banbury, C., Reddi, V.J., Torelli, P., Holleman, J., Jeffries, N., Kiraly, C., et al.: MLPerf Tiny Benchmark: 2021 (2021)
9. Sze, V., Chen, Y.-H., Yang, T.-J., Emer, J.S.: Efficient processing of deep neural networks: a tutorial and survey. Proc. IEEE **105**(12), 2295–2329 (2017)

10. Shi, W., Cao, J., Zhang, Q., Li, Y., Xu, L.: Edge computing: vision and challenges. IEEE Internet Things J. **3**(5), 637–646 (2016)
11. Google: Teachable Machine-Train a computer to recognize your own images, sounds, and poses. Online tool, Google. Available at: https://teachablemachine.withgoogle.com/ (Accessed 19 Nov 2025)
12. TensorFlow: TensorFlow Lite for Microcontrollers. TensorFlow Documentation. Available at: https://cms.tinyml.org/wp-content/uploads/2020/09/bay-area-situnayake.pdf?ref=blog.roboflow.com. (Accessed 19 Nov 2025)
13. STMicroelectronics: STM32Cube.AI-STM32Cube expansion package for Artificial Intelligence (X-CUBE-AI). Product/solution page, STMicroelectronics. Available at: https://www.st.com/en/embedded-software/x-cube-ai.html. (Accessed 09 Nov 2025)
14. STMicroelectronics: STM32H747I-DISCO-Discovery kit with STM32H747XI MCU. Product page, STMicroelectronics. Available at: https://www.st.com/en/evaluation-tools/stm32h747i-disco.html. (Accessed 07 Nov 2025)
15. STMicroelectronics: B-CAMS-OMV-Camera module for STM32 boards. Product page, STMicroelectronics. Available at: https://www.st.com/en/imaging-and-photonics-solutions.html#tools-software. (Accessed 07 Nov 2025)
16. STMicroelectronics: AI: How to use Teachable Machine to create an image classification application on STM32. STM32 MCU Wiki, STMicroelectronics. Available at: https://wiki.st.com/stm32mcu/wiki/AI:How_to_use_Teachable_Machine_to_create_an_image_classification_application_on_STM32. (Accessed 07 Nov 2025)

Data-Supported Visual Supervision for Reducing Production Losses in a Semi-Automated Assembly Process: An Automotive Case Study

Julia Olesińska and Maria Rosienkiewicz(✉)

Department of Production Engineering and Management, Wroclaw University of Science and Technology, Wroclaw, Poland
maria.rosienkiewicz@pwr.edu.pl

Abstract. This study examines production losses caused by dropped components in a semi-automated automotive assembly process and evaluates a simple, data-supported supervision approach to reduce these events. One year of loss records was combined with structured Root Cause Analysis and on-site observations to understand when and why drops occur, linking quantitative patterns with specific ergonomic conditions and operator routines. Based on these findings, several targeted improvements were introduced, including adjustments to part feeders, clearer handling standards, small ergonomic changes and a lightweight dashboard showing drop-related indicators. After implementation, the assembly lines recorded a clear reduction in both the number and financial impact of drop events. The results show that these losses follow repeatable behavioural and task-related mechanisms and can be reduced through focused, low-cost measures. The study also shows that drop losses represent a form of operator-dependent waste that is not captured in aggregated performance metrics. By combining simple supervisory tools with established analytical methods, the proposed approach offers a practical way to improve stability in manual and hybrid assembly environments without requiring extensive digital infrastructure.

Keywords: Manual assembly · human error · micro-losses · data-driven supervision · ergonomic improvement · root cause analysis · automotive manufacturing

1 Introduction

Automotive manufacturing continues to face rising demands for product variety, shorter delivery times and high quality, which increase the complexity of planning and control in high-mix production environments [1]. Despite rapid advances in automation, many tasks still rely on manual assembly because they require human dexterity, judgement and fine-grained handling of parts [2]. As a result, manual workstations remain central to value creation in processes involving numerous product variants and tight quality

J. Trojanowska et al. (Eds.): MANUFACTURING 2026, LNME, pp. 267–279, 2026.
https://doi.org/10.1007/978-3-032-21654-0_20

requirements. However, this reliance on human input introduces variability, making the process vulnerable to fluctuations in operator performance. Research shows that factors such as fatigue, time pressure and limited workspace influence the reliability of operator actions and increase the likelihood of errors [3]. Many of these errors are minor and transient; they do not cause immediate disruptions but accumulate over time, undermining overall process stability. Traditional reporting systems often group all defects into broad categories. As a result, behaviour-related micro-losses, such as dropped components or handling slips, frequently go unnoticed in routine performance analysis and receive little managerial attention. These challenges increase the need for more systematic and continuous forms of supervision.

Intelligent supervision refers to a data-driven approach to monitoring manual assembly processes in real time. It uses operational data, digital visualisation, and basic analytical reasoning to detect disruptions early. Modern manufacturing environments generate vast amounts of information through MES, ERP, and quality monitoring systems. By analysing these datasets, companies can identify patterns of human error and understand how workstation conditions influence operator behaviour. Prior studies indicate that such insights can help predict operator-induced disturbances and contextualise their occurrence [4]. These capabilities enhance rather than replace traditional methods such as ergonomics, structured error-prevention approaches and classical RCA. Whereas conventional tools provide root cause insights through periodic review, intelligent supervision enables continuous monitoring of process behaviour, highlighting small irregularities that may not appear in aggregated indicators but still degrade performance over time. In this paper, intelligent supervision is defined as the use of simple data collection tools, visual feedback mechanisms, and lightweight analytic techniques such as trend detection and deviation flagging to support control of manual assembly.

Current research provides valuable insights into human error mechanisms and ergonomic constraints in manual assembly [2, 3]. However, most studies focus on high-impact failures or major defects. Little attention is paid to frequent, small-scale disturbances such as dropped components, which remain under-analysed and are not treated as a distinct waste category. Moreover, although digitalisation in high-mix manufacturing is well documented, existing studies rarely address how low-cost digital supervision can be used to detect and prevent such micro-errors in real time. As a result, existing approaches remain fragmented. They typically address either human error, or ergonomics, or monitoring in isolation but do not integrate all three into a unified framework.

This study addresses these gaps by proposing a simple, data-driven supervision model tailored for manual assembly in automotive production. The aim is to combine human-error analysis, ergonomic principles, and real-time process monitoring to identify and reduce micro-losses. The research investigates whether lightweight digital supervision-based on basic tools and operator feedback, can stabilise manual tasks.

2 Literature Review

Manual assembly remains widely applied in manufacturing sectors characterised by high product complexity and variant diversity, such as automotive production. In these environments, many tasks require operators to interpret information, perform spatial judgements, and apply controlled physical actions, making full automation a less flexible or viable alternative [2]. However, manual work is strongly affected by human performance. Research shows that operator performance varies during the workday and is influenced by factors such as fatigue and workload, which impact both productivity and process stability [5]. Additionally, the increasing variety of products places growing demands on operator resilience and capability, intensifying the risk of human error in manual assembly [4]. Operational conditions further shape error occurrence. Experimental results show that time pressure significantly increases human error rates, as workers accelerate task execution and experience higher mental strain. Likewise, workstation configuration influences performance: flat component bins result in more mistakes compared to tilted bins, which are designed to simplify component retrieval [6]. Human Factors and Ergonomics (HF/E) literature also highlights that manufacturing systems with insufficiently developed HF/E capabilities can impose additional cognitive demands on operators, making task execution more challenging in increasingly complex production environments. [7]. Ergonomic conditions are another major determinant of assembly quality. Poor physical ergonomics lead to higher physical strain and contribute to execution errors such as misalignments or damage to components [2]. Evidence from automotive assembly confirms that high physical workload is associated with substantially higher error rates, reinforcing the need for ergonomic workstation design [8]. Documented manual assembly errors include part mis selection, orientation mistakes, and force-related damage [2]. Overall, the literature shows that manual assembly remains essential but is vulnerable to human error driven by performance variability, time pressure and ergonomic conditions.

Lean manufacturing defines waste as any activity that does not add value. Recent research shows that waste may also result from organizational and human factors, which influence the overall performance of production systems [9]Well-being waste and stress-related inefficiencies arise when operators face physical or cognitive exhaustion, unclear responsibilities or decision-making pressure [9]. Even with these broader views, defect data in Lean studies are often reported only at a general level. Many industrial cases summarize defect outcomes without distinguishing between behavioural, material or process-based mechanisms [10]. A similar limitation appears when scrap is monitored as a single Lean Six Sigma indicator, even though the root-cause analysis may decompose influences into operator-, material- and method-related factors [11]. With such limited granularity, some inefficiencies remain unrecognized in conventional reporting. Stress-related inconsistencies, unclear KPIs and rigid procedures can generate non-core manufacturing and well-being waste that is difficult to measure within classical Lean categories [9]. Smart tools and digital dashboards can enhance the visibility of such disruptions by integrating real-time data and supporting lean green smart decision-making [12].

Root Cause Analysis (RCA) is a central method used in manufacturing to identify why defects and process failures occur. It provides a structured way to move beyond visible symptoms and understand the deeper factors that influence quality and performance. Two traditional RCA tools remain especially common: the Ishikawa (fishbone) diagram and the 5 Whys technique. This visual structure helps teams systematically examine different aspects of production and identify where problems may originate. It supports both early exploratory analysis and more detailed investigations, especially when multiple factors must be considered at once [13, 14]. The 5 Whys technique complements the fishbone diagram by guiding users through a sequence of "why" questions to trace a defect back to its root cause. Despite their extensive use, conventional RCA tools also has important limitations. Manual RCA heavily depends on expert knowledge and judgment, which can lead to inconsistent outcomes. This subjectivity is often highlighted as a major obstacle to effective and reliable decision-making in industrial settings. Moreover, traditional RCA tools are not well-suited to complex production systems with high product variety, large data volumes, and dynamic process interactions. In practice, companies face challenges such as time-consuming manual investigations, a lack of unified data structures, and difficulties in quantifying the influence of individual causes [15]. To address these constraints, Lean Six Sigma (LSS) introduces structured improvement cycles such as DMAIC, which integrate RCA with tools like Value Stream Mapping, Kaizen and Pareto analysis. These approaches provide staged procedures for mapping current conditions, identifying bottlenecks and implementing corrective actions. A case study on bogie assembly demonstrated that combining DMAIC with tools such as Kaizen, SMED, 5S and VSM leads to shorter cycle times, reduced non-value-added activities and fewer defects, demonstrating that structured improvement can significantly strengthen RCA outcomes [16]. A second direction in recent research is the shift toward data-driven and automated RCA. One proposed system uses large datasets from ERP systems, audit records and production logs to create a detailed feature space for analysis. Techniques such as principal component analysis and machine learning classifiers are then applied to identify likely root causes more consistently and with less reliance on expert judgment [15]. This modular approach offers better scalability and supports more precise decision-making. Interpretability is also a key concern, especially in industries where operators must understand and validate diagnostic results. For this reason, new RCA approaches based on Association Rule Mining (ARM) have been developed. ARM provides transparent, rule-based explanations while adjusting to changes in production conditions. Tests in automotive final assembly show that ARM can detect evolving causal patterns and remain understandable to domain experts [17]. Further extensions include RCA approaches integrated into resilient production systems, where automated monitoring supports detection of disruptions and deviations, accelerating the RCA cycle [18]. Network-based RCA extends this further by modelling non-linear causal relationships in high-precision manufacturing, helping organizations understand how technical and procedural factors jointly affect Overall Equipment Effectiveness (OEE) [19]. Real time anomaly detection methods also support RCA by identifying deviations early, allowing faster investigation and correction [20]. In summary, traditional RCA is still valuable but no longer adequate on its own. Manufacturing increasingly uses hybrid approaches that mix classical tools with Lean Six Sigma, data analytics and real-time monitoring to improve diagnostic accuracy and system resilience.

Manufacturing companies are moving from traditional supervision methods to more intelligent, data-based systems. Paper-based boards have been replaced by digital dashboards connected to real-time systems, enabling the monitoring of Overall Equipment Effectiveness (OEE), production status, and resource availability via MES and IoT integration [21]. Visual management patterns integrate digital and physical interfaces, such as e-Andon and smart Kanban systems, enabling dynamic and distributed supervision [22]. Dashboards show key indicators such as quality, downtime, and availability, and help detect problems earlier [23]. Distinguish between users (operators, supervisors, and managers) to see the data they need, helping align monitoring with company goals [23]. Lightweight systems, such as open-source dashboards based on AppSheet and Google Sheets, offer low-cost supervision for small or resource-constrained manufacturers [21]. These solutions are simple but useful, especially when resources are limited. The integration of intelligent tools with visual supervision enhances the speed and clarity with which production teams can interpret operational data and respond to disruptions. This shift reflects a transition from reactive monitoring to proactive process control, driven by improved data accessibility and user-oriented visualization.

systems [22].

The reviewed literature confirms that manual assembly remains essential in complex production environments, yet it is highly vulnerable to operator variation and ergonomic stressors. Factors such as fatigue, workload, and poor workstation layout contribute directly to quality issues, making manual tasks particularly prone to instability and human error. Fatigue, workload and suboptimal workstation layout contribute directly to quality issues, while Lean studies typically report defects only in aggregated forms such as scrap. As a result, small operator-induced losses such as dropped components or handling mistakes remain largely invisible in conventional reporting. Although RCA tools like the Ishikawa diagram and 5 Whys are widely used, the reviewed studies do not document their application to frequent, small-scale manual errors. RCA is generally applied to high-impact system failures and often lacks the granularity needed for diagnosing micro-level disruptions. At the same time, Industry 4.0 research focuses on advanced supervisory infrastructures such as real-time dashboards and cyber-physical systems, with no discussion of lightweight, low-cost solutions suitable for small or resource-limited assembly settings. Overall, the literature reveals three gaps: a lack of data-driven analyses addressing drop-related losses in manual assembly; an absence of integrated frameworks combining ergonomics, RCA and visual supervision for such micro-losses; no reported case studies demonstrating measurable improvements. This study addresses these gaps by proposing a low-cost, data-supported supervision model for reducing drop losses in manual automotive assembly.

Research on losses in manual assembly consistently indicates that quality-related outcomes are shaped by measurable interactions between process complexity, workstation design, and operator-related factors. These interactions demonstrate that operator-dependent losses are systematic in nature and can be analytically characterized rather than treated as random variability [24]. Existing modeling approaches show that task structure, spatial constraints, and human factors significantly influence defect propensity in automotive assembly environments. However, this body of work primarily conceptualizes losses in terms of discrete quality defects or rework events, offering limited insight into small, repetitive disturbances occurring during normal production flow. Complementary findings emerge from ergonomics-oriented studies, which show that quantified ergonomic risk indicators are closely associated with production performance metrics such as process instability, execution errors, and cycle time variability [25]. These results confirm that unfavorable ergonomic conditions contribute to degraded operational performance. Nevertheless, current studies do not explicitly address how ergonomics-driven micro-errors, such as handling inaccuracies, translate into cumulative material and cost losses. As a result, a research gap remains in linking ergonomic handling risks with operator-dependent micro-loss mechanisms in manual and semi-automated assembly processes.

3 Research Methodology

This study was designed as a data-driven case study carried out in a large automotive manufacturing plant located in Poland. The research focused on one assembly process in which most operations are automated, but selected handling steps are still performed manually. These manual steps were identified as a potential source of production losses, especially in relation to dropped components.

The methodological approach consisted of three stages. First, twelve months of loss data were reviewed to define the scale and distribution of component drops. Next, a Discovery Workshop was conducted in which four process engineers observed the relevant workstations across several shifts to identify typical handling patterns and situations where drops occurred. Finally, root-cause analysis tools were applied to link the quantitative findings with direct shop-floor observations. The study follows a before–after case study design, in which production loss data collected from January to December 2024 serve as the baseline period, the improvement actions were implemented in December 2024, and the period from January to April 2025 represents the post-intervention stabilization and evaluation phase.

3.1 Data Collection

The dataset used in this study consisted of internal scrap and loss records covering the period January–December 2024. These records are generated as part of the standard production reporting process and document each registered loss event. For every entry, the system logs the loss category, the component involved, the workstation where the event occurred, and the associated cost.

All recorded losses were initially reviewed to determine the overall structure of waste in the assembly process. Events classified as dropped components were selected for further analysis, as they represented a substantial share of the total loss value and were directly linked to manual steps. To support further diagnostic work, the data were grouped by assembly lines to identify which sections of the process were most affected.

3.2 Root Cause Analysis

A structured Root Cause Analysis (RCA) was conducted to identify the factors contributing to dropped components in the assembly process. The analysis combined two complementary tools: an Ishikawa (5M) diagram and the 5 Why technique, with direct observations gathered during the Discovery Workshop.

An Ishikawa diagram shown in Fig. 1 was used to organize potential causes into the five classical categories: Man, Machine, Material, Method, and Measurement.

The analysis showed that manual errors during refilling trays, difficulties caused by high feeder positioning, the tendency of elastic rubber parts to slip, and the lack of a clear refilling standard were recurring issues. Observations also indicated that operators frequently worked under time pressure, which increased the likelihood of mistakes, particularly during small-part handling.

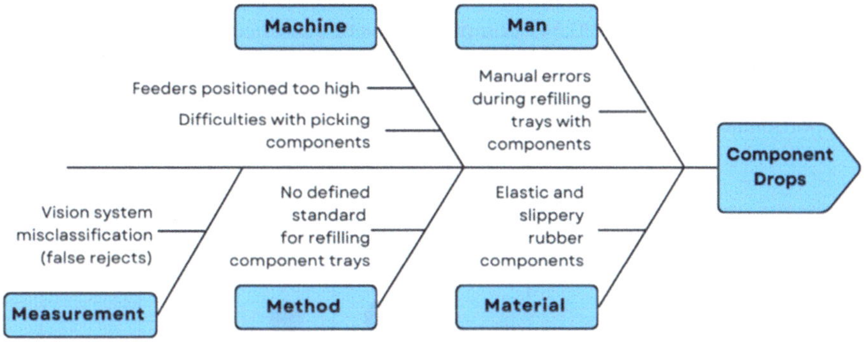

Fig. 1. Ishikawa diagram for component drops.

The 5 Why method was applied to the branches of the diagram that appeared most significant. This analysis helped distinguish immediate handling issues from the underlying factors that led to them. For example, what initially seemed to be a simple slip during gripping was often linked to deeper process characteristics, such as the way components were presented in the feeder or the limited working space at certain stations.

3.3 Implementation of Corrective Actions and Supervisory Controls

Based on the root causes identified in Sect. 3.2, a series of corrective actions was implemented to address the main factors contributing to component drops. The height and orientation of feeders were adjusted to improve the stability of part presentation, and small guiding rails together with protective meshes were added in areas where components frequently slipped, bounced or fell through gaps. To limit unintentional slipping of small elastic parts such as grommets, caps and O-rings, operators were equipped with PU-coated gloves, which provided better grip and improved control during grasping. Ergonomic refinements focused on simplifying the layout of the operator work zones and reducing unnecessary movements, particularly at stations with the highest number of drop events. These adjustments aimed to stabilize manual handling and reduce the physical effort required during tray refilling or part transfers. In parallel, standardized procedures for refilling trays and moving components between stations were introduced to reduce handling variability and ensure consistent practices across shifts.

To support these improvements, a simple supervisory control layer was established. Two indicators: drop count and estimated cost per drop were monitored on a routine basis and displayed in a visual dashboard accessible to operators on the shop floor. The combination of data monitoring, visual feedback and standardized handling created a supervision framework that supported correct practices and helped identify early signs of process instability.

4 Results and Discussion

Loss records for the baseline period (January–December 2024) show that dropped components accounted for a significant share of the total waste generated during the analysed assembly process. In total 290,547 components were classified as dropped during this period, which resulted in annual losses reaching several hundred thousand PLN. The distribution of events was uneven across assembly lines, with several stations showing disproportionately high counts.

Station / Components	Station 1	Station 2	Station 3	Station 4	Station 5	Station 6	Station 7
Reservoirs				●	●		
Pump housing			●		●		
Pistons		●					
Spring assemblies		●					
Seals	●						
TMC+RSV assemblies					●	●	●
Caps							●
O-rings						●	
Pin/ Screws							●

Fig. 2. Distribution of Component Drop Events by Station

These stations shared two characteristics: multiple manual handling steps and the presence of high-risk components defined by low friction or deformability. To visualize the distribution of losses across workstations and component types, a loss map was constructed (Fig. 2). The map illustrates where drop events occurred most frequently and which components generated the highest cost impact. A deeper inspection revealed that a small group of parts accounted for the majority of drops. The most affected items were springs (6,793 drops/month on average), grommets (4,058/month), and sealing caps (1,205/month), followed by pins/screws (1,625/month), O-rings, pistons, reservoirs, and assembled pump-and-reservoir units. Although the mechanisms differed across materials, all high-risk components shared a dependency on stable operator control and predictable presentation conditions, which is consistent with the literature on handling errors involving small, low-friction parts [26].

The post-intervention period (January–April 2025) demonstrated a reduction of more than 35% in total drop-related losses, with the largest improvements observed for grommets, caps and O-rings. Medium-level reductions were observed for springs, pins/screws and pump housings, while pistons showed only minor improvements. Components with high unit cost showed the most pronounced financial impact, indicating that the interventions targeted the most consequential failure modes.

Although the implemented improvement actions were introduced as a combined intervention package, qualitative observations during the post-intervention period allow for a partial attribution of effects to specific countermeasures. Adjustments to part feeders had the most direct impact on drop events involving small and flexible components, such as grommets, caps, and O-rings, by improving part presentation and reducing unstable handling situations. The introduction of PU-coated gloves contributed to improved grip stability during manual handling, particularly for lightweight components with smooth surfaces, reducing unintended slips during transfer operations. Standardization of handling sequences and workstation layout reduced variability between operators and shifts, leading to more consistent execution of manual tasks. Finally, the visual supervision dashboard did not directly prevent drop events but acted as a reinforcing mechanism by increasing operator awareness and sustaining attention to handling-related losses. These qualitative observations support the interpretation of the observed reduction in drop-related losses as the result of complementary and mutually reinforcing interventions rather than a single dominant factor.

The reduction in drop-related losses after the intervention indicates that these events are systematic rather than random, shaped by interactions between operator behaviour, workstation ergonomics and component properties. Workstation visits revealed repeated slips during tray refilling and handling instability related to feeder height. These patterns point to micro-errors driven by local misalignment between task conditions and operator capabilities strain [8]. This interpretation is consistent with earlier research highlighting the influence of workload, task complexity and physical constraints on manual assembly quality [7]. These findings demonstrate that low-intensity ergonomic refinements can stabilize micro-behaviours linked to drops, complementing evidence that well-designed workstations reduce execution errors and cognitive strain [8].

The present findings reinforce earlier research on the role of human factors in manual assembly, particularly regarding slips, misalignments and errors induced by time pressure or poor part presentation [6]. However, the concentration of losses in a narrow set of components diverges from many Lean studies, where defect indicators are typically aggregated and do not isolate micro-scale handling errors. This suggests that drop losses represent a distinct class of micro-waste not captured in traditional reporting frameworks. Regarding diagnostics, the study applies classical RCA tools to small, frequent events. This contrasts with typical RCA applications documented in the literature, which focus on major failures or systemic deviations [14]. By integrating RCA with direct shop-floor observation, the analysis delivers causal insights on par with more advanced data-driven methods, including PCA-based or classifier-supported systems [15].

The results extend the theoretical understanding of micro-loss mechanisms in semi-automated assembly. They reveal that drop losses form a specific, operator-dependent waste category that is largely invisible in aggregated Lean metrics. This supports earlier research showing that traditional reporting systems overlook subtle, behaviour-related inefficiencies [27]. Furthermore, the effectiveness of the lightweight dashboard shows that simple digital supervision can reinforce behavioural stability without requiring full Industry 4.0 infrastructure. This complements earlier work on visual management and digital dashboards [28].

The findings indicate that targeted ergonomic modifications, clear handling standards and minimal visual supervision are sufficient to achieve meaningful reductions in micro-losses. This is especially relevant for manufacturing environments where automation is partial and operators handle small deformable components. The results also suggest that high-risk components should be prioritized in continuous improvement cycles due to their disproportionate impact on both frequency and cost.

5 Conclusions

This study examined production losses caused by dropped components in automotive assembly process and demonstrated that a low-cost, data-supported supervision model can effectively reduce such events. The combination of loss analytics, root-cause analysis and workstation observations made it possible to identify micro-level handling issues. These issues were linked to clear ergonomic and behavioural factors. The findings confirm that drop losses are not random but follow repeatable operator–task patterns that can be addressed through focused interventions.

The results show that lightweight supervisory practices, supported by a small set of indicators and simple ergonomic adjustments, can stabilize manual handling. This applies in particular to environments where operators work with small or deformable components. The findings also indicate that micro-losses form a distinct, operator-dependent waste category that often remains hidden in aggregated Lean metrics. In addition, the study demonstrates that classical RCA, when combined with systematic shop-floor observation, can deliver clear and actionable causal insights without relying on advanced analytical systems.

From a theoretical perspective, the study advances current understanding in several ways. First, it shows that drop losses represent a specific form of micro-waste related to human factors. This type of waste is not visible in standard Lean reporting, which often aggregates defect data to a level that masks behaviour-driven inefficiencies. Second, the findings demonstrate that classical RCA tools, when combined with structured shop-floor observation, can produce causal insights comparable to those achieved with more advanced analytical systems. This is especially relevant in settings where the frequency of disturbances is high but their individual impact is small. Third, the study confirms that lightweight visual supervision, based on a small number of relevant indicators, can reinforce behavioural consistency without requiring large digital infrastructures. The results also carry practical implications. Manufacturing organizations operating under partial automation can achieve meaningful reductions in micro-losses without major investments. Improvements are possible when companies target specific combinations of ergonomic constraints, operator routines and component characteristics. The proposed framework may be valuable in settings where operators must handle small or deformable parts and where conventional reporting systems underestimate the cumulative impact of micro-disturbances. The findings further suggest that focusing on high-risk components can increase the efficiency of improvement cycles, especially when resources are limited.

Several limitations should be noted. The study was conducted in one automotive plant and focused on a single product family across several semi-automated brake master cylinder assembly lines. This narrow scope may limit the transferability of the findings to plants with different process layouts, material types or operator tasks. The interventions were also restricted to low-cost ergonomic changes and simple supervisory controls. Broader organizational influences, such as training practices or shift structures were not examined. Future research should analyse similar micro-loss mechanisms in other manual or hybrid assembly processes and across different product families. It would also be valuable to study how lightweight supervision can be combined with more advanced diagnostic tools. Future studies could analyse the long-term persistence of these behavioural improvements and the ability of feedback-based monitoring to sustain process stability.

References

1. Gan, Z.L., Musa, S.N., Yap, H.J.: A review of the high-mix. Low-Volume Manuf. Ind. (2023). https://doi.org/10.3390/app13031687
2. Torres, Y., Nadeau, S., Landau, K.: Classification and quantification of human error in manufacturing: a case study in complex manual assembly. Appl. Sci. (Switzerland) **11**, 1–23 (2021). https://doi.org/10.3390/app11020749
3. Klages, B., Jordan, P., Kroeger, S., Zaeh, M.: Towards a methodology for the selection of preventive measures to mitigate error influences in manual assembly lines. In: Procedia CIRP. pp. 1096–1103. Elsevier B.V. (2024). https://doi.org/10.1016/j.procir.2024.10.212
4. Klages, B., Zaeh, M.: Concept of a data-based approach for the prediction and reduction of human errors in manual assembly. In: Procedia CIRP. pp. 209–214. Elsevier B.V. (2023). https://doi.org/10.1016/j.procir.2023.02.036
5. Vilela, F.F., Leal, F., Montevechi, J.A.B., Piedade, D.D.C.: Effect of human factor performance on the productivity of a manual assembly line. Int. J. Simul. Model. **19**, 365–374 (2020). https://doi.org/10.2507/IJSIMM19-3-508
6. Saptari, A., Leau, J.X., Mohamad, A.: The effect of time pressure, working position, component bin position and gender on human error in manual assembly line
7. Reiman, A., Kaivo-oja, J., Parviainen, E., Takala, E.P., Lauraeus, T.: Human factors and ergonomics in manufacturing in the industry 4.0 context–A scoping review. Technol. Soc. **65** (2021). https://doi.org/10.1016/j.techsoc.2021.101572
8. Fritzsche, L., Wegge, J., Schmauder, M., Kliegel, M., Schmidt, K.H.: Good ergonomics and team diversity reduce absenteeism and errors in car manufacturing. Ergonomics **57**, 148–161 (2014). https://doi.org/10.1080/00140139.2013.875597

9. Purushothaman, M.babu, Seadon, J., Moore, D.: Waste reduction using lean tools in a multicultural environment. J. Clean. Prod. **265** (2020). https://doi.org/10.1016/j.jclepro.2020.121681
10. Mukti, A.S.: Defect reduction in the manufacturing industry: systematic literature review. Int. J. Ind. Eng. Eng. Manage. **5**, 73–83 (2023). https://doi.org/10.24002/ijieem.v5i2.7495
11. Enache, I.C., Chivu, O.R., Rugescu, A.M., Ionita, E., Radu, I.V.: Reducing the scrap rate on a production process using lean six sigma methodology. Processes **11** (2023). https://doi.org/10.3390/pr11041295
12. Fiorello, M., et al.: Towards a smart lean green production paradigm to improve operational performance. J. Clean. Prod. **413** (2023). https://doi.org/10.1016/j.jclepro.2023.137418
13. Ciecińska, B.: Identification of Defects Causes: Ishikawa Diagram and 5 Whys in Theoretical and Practical Terms
14. Cox, M., Sandberg, K.: Modeling causal relationships in quality improvement. Curr. Probl. Pediatr. Adolesc. Health Care **48**, 182–185 (2018). https://doi.org/10.1016/j.cppeds.2018.08.011
15. Ma, Q., Li, H., Thorstenson, A.: A big data-driven root cause analysis system: application of machine learning in quality problem solving. Comput. Ind. Eng. **160** (2021). https://doi.org/10.1016/j.cie.2021.107580
16. Daniyan, I., Adeodu, A., Mpofu, K., Maladzhi, R., Kana-Kana Katumba, M.G.: Application of lean Six Sigma methodology using DMAIC approach for the improvement of bogie assembly process in the railcar industry. Heliyon **8** (2022). https://doi.org/10.1016/j.heliyon.2022.e09043
17. Pietsch, D., Wieland, U., Stoll, M., Ihlenfeldt, S.: Interpretable and adaptive data-driven root cause analysis in manufacturing through association rule mining. In: Procedia CIRP. pp. 31–36. Elsevier B.V. (2025). https://doi.org/10.1016/j.procir.2025.03.006
18. Ito, A., et al.: Improved root cause analysis supporting resilient production systems (2022). https://doi.org/10.1016/j.jmsy.2022.07.015
19. Saretzky, F., Engel, T., Ansari, F.: Network-based root cause identification to improve OEE in high-precision manufacturing. In: IFAC-PapersOnLine. pp. 2415–2420. Elsevier B.V. (2025). https://doi.org/10.1016/j.ifacol.2025.09.406
20. Gram, J., Drees, J., Stambera, S., Jestädt, M., Cagno, M. Di, Bauernhansl, T.: Real-Time anomaly detection in manufacturing: a data-driven approach to enhancing production efficiency through online KPI monitoring. In: Procedia CIRP. pp. 145–150. Elsevier B.V. (2025). https://doi.org/10.1016/j.procir.2025.03.047
21. Steenkamp, L.P., Hagedorn-Hansen, D., Oosthuizen, G.A.: Visual management system to manage manufacturing resources. In: Procedia Manufacturing. pp. 455–462. Elsevier B.V. (2017). https://doi.org/10.1016/j.promfg.2017.02.058
22. Fenza, G., Loia, V., Nota, G.: Patterns for visual management in industry 4.0. Sensors **21** (2021). https://doi.org/10.3390/s21196440
23. Brundage, M.P., Bernstein, W.Z., Morris, K.C., Horst, J.A.: Using graph-based visualizations to explore key performance indicator relationships for manufacturing production systems. In: Procedia CIRP. pp. 451–456. Elsevier B.V. (2017). https://doi.org/10.1016/j.procir.2016.11.176
24. Krugh, M., Antani, K., Mears, L., Schulte, J.: Statistical modeling of defect propensity in manual assembly as applied to automotive electrical connectors. In: Procedia CIRP. pp. 441–446. Elsevier B.V. (2016). https://doi.org/10.1016/j.procir.2016.02.335
25. Krajný, P., Janeková, J., Badida, M.: Monitoring of ergonomics score impact on production processes. Processes **13** (2025). https://doi.org/10.3390/pr13082626
26. Di Pasquale, V., Miranda, S., Neumann, W.P., Setayesh, A.: Human reliability in manual assembly systems: a systematic literature review. In: IFAC-PapersOnLine. pp. 675–680. Elsevier B.V. (2018). https://doi.org/10.1016/j.ifacol.2018.08.396

27. Falck, A.C., Rosenqvist, M.: A model for calculation of the costs of poor assembly ergonomics (part 1). Int. J. Ind. Ergon. **44**, 140–147 (2014). https://doi.org/10.1016/j.ergon.2013.11.013
28. Tokola, H., Gröger, C., Järvenpää, E., Niemi, E.: Designing manufacturing dashboards on the basis of a key performance indicator survey. In: Procedia CIRP. pp. 619–624. Elsevier B.V. (2016). https://doi.org/10.1016/j.procir.2016.11.107

Assumptions of the Production Scheduling Model in the Smart Factory Demonstrator

Piotr Kaźmierczak(✉) and Krzysztof Żywicki

Poznan University of Technology, Poznan, Poland
piotr.kazmierczak@put.poznan.pl

Abstract. The article is focused on the characteristics of the production scheduling and rescheduling model in a laboratory smart factory demonstrator at the Poznan University of Technology. As a key element of the Industry 4.0 concept, smart factories require advanced IT systems capable of interpreting data and making decisions in real time. Traditional production scheduling is insufficient in a dynamically changing market demanding product customization. The paper presents the architecture of the 4Factory IT system, along with the conceptual implementation of supporting technologies such as IoT and RFID, enabling real-time data collection and processing. The process of generating production schedules and the implementation of a rescheduling strategy are described, allowing adaptive modifications to the production plan in response to disruptions such as workstation failures, urgent orders, or changes in priorities. The key conclusion is that effective rescheduling requires access to high-quality, up-to-date data from the production infrastructure. The added value of the article is the development of a coherent description of the production scheduling and rescheduling model, which is implemented in the smart factory demonstrator. The creation of a real research environment allows for conducting research in various simulation conditions, which will form the basis for further research.

Keywords: Smart Factory · Production Scheduling · Industry 4.0

1 Introduction

Production systems are currently undergoing a dynamic transformation related to the Industry 4.0 concept. Digital transformation brings both opportunities and significant complications for traditional approaches to production process management. The smart factory, a key element of the Industry 4.0 concept, focuses on automated and adaptive production management based on the direct flow of information from the physical layer to decision-making systems.

Key technologies used in smart factories, such as RFID systems, the Internet of Things, Artificial Intelligence, and Digital Twin, enable real-time data collection and processing, which opens up new possibilities for production management. However, the integration of these technologies requires not only the implementation of advanced hardware infrastructure, but above all the development of intelligent IT systems. These

J. Trojanowska et al. (Eds.): MANUFACTURING 2026, LNME, pp. 280–289, 2026.
https://doi.org/10.1007/978-3-032-21654-0_21

systems should enable data interpretation and decision-making. For efficient production management, decisions regarding production scheduling are crucial and must evolve towards flexible solutions that allow for ongoing adaptation to unpredictable events and changing conditions.

The subject of this article is the characteristics of the production scheduling and rescheduling model in a laboratory environment of a production system that is a demonstrator of the smart factory concept.

2 The Role of Production Scheduling in the Smart Factory Concept

2.1 Smart Factory as Part of the Industry 4.0 Concept

The smart factory is one of the many key achievements of the fourth industrial revolution, known as Industry 4.0. The term smart factory refers to a highly integrated production system characterized by a high level of integration of digital technologies with physical processes. The foundation of the smart factory is the Industry 4.0 concept, which is characterized by the integration of advanced technologies such as the Internet of Things (IoT), artificial intelligence (AI), and big data analytics. The use of various modern technologies and their mutual integration is aimed at creating smart factories where the flow of information is smooth, resulting in increased efficiency and flexibility of the entire process [1–3].

These advanced technologies are the cornerstone of the smart factory, one of which is the Internet of Things, which is based on real-time operation. Data is collected from IoT sensors, then processed and analyzed by a central system that allows for real-time response and decision-making based on the current status provided by the IoT [4]. Other related technologies used in smart factories include Digital Twin, which can be defined as a virtual representation of an object or system, provided in real time by data coming from its physical counterpart [5]. RFID (Radio Frequency Identification) and RTLS (Real-Time Locating System) technologies also support the functioning of smart factories by monitoring the flow of materials and products [6, 7].

The concept of a modern approach to production management within a smart factory focuses on the information flow zone, which is a complex, multi-layered system integrating technologies with infrastructure. These technologies, in the form of sensors, are necessary for collecting data and using it for the effective operation of manufacturing companies [8]. Intelligent sensors used within the IoT convert information in real time into digital data, which enables real-time prediction and monitoring of processes, as well as immediate corrective action when necessary [9, 10].

2.2 Production Scheduling in the Smart Factory

Production scheduling in the Industry 4.0 and smart factory concept is its main element, playing a key role in modern production management. Better organization of the entire manufacturing enterprise, along with each of its components, enables dynamic adaptation to changing market conditions. Current market demand, characterized by high and mass customization and personalization of products, requires adaptation to a wide variety of products manufactured in smaller quantities [11]. This situation means that classic scheduling is not sufficient and production must remain flexible at all times and adapt on an ongoing basis to variability and unexpected events such as due date fluctuations, order cancellations, new orders and changes that may result from customer preferences, material shortages, unexpected machine breakdowns, and (un)scheduled maintenance activities [12, 13].

In volatile and uncertain industrial conditions, scheduling faces new challenges that require dynamic adaptation of production processes in response to current needs and circumstances. To this end, the smart factory integrates advanced cyber-physical systems (CPS), Internet of Things (IoT) technologies, radio frequency identification (RFID) technology, and digital twins (DT), which enable real-time monitoring, control, and adaptive management of production schedules. Full system integration will allow for flexibility, optimal scheduling, and decisions on rescheduling due to random events will be made immediately, with a higher level of certainty thanks to up-to-date data from production processes [14, 15].

Intelligent scheduling within a smart factory also has challenges and problems that need to be highlighted. These include challenges related to the collection, processing, and analysis of large data sets in real time. Simply connecting a sensor is not enough; the data flowing from the production system must be relevant, reliable, and collected from multiple sources, and then processed appropriately so that the right scheduling decision can be made. Another challenge, not only related to smart scheduling but to the entire adaptation to Industry 4.0, is the high cost of implementing the technical infrastructure that supports the factory. Companies need significant financial resources to purchase digital technologies, communication devices, and software applications, and additional resources must be allocated for employee training. Overall, this can be a challenge for small and medium-sized companies, which often do not have such resources or are afraid to take the risk associated with the investment [16, 17].

3 Assumptions for Dynamic Production Scheduling in the Smart Factory Demonstrator

3.1 Laboratory Production System

The production system in the SmartFactory laboratory at the Poznań University of Technology, shown in Fig. 1, is being built for research and teaching purposes for various technical and organizational solutions in line with the concept of a smart factory. The laboratory equipment allows for the simulation of the main processes occurring in real production systems.

The main element of the demonstrator is an automatic assembly line. It consists of three transport loops with assembly stations located next to them. The production process involves assembling products from parts in the form of Lego blocks. These have been adopted as the basic building blocks for finished products. This ensures full flexibility in product design in line with the idea of production customization, which is the fundamental goal of the Industry 4.0 concept. The flow of products on the production line is based on the one-piece flow principle, where the carrier is a transport pallet. The assembly line is controlled by a control cabinet equipped with power supply components, safety systems, a PLC controller, and a safety system module. The system uses two industrial networks, AS-interface and ProfiNet, to control the devices.

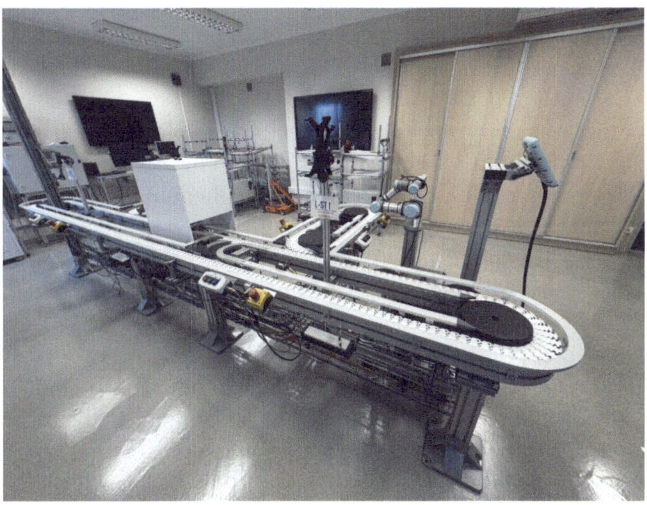

Fig. 1. SmartFactory laboratory at the Poznań University of Technology.

The production system is controlled by an IT system called 4Factory. It is composed of modules whose functionalities enable production scheduling, material flow monitoring, and production line control.

3.2 Production Scheduling in Smart Factory

Production management is supported by the 4Factory IT system. This integrated IT platform enables central control of all processes carried out in the smart factory demonstrator.

One of the key modules of this IT system is the production scheduling module (shown in Fig. 2). The effective and optimal functioning of this module requires the definition of a number of necessary pieces of information. These include, first and foremost, the structures of finished products, which make it possible to determine the dependencies and relationships between individual components. Each finished product is described by a BOM (Bill of Materials) list, which defines the set of necessary components and their quantities required to complete and manufacture the finished product. Another key element is the technological data, which includes detailed information on the technological process for a given product. This data includes a set of technological operations along with their designated execution time and preparation and completion time. Dedicated production stations are assigned to individual operations and, if necessary, alternative stations are assigned to maintain production continuity in the event of disruptions. Production orders are also entered into the system, containing information about the type and number of products to be manufactured, as well as, if necessary, the assigned priority of the order.

In order for the production process to be effectively implemented in the smart factory demonstrator environment, it is necessary not only to have access to data on finished products and production orders, but also to integrate them into the scheduling module. This integration makes it possible to generate production schedules based on orders entering the system. The implemented IT system allows the user to select scheduling strategies, methods, and priority rules. Based on individual preferences, the user can determine how a given pool of production orders is to be executed by choosing from among the implemented priority rules. The use of rules such as FIFO (First-In-First-Out), LIFO (Last-In-First-Out), and others enables production to be carried out according to different strategies and also allows the user to conduct comparative experiments and analyze the impact of individual rules on the production process.

The result of the scheduling process is a production schedule, which forms the basis of the work plan for the entire integrated production system within a specified time frame.

After generating the production schedule, the system moves on to the implementation stage. Thanks to the possibility of integrating the smart factory infrastructure with the IT system and the scheduling module, it is possible to obtain real data directly from the production process.

Data acquisition is carried out using advanced Internet of Things (IoT) infrastructure. Sensors located within the smart factory enable real-time recording of the status of each production station, including information about the work being performed, waiting for material delivery, and any time deviations from the original schedule. The system also collects the actual times of individual operations, allowing them to be compared with planned values, and monitors the flow of materials and semi-finished products between production stations.

The integration of all components into a single, coherent, IT system enables a modern approach to scheduling in a smart factory environment. This solution eliminates human errors resulting from inaccuracy, fatigue, and other random factors. Comprehensive information about production processes is available in a single system, which allows the planner to react quickly in the event of unforeseen situations, based on the analysis of

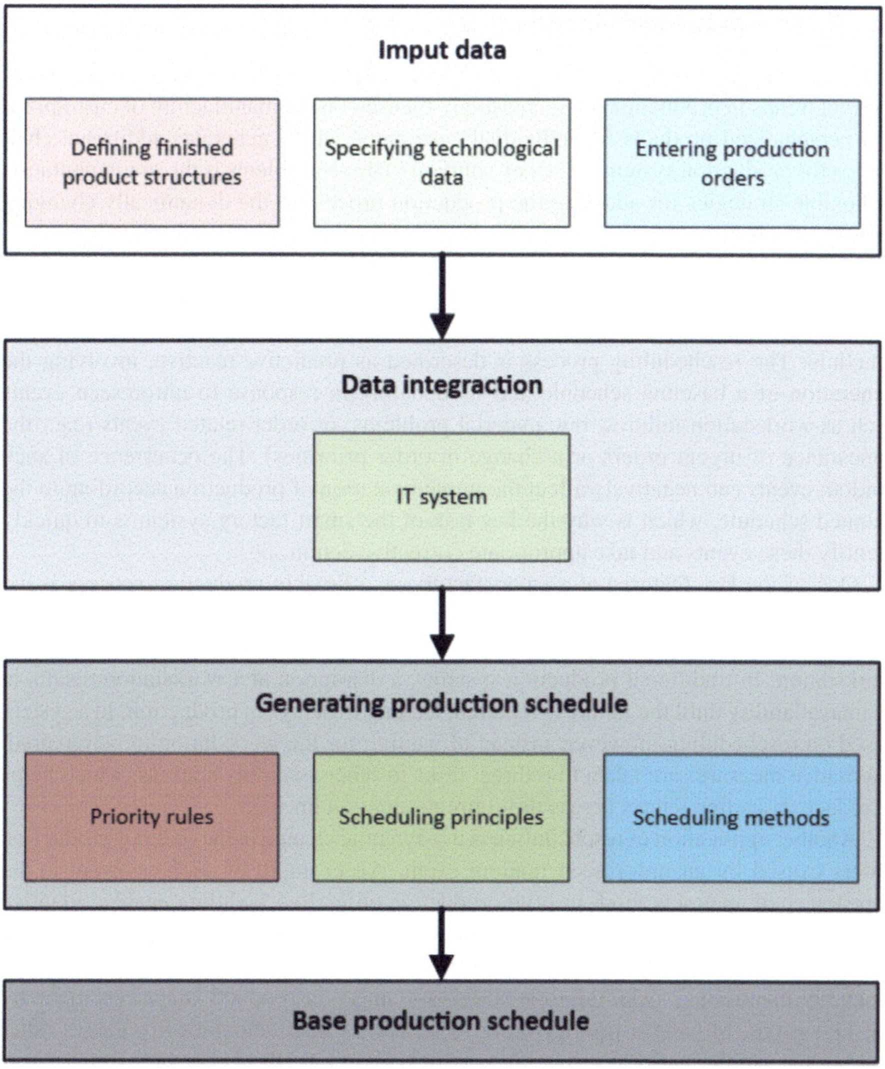

Fig. 2. Production scheduling process in smart factory.

real data presented through visualization on the monitoring dashboard, providing full insight into current production in the smart factory. The production monitoring system also includes workstations equipped with operator consoles that display the current status of operations in progress, along with detailed information on the expected completion time and progress of work, both completed and pending.

3.3 Production Rescheduling Model

Modern smart factories are not limited to traditional production scheduling methods. Current trends in production are increasingly focused on the manufacture of customized and personalized products for individual customers, which generates additional challenges for production systems. The key solution to these problems is the implementation of flexible strategies for adapting the production process to the dynamically changing current situation.

One of the solutions supporting production flexibility is rescheduling, which, within the smart factory demonstrator (show in Fig. 3), can be implemented as an extension of the computer program responsible for making changes to the current production schedule. The rescheduling process is described as predictive-reactive, involving the generation of a baseline schedule and its updating in response to unforeseen events such as workstation failures, raw material problems, or order-related events (e.g., the appearance of urgent orders or a change in order priorities). The occurrence of such random events can negatively affect the implementation of production according to the planned schedule, which is why the key task of the smart factory system is to quickly identify these events and take appropriate corrective action.

One of the key features of a smart factory as a flexible production process is the use of rescheduling, which allows production or assembly operations to be moved from one workstation (e.g., affected by a failure or unplanned event) to another, alternative workstation. In traditional production systems, a disruption at a workstation results in its unavailability until the failure is repaired, leading to delays in production. In a system based on rescheduling, however, instead of waiting for the workstation to be repaired, alternative measures are taken to redirect tasks to other assembly stations, which helps to reduce cascading delays across the entire production line.

Another application of rescheduling is the dynamic change in the order of production orders caused by an unforeseen random event. An example of such an event is the appearance of an urgent, high-priority production order. Rescheduling enables adaptive modification of the schedule, allowing for the quick introduction of a new order into the current production plan, with simultaneous automatic correction of other orders and assembly operations, including time shifts or changes in production stations made by the IT system. Moving a priority order to an earlier date automatically delays other operations, which can result in timeliness issues. Therefore, the rescheduling model will include mechanisms to analyze the impact of the changes on the entire order queue and minimize their impact on the original production plan.

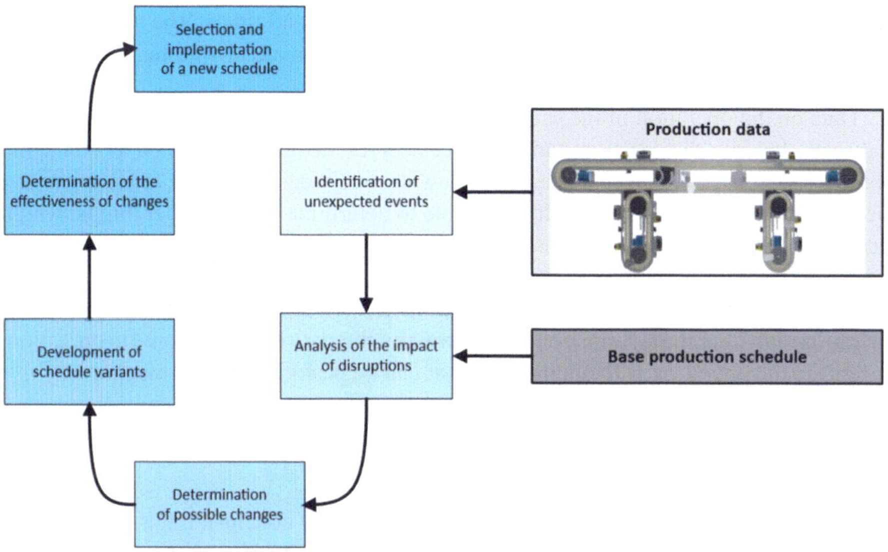

Fig. 3. Production rescheduling process in smart factory.

A flexible approach to production management ensures continuity of the production process and reduces the risk of a complete shutdown of the production line caused by a single workstation failure. Nevertheless, it is important that the rescheduling process is carried out in an optimal manner, as excessive frequency of schedule changes can lead to increased instability and nervousness in the production system. This results in increased operating costs and reduced production efficiency.

For the rescheduling model to function effectively and support the decision-making process, the system must operate on accurate, up-to-date, and complete data that reflects the current state of the smart factory. The technologies discussed in previous chapters, such as IoT and RFID, enable the acquisition of real-time data, which forms the foundation of the rescheduling process. An important feature of modern systems is their ability to transmit information instantly, allowing decisions to be made in real time and eliminating the need to manually collect production data.

From the perspective of the rescheduling process, the quality and nature of the data are of significant importance, as they determine the ability to properly adjust the plan in response to disruptions. Information on the start and end times of operations at individual workstations, transmitted to the system via RFID technology, allows for precise determination of the actual times of operation completion. Thanks to this, the system is able to identify which operations are performed faster and which slower than anticipated in the plan. In addition, it is important to extract data on operational delays. The system should monitor and report any delays in production, as well as record this data as historical information, enabling the identification of potential delay trends for specific types of operations and workstations.

Other types of data, such as information on breakdowns, downtime, interruptions at assembly stations, and any other events causing production delays, should be recorded in the system with a precise indication of their start and end times. The availability of high-quality historical data enables the IT system to estimate downtime, which is the basis for an accurate rescheduling process.

Data on the operation of the smart factory demonstrator (both actual and historical) is directly fed into the decision-making process in rescheduling. Based on this data, the 4Factory system performs analyses that allow rational decisions to be made regarding the implementation of processes. It is possible to determine, among other things: which technological operations are at risk due to the unavailability of production stations, forecasts of station downtime, and alternative production stations. Based on this analysis, the system decides whether to move operations to an alternative station, postpone the operation until after the repair is completed, wait for the station to be restored to operation, or replace the operation with another one that can be performed in the meantime.

4 Summary and Conclusions

An analysis of the issues raised in this article leads to the conclusion that the future of production management lies in the ability of IT systems to quickly process real-time data and make optimal adaptive decisions. Production scheduling within a smart factory cannot be treated as a one-time activity, but as a continuous process resulting from variable random factors that arise in the production process. For rescheduling to be effective, the quality and timeliness of data flowing from the production infrastructure must be adequate. Technologies such as the Internet of Things and RFID are an essential foundation for providing real-time information on the status of each workstation, operation times, and unforeseen disruptions. Without access to reliable data, the IT system supporting the smart factory is unable to make accurate decisions about operational changes. It is important that these changes do not disrupt the overall stability of the production system. Although rescheduling allows for a quick response to a random event, excessive frequency of changes can lead to an overall decline in efficiency.

The future of smart factories is based on the development of comprehensive IT systems capable of integrating data from numerous sources and using it effectively in production scheduling processes. Such systems are a key element enabling manufacturing companies to gain a real competitive advantage by increasing production flexibility and the ability to immediately adapt to unforeseen situations related to changing market conditions.

Acknowledgments. The presented results derive from a scientific statutory research conducted by Chair of Production Engineering, Faculty of Mechanical Engineering, Poznan University of Technology, Poland, supported by the Polish Ministry of Science and Higher Education.

References

1. Alade, B., Ocheni, E. S., Onyekachukwu, M., Olufinmilayo, S.: Big data for predictive maintenance in Industry 4.0: enhancing operational efficiency and equipment reliability. Int. J. Comput. Appl.Technol. Res. **13**(10), 37–51 (2024)

2. Ivanov, D., Sokolov, B., Dolgui, A.: Introduction to scheduling in Industry 4.0 and cloud manufacturing systems. In: Ivanov, D., Sokolov, B., Dolgui, A. (eds.) Scheduling in Industry 4.0 and Cloud Manufacturing, pp. 1–9. Springer, Cham (2020)
3. Karic, H., Schilberg, D., Arockiadoss, A.: Smart factory and Industry 4.0: a survey on advancements, technologies, methods and perspectives of digital transformation in manufacturing. In: Karwowski, W., Mrugalska, B., Trzcielinski, S. (eds.) AHFE 2025, vol. 184, pp. 1–15. AHFE International, USA (2025)
4. Chang, T.W., Cho, E., Jun, J.-H., Ahn, H.: Implementation of smart factory for SME: focusing on data acquisition and monitoring. ICIC Express Lett. Part B: Appl. **10**(6), 551–558 (2019)
5. Pronost, G., Mayer, F., Camargo, M., Dupont, L.: Digital Twins along the product lifecycle: a systematic literature review of applications in manufacturing. Digital Twin **1**(2) (2024)
6. Guo, Z.X., Ngai, E.W.T., Yang, C., Liang, X.: An RFID-Based intelligent decision support system architecture for production monitoring and scheduling in a distributed manufacturing environment. Int. J. Prod. Econ. **159**, 16–28 (2015)
7. Soori, M., Arezoo, B., Dastres, R.: Internet of things for smart factories in Industry 4.0, a review. Internet of Things Cyber-Phys. Syst. **3**, 192–204 (2023)
8. Duan, L., Da Xu, L.: Data analytics in Industry 4.0: a survey. Inf. Syst. Front. **26**(6), 6783 (2024)
9. Kalsoom, T., Ramzan, N., Ahmed, S., Ur-Rehman, M.: Advances in sensor technologies in the era of smart factory and Industry 4.0. Sensors **20**(23), 2287–2303 (2024)
10. Syafrudin, M., Alfian, G., Fitriyani, N.L., Rhee, J.: Performance analysis of IoT-Based sensor, big data processing, and machine learning model for real-time monitoring system in automotive manufacturing. Sensors **18**, 2946 (2018)
11. Mourtzis, D.: Advances in adaptive scheduling in Industry 4.0. Front. Manuf. Technol. **2** (2022)
12. Penchev, P., Vitliemov, P., Georgiev, I.: Optimization model for production scheduling taking into account preventive maintenance in an uncertainty-based production system. Heliyon **9**(7) (2023)
13. Prashar, A., Tortorella, G.L., Fogliatto, F.S.: Production scheduling in Industry 4.0: morphological analysis of the literature and future research agenda. J. Manuf. Syst. **65**, 33–43 (2022)
14. Wang, H., Yu, J.: Smart factory construction: a path to efficiency and innovation. Int. J. Eng. Inventions **12**(8), 47–52 (2022)
15. Khadiri, H., Sekkat, S., Herrou, B.: An intelligent method for the scheduling of cyber physical production systems. Manage. Prod. Eng. Rev. 44–51 (2022)
16. Gunasekaran, A., Kamble, S., Ghadge, A., Kumar, V.: Investments in Industry 4.0 technologies and supply chain finance: approaches, framework and strategies. Int. J. Prod. Res. **62**(22), 8049–8055 (2024)
17. Bagheri Rad, N., Behnamian, J.: Real-Time multi-factory scheduling in Industry 4.0 with virtual alliances. Eng. Appl. Artif. Intell. **125**, 106636 (2023)

A Concept of a Production Flow Control IT System for Discrete Production of Mechanical Parts–Communication and Database Aspects

Jaroslaw Chrobot(✉)

Wroclaw University of Science and Technology, Wroclaw, Poland
jaroslaw.chrobot@pwr.edu.pl

Abstract. Requirements for product traceability in selected industrial sectors have led to an increasing demand for Production Flow Control (PFC) systems as an integral component of modern manufacturing enterprises. The implementation of such systems provides significant value by enhancing product quality assurance and minimizing the risk of delivering defective products to customers. This is achieved through the automatic identification and blocking of defective components once the root cause of a defect has been detected. The present paper introduces a proposed basic functionality of a PFC system designed to satisfy fundamental industrial requirements within the domain of discrete mechanical manufacturing. Particular attention is devoted to the system's database structure and its functional mechanisms as well as communication method with devices, which control production stations, including PLCs and industrial PCs. Furthermore, the article presents and analyses selected results obtained from research on the system's operational effectiveness, demonstrating its potential to improve production reliability and traceability.

Keywords: Production Flow Control · Traceability · Discrete Production · Mechanical Products

1 Introduction

1.1 Product Traceability as an Industrial Requirement

The increasing requirements for product traceability in various industrial sectors, such as the food [1–5], pharmaceutical [6–8], electronics, aerospace, and automotive industries [9–12], have made the implementation of traceability systems an essential component of modern industrial operations. In these sectors, where the safety and health of end users depend directly on product quality, traceability systems serve as critical tools for ensuring reliability and compliance. In the food industry, traceability is not only mandated by regulatory authorities and required by third-party audit standards but also represents a key element of risk management and safety assurance. Traceability can be defined as a structured and chronological collection of information concerning uniquely identified products, documented throughout the production process. Such systems enable detailed

J. Trojanowska et al. (Eds.): MANUFACTURING 2026, LNME, pp. 290–304, 2026.
https://doi.org/10.1007/978-3-032-21654-0_22

tracking of each significant production step, from raw material acquisition (including origin and composition), through the processing of semi-finished components and technological operations (along with corresponding process parameters), to the completion of finished goods. The systematic recording of production data allows continuous monitoring of process execution, verification of product history, and timely prevention of defective components from advancing to subsequent manufacturing stages. By automatically collecting and storing process data in a database system, traceability solutions reduce the need for repetitive testing and manual inspections. This, in turn, facilitates comprehensive data analysis, enabling the identification of frequent errors, the assessment of their extent, and the determination of their root causes, thereby mitigating product liability risks. Moreover, process transparency achieved through traceability contributes to improved product quality, enhanced production planning, and increased compliance with both customer expectations and relevant standards. In essence, traceability systems are designed to determine the characteristics and parameters of products at each production stage and compare them against mandatory specifications. As highlighted by [13], traceability also extends beyond individual manufacturing plants, supporting control and visibility across entire supply chains. Consequently, advanced IT-based traceability systems have evolved into crucial management information tools within production enterprises. They enable effective supply chain oversight, minimize production risks, and strengthen customer trust in the quality and reliability of manufactured products. Contemporary product traceability systems represent a crucial component of digital process management, enabling a comprehensive understanding of production workflows, complete quality control, and effective handling of customer complaints and defective products. Moreover, they contribute to the identification and elimination of inefficiencies within manufacturing processes and facilitate the optimal distribution of responsibilities across the supply chain [14].

1.2 Production Flow Control Systems and Runtime Defect Prevention

Production Flow Control (PFC) systems extend beyond the conventional scope of traceability solutions. These systems represent advanced IT tools designed to record discrete production data in a database at various stages of industrial processes. The primary sources of such data are Programmable Logic Controllers (PLCs) and industrial PCs integrated into production lines. In this capacity, PFC systems inherently fulfill the role of traceability systems by enabling the monitoring and tracking of both products and technological processes. However, due to their dedicated communication protocols and advanced database functionalities, PFC systems provide additional capabilities that distinguish them from standard traceability solutions. In particular, they enable the automatic identification and blocking of specific products that have been incorrectly processed in earlier stages of production or that deviate from the predefined technological path. This mechanism serves as an essential preventive measure, reducing the risk of defective products progressing through subsequent production phases. For effective operation, PFC systems require production stations equipped with PLC or PC control units, or alternative interfaces that control the stations and allow data acquisition from manufacturing processes. In some configurations, data can also be obtained through independent PC-based systems that are not directly integrated with the hardware of production equipment. By combining data collection, process monitoring, and preventive control functions, PFC systems contribute to higher production reliability, improved quality assurance, and enhanced operational efficiency within discrete manufacturing environments.

PFC systems facilitate the comprehensive registration of discrete product histories by recording individual production steps together with detailed process data, such as date and time stamps, numerical process parameters, component identifiers, operation results, and workstation identifiers. This information uniquely characterizes each production step, enabling in-depth analysis of the manufacturing process and the detection, and subsequent blocking, of defective products before they proceed to subsequent production stages, especially in case of a discrete production of mechanical parts. In PFC systems, the blocking mechanism prevents further processing by inhibiting part release, where available via physical interlocks, once a defect has been identified. Such a defect may result from an incomplete product assembly, an improperly executed technological operation, or the occurrence of duplicate product identifiers. Defect detection is achieved through automated analysis of database records, particularly data obtained from preceding workstations, and through the application of system configuration rules, including product-specific workstation sequences, required component sets, and permissible alternative workstations. By integrating these analytical and preventive functionalities, PFC systems contribute to the reduction of defective outputs and help prevent potential customer complaints, or at least significantly decrease their frequency. In addition to improving product quality, PFC systems provide measurable economic benefits to manufacturers by reducing the financial resources needed to resolve customer claims and by enhancing overall customer satisfaction. The resulting increase in trust and brand reliability supports higher competitiveness and facilitates the acquisition of new clients.

1.3 Limitations of Existing Traceability Solutions and Research Gap

Recent scientific publications present advanced models and concepts of traceability systems [15, 16]. In parallel, several commercial traceability solutions are available on the market, such as the DELMIAWorks Manufacturing Track & Trace System and XPRIMER.TRC system. Both research and industrial implementations, however, primarily focus on the registration and tracking of product or process data. While existing research and commercial solutions predominantly focus on data registration, tracking, and post-process analysis, relatively little attention has been devoted to system-level mechanisms that actively enforce production correctness during runtime. In practice, many traceability systems operate as passive repositories of production history, leaving decision-making and defect handling largely dependent on human operators or external quality systems. As a result, defective products may continue through subsequent production stages before being detected, increasing rework costs and the risk of customer complaints. To date, no documented concept enabling automatic stoppage in the case of defective products has been identified in the literature, except [17], nor is such functionality offered by existing IT solutions. PFC systems are typically developed as in-house solutions by manufacturing companies and are thus tailored to their specific operational requirements. Consequently, there is a clear need for a universal and standardized approach to traceability and defect management that could be applied across different industrial contexts.

1.4 Contribution and Scope of the Paper

This study presents a conceptual basic framework for a PFC system, this time with an emphasis on the database structure and its functionality, as well as on communication algorithms with PLC and industrial PC, developed to address the core industrial requirements characteristic of discrete mechanical manufacturing environments. Production Flow Control (PFC) systems extend conventional traceability by combining real-time communication with shop-floor control devices, centralized production data management, and runtime enforcement of production rules. Instead of introducing new control or optimization algorithms, PFC provides an operational framework that coordinates data acquisition, verifies technological routes, and automatically blocks products that violate predefined constraints. Its effectiveness depends on a clear architecture, reliable communication, and a database structure that enables fast verification and decision-making. This paper contributes at the system integration and architectural level by proposing and validating a reference PFC architecture for discrete mechanical manufacturing. It focuses on: the PFC database structure and functionality, communication between the PFC system and PLCs/industrial PCs, as well as on database-driven blocking strategies based on historical data and enforced at runtime. Laboratory experiments demonstrate the feasibility, performance, and scalability of the framework under realistic industrial conditions, providing a practical foundation for future advanced analytical or algorithmic solutions.

1.5 Benchmarking Against Existing Industrial IT Systems

To place the proposed Production Flow Control (PFC) concept among industrial IT solutions, a qualitative functional benchmark was conducted, focusing on execution-level needs in discrete mechanical manufacturing: real-time enforcement of production correctness, interaction with PLC-controlled stations, openness of communication, and adaptability to plant-specific constraints. Because commercial systems are proprietary and low-level performance data are rarely available, the comparison relies on functionalities reported in the literature and vendor materials. Commercial traceability tools (e.g., DELMIAWorks Track & Trace, XPRIMER.TRC) mainly register, store, and report production/quality data, offering strong post-process traceability but limited active flow enforcement. Defects are typically handled after operations and often depend on operators or external quality systems. Similarly, common MES platforms (e.g., Siemens Opcenter Execution, SAP Digital Manufacturing, Rockwell FactoryTalk ProductionCentre) support workflows, routes, and rework at the MES level, yet correctness enforcement is usually indirect (workflow logic, alarms, post-checks) rather than mandatory PLC-level authorization driven by real-time database verification. Shop-floor connectivity is often mediated through proprietary interfaces or middleware, with limited coupling between database state and machine execution. In contrast, the proposed PFC targets the execution layer between traceability systems and machine control by combining centralized, database-driven verification with real-time communication to PLCs/industrial PCs. Each step is authorized before execution based on the current and historical product state in the PFC database, and blocking is enforced at runtime to prevent incorrectly processed products from advancing. The concept also emphasizes openness and adaptability: Ethernet/TCP/IP communication (optionally via open standards such as OPC) and customizable database logic support repair modes, controlled rework, and commissioning scenarios. Overall, the benchmark suggests that while existing traceability and MES solutions excel at data collection and coordination, they generally lack systematic, real-time, database-driven PLC-level enforcement, an execution-layer gap the proposed PFC framework is designed to fill.

2 Research Problem

Modern production environments increasingly rely on integrated IT systems to ensure efficient control and monitoring of production flow. In such systems, the efficiency of communication between networked components is a key factor affecting overall system performance and reliability. After scanning a product's identifier, a worker should receive immediate feedback on whether a given product has been approved for production. A noticeable delay poses a problem for the efficiency of the production process itself. This research addresses the problem of evaluating the communication efficiency within an IT-based PFC system, consisting of a central database connected to an intermediary PC through an Ethernet network. The intermediary PC acts as a bridge between the database, PLCs, and industrial PCs, all communicating over the same Ethernet infrastructure.

3 Main Components of the PFC System

The core functionality of the PFC system consists of the following main components:

- PFC Database - at each stage of the production process, i.e., at every workstation, the PFC system records process-related information within an SQL database. The database structure comprises a set of tables, functions, and stored procedures responsible, among other tasks, for the decision-taking regarding automatic blocking of defective parts.
- PFC System Protocol - this communication protocol is specifically designed for Ethernet-based data exchange between the PFC system and the PLC or industrial PC units controlling production stations. It provides the foundation for real-time data transfer and automated decision-making processes within the system.
- Main PFC System Application - this software application, installed on a dedicated PC, serves as the central operational interface of the system. It enables data exchange with the database, facilitates event visualization, provides information regarding product blocking causes, and communicates with production workstation controllers (PLCs and industrial PCs) through a dedicated internal communication protocol.

An overview of the system architecture is presented in Fig. 1, which illustrates the interconnection of components via an Ethernet network. The PFC database is hosted on a dedicated high-performance server that communicates thorough the Main PFC System Application with both PLC and PC controllers through the PFC System Protocol. The overall system structure reflects typical industrial requirements and functional expectations, corresponding to the architecture of modern production environments that include both PLC/PC-controlled workstations. The proposed concept is specifically adapted to discrete manufacturing processes involving identifiable mechanical products, which may consist of both traceable components (e.g. electrical or manual actuators) and non-traceable elements (e.g. springs, screws, rivets).

3.1 PFC Database

At each stage of production, corresponding to individual production stations critical from the perspective of traceability, the PFC system records relevant data in an SQL database. This database comprises structured tables, functions, and stored procedures responsible not only for data storage but also for the automatic blocking of defective products. Upon the arrival of a product at a subsequent production station, the PFC system queries the database to verify whether the previous operation was successfully completed, if the product is following the prescribed technological route, and whether the correct component is to be assembled into the main product. These queries are executed through internal database functions and procedures that process the collected data. In cases where the verification yields a negative result, a function returns a specific value indicating the reason for rejecting further processing of the product. The database tables store detailed information on individual technological steps, including parameters such as e.g. tightening torque, tightening angle, number of rivets, and component identifiers, alongside operation outcomes (pass/fail). Furthermore, data related to individual products in progress, including expected subsequent technological steps and completed products, are maintained. Database functions are responsible for querying the status of prior technological operations as a product enters a new station, while stored procedures manage the recording of results upon completion of each technological operation. To ensure optimal system performance, particularly under conditions involving numerous production stations, the PFC system is designed to minimize conflicts during intensive data exchanges by allocating dedicated database connection channels for each production station. This approach mitigates query collisions and supports efficient, concurrent operation across the manufacturing process. The database manages blocking strategies related to defects originating from preceding production stages, as well as multiple attempts to process a product at a production station that has already completed the corresponding operation.

The PFC database consists of 3 main tables. The "part_data" table includes data regarding a part including unique identifier and result of last process. This table serves for checking if the product is allowed to be further processed. The "final_product_data" table includes data about the finished products (i.e. unique identifier, components' identifiers). The "step_data" table includes data regarding each technology step executed on a product (i.e. assembly, machining etc.) with values registered. Figure 2 shows an example of a database function algorithm for adding disassembly operation to product's history.

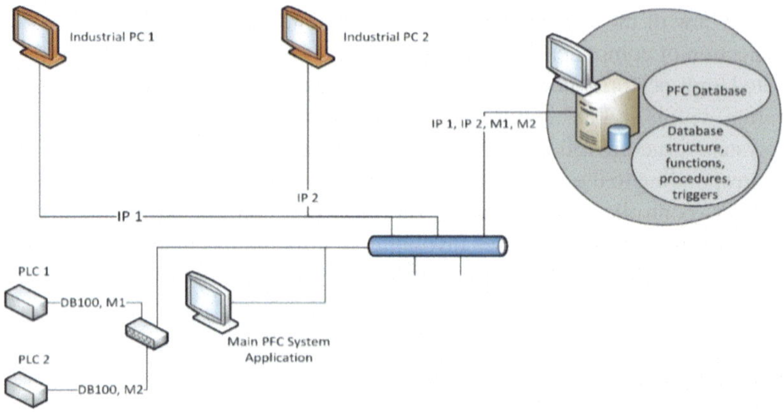

Fig. 1. Interconnection of PFC system components [source: own development]

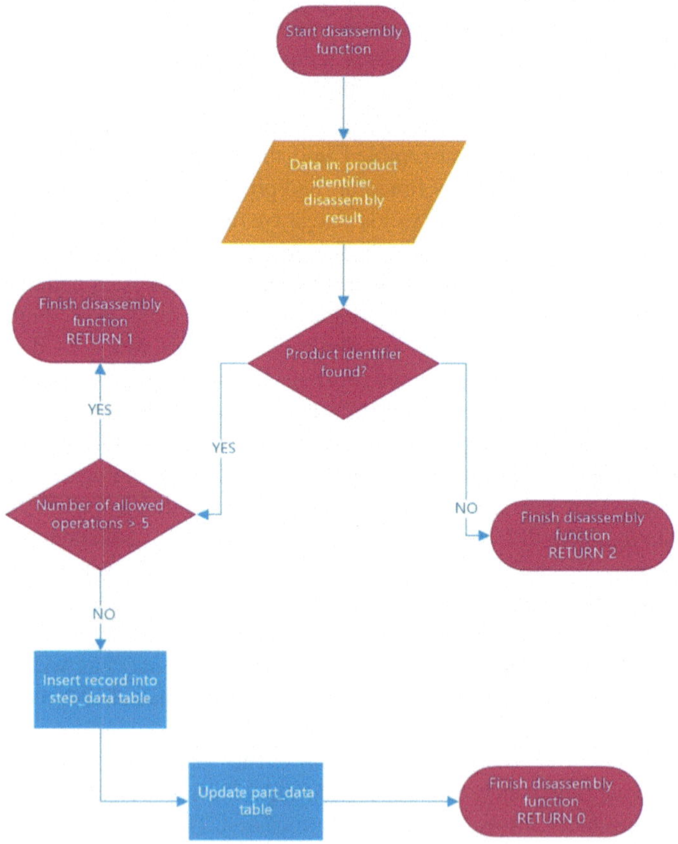

Fig. 2. PFC system's algorithm of an example database function [source: own development]

3.2 PFC Protocol

The communication protocol has been specifically designed to facilitate data exchange over an Ethernet network between the Main PFC System Application and devices controlling production stations, including PLCs and industrial PCs. Its primary functions include:

- verifying the availability and operational status of the communicating device,
- acquiring data related to products entering a production station,
- transmitting authorization commands to permit or block the continuation of product manufacturing,
- retrieving results of technological operations upon completion, and
- confirming the receipt of transmitted messages.

The protocol supports implementation based on the OPC standard, ensuring interoperability of the PFC system with diverse industrial equipment. Alternatively, a proprietary communication algorithm, initially limited to PLCs from selected vendors, can be applied. Effective operation of the protocol requires programming of both the Main PFC System Application and the PLCs, utilizing a specially designed data block (DBL), as well as configuration of the industrial PCs, which control production stations, in accordance with specified programming guidelines. Functionally, the protocol operates as a telegram-based communication system between the controlling PLC/PC and the main application. It relies on the TCP/IP network protocol to ensure reliable data transmission. All relevant data are prepared within the PLC or PC and subsequently collected by the Main PFC System Application into the PFC Database. When communicating with PLCs, each PLC is assigned a dedicated section within the DBL containing all necessary data to guarantee complete product traceability. The main application establishes communication with the PLC by connecting to a predefined IP address and port, reading data from the specified DBL, and writing response data back into the same DBL.

The DBL is logically divided into two areas:

- the first area, exclusively written by the PLC, contains current operational data such as product identifiers and results of screwing operations; the main application does not modify this section.
- the second area is designated for responses sent by the Main PFC System Application following receipt of data from the PLC.

In the case of communication with industrial PCs, data exchange occurs through specially formatted text messages between the PC at the production station and the Main PFC System Application. These messages encapsulate all necessary information required to register products and record process results. Figure 3 illustrates the proposed communication protocol structure for data exchange between the main application and a production station PC.

In the case of communication with a PLC, the interface is based on a standardized data structure, referred to as e.g. DB100 data block, alongside dedicated data population procedures. When a single PLC manages multiple workstations, an internal buffer within the PLC is implemented as a matrix of several data blocks, numbered in a unique way, to store operation results corresponding to each workstation.

COMMUNICATION SPECIFICATION:
production station PC ←→ PFC (main application)

First scan of product code, when products enters station:

Status query (production station PC → PFC):
Q:label_in:line_in:station_in;

Answer (PC ← PFC):
Q:status:reference:operation_number_out;

Comments:
1. status=1 means operations allowed.
2. status=0 means operations not allowed

Operation finish:

Finish request (production station PC → PFC):
F:label_in:line_in:station_in:assembling_status_in:assembling_datetime_in:treatment_type_in:operation_number_in:operation_status_in:operator_id_in:final_label_in:{scanned_label_component_in}:{value_name_in}: {value_value_in}:{value_status_in}:

Comments:
1. assembling_status_in - value 1-OK, 0-NOK
2. scanned_label_component_in – labels of possible components shall be put in quotation marks and separated by semicolons, for example {"1234";"5454";"5454"}
3. value_value_in - numeric values in floating point format with a dot, separated by a semicolon, for example {12.4;43.2;12.7}, if a value has status only please put as empty, for example {12.4;;12.7}
4. value_status_in - value 0-NOK, value 1-OK separated by semicolon, for example {1;1;0}

Answer (production station PC PFC):
F:status

Comments:
1. status=1 means OK.
2. status=0 means "data storing error"

Fig. 3. Communication protocol between an industrial PC and PFC, based on [17]

Upon completion of an operation on a product, the machine's PLC writes the resulting data into the corresponding DB100 and sets the PLC variable "Request number" to the value 3, indicating that data is ready for retrieval by the PFC main application. Subsequently, the PFC main application reads the contents of the DB100 from the PLC and integrates this information with the product's historical data stored in the PFC database. As feedback, the PFC main application updates specific variables within the DB100, including:

- the "PFC result" variable, which is assigned a value of 30 to indicate success or 31 to denote failure,
- the "Request number" variable, which is set to 103 to acknowledge data reception.

Upon successful receipt of complete data, the PLC resets the "Request number" variable to 0, as illustrated in Fig. 4.

If the PLC needs to request authorization from the PFC main application to proceed with the product, it sets the "Request number" variable to 2, representing a cross-check request. The PFC main application responds by setting the "Request number" variable to 102. After verifying the product's history within the database, the PFC main application updates the "PFC result" variable with either 20 (authorization granted) or 21 (processing blocked).

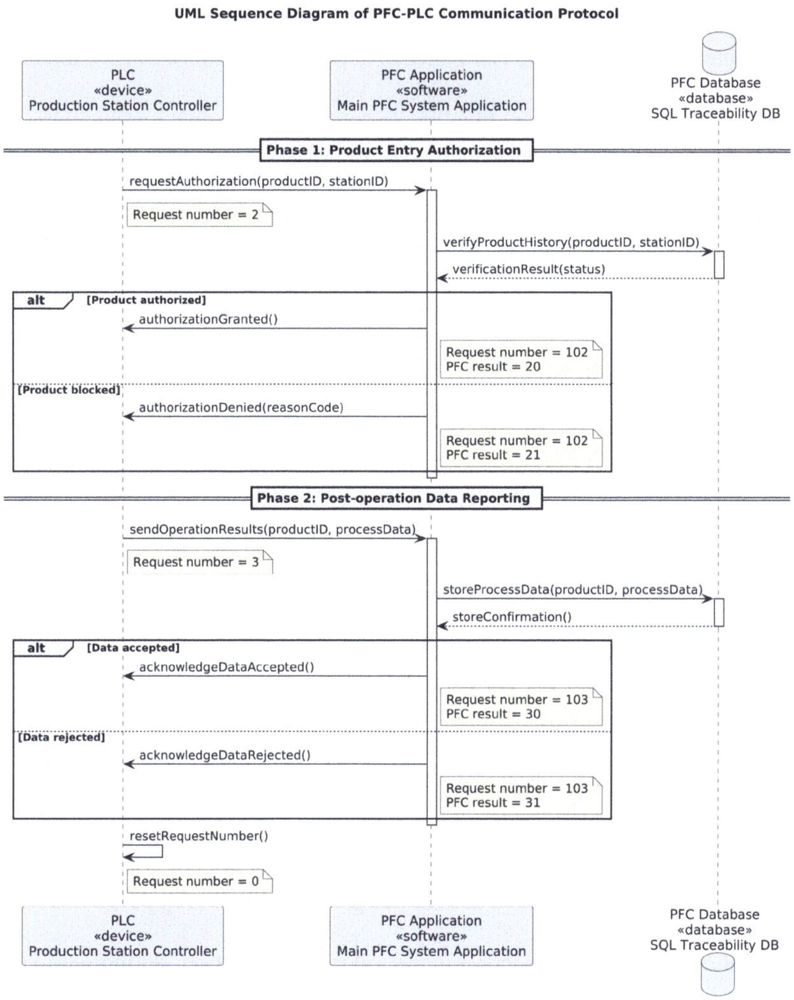

Fig. 4. PFC and PLC communication algorithms [source: own development]

3.3 Main PFC System Application

The PFC main application runs on a single PC (either on the shop floor or remotely) and provides:

- Real-time communication with station controllers (PLCs/industrial PCs): bidirectional Ethernet exchange of product/component IDs and process parameters, plus acknowledgments and commands to authorize or block production.
- Database Interaction: SQL-based read/write to a central PFC database, where all events are stored and linked to a unique product ID (barcode/Data Matrix/RFID) to ensure traceability.

- Blocking logic: runtime authorization or blocking based on defects from prior steps, route deviations, wrong components, duplicate/unauthorized IDs, and repeated processing at stations.
- Operational modes per station: blocking (stop defective items), non-blocking (allow passage during commissioning), and repair (repeat an operation a limited number of times).
- GUI monitoring: live status of connected stations and modes, an event/telegram log, and detailed explanations including blocking reasons.
- High configurability: user-friendly definition of stations, network addresses, product types and routes, allowed modes, and rework limits (via configuration files).

3.4 Potential for Future Algorithmic Extensions

Although the PFC system is deliberately rule-based and database-driven, it provides a solid base for future algorithmic upgrades. Time-stamped, structured data collected at each workstation and stored centrally, together with consistent communication, creates good conditions for decision-support methods. A key extension is predictive blocking, where historical data and process trends are used to anticipate defects (e.g., detecting parameter drift or recurring deviations tied to specific components or stations) rather than reacting only to explicit rule violations. Another direction is AI-based anomaly detection to uncover previously unknown defect patterns or rare failure modes from large-scale process data. Thanks to the modular architecture separating communication, database logic, and operational control, such enhancements can be added without major changes, making the framework a practical platform for combining deterministic rules with data-driven intelligence in future smart manufacturing.

4 Report on Test of Connecting the PFC Database with PLC and PC Through a Data Collecting PFC Main Application

The concept discussed in this paper requires practical validation. Preliminary verification was carried out in a laboratory environment, where prototype software components were tested. During the experiments, multiple PLC and PC stations were simulated, and a dedicated database server was implemented. System's response time was measured in order to evaluate the efficiency and robustness of the proposed solution. All tests were conducted using a custom-developed communication protocol on the side of the PFC main application. To simulate an industrial PC, a multithreaded application was launched. Each thread represents an industrial PC controlling a production station and communicating via Ethernet. Multiple PCs were used, each with a multithreaded application installed and running to address processor load. OpenPLC, a free, open-source PLC simulator, was used to simulate the PLC. The goal of the tests was to verify the performance and reliability of communication between the components of a production system based on a PostgreSQL database, Siemens S7 PLCs (OpenPLC in simulation mode), and industrial PCs. The test process replicated a real-world 2 phases production situation, where a product enters the production station (phase 1), the data-matrix code gets scanned, and the database query result influences the subsequent production process.

After a couple of minutes the same production station, where the production process is just completed, generates after-production data (phase 2). Communication between all components occurred via an intermediary PC (running PFC main application), which exchanged data using the TCP/IP protocol. The tests were specifically aimed at:

- measuring the system response time from the moment a query is sent from the PLC/industrial PC to the moment the response is received from the database,
- verifying the correctness of data writing and reading in the PostgreSQL database,
- determining the impact of the number of active devices and the size of the data table on system response times,
- assessing the stability of connections and data consistency during information exchange.

The test environment consisted of the following components:

- database - PostgreSQL 15, Windows 10, Intel i7–10700 CPU, 16 GB RAM, SSD drive,
- database tables - part_data, step_data, final_product_data - 20 text fields, indexed by scanned_label column,
- computer with the main application - Windows 10, Intel i7–10700 CPU, 16 GB RAM, SSD drive,
- PLCs - Siemens S7–1500 instances (simulated using OpenPLC over Modbus TCP),
- industrial PCs - TCP client applications,
- communication network - 1 Gbps Ethernet, dedicated test VLAN, latency < 1 ms,
- network communication protocol - TCP/IP (socket client-server),
- PLC-PC communication protocol - own development.

Applied were 4 research scenarios:

- S1 - 1 PLC instance (PLC), 1 PC application (PC), 10 000 database records,
- S2 - 5 PLCs, 5 PCs, 100 000 database records,
- S3 - 10 PLCs, 10 PCs, 1 000 000 records,
- S4 - 20 PLCs, 20 PCs, 1 000 000 records.

Analysis of the obtained results shows that:

- the system maintains stability and data consistency in all test scenarios,
- scaling the number of devices causes a linear increase in response time, confirming the good performance of the TCP/IP architecture and PostgreSQL performance,
- no packet loss or data write errors were observed,
- performance at low load (1–5 instances of each type and less than 100 000 records) is very good - response times do not exceed 100 ms,
- scaling beyond 10 or more instances of each type and setting one million of records causes an increase in response time,
- exceedances of the 250 ms threshold begin to occur with the number of instances exceeding 20 of each type.

Table 1 shows measured response times of the PFC system under different load scenarios.

Table 1. Measured response times of the PFC system under different load scenarios.

Scenario	PLCs	PCs	Db records	Avg.resp.	Min[ms]	Max[ms]	95th percentile[ms]
S1	1	1	10 000	42	31	68	59
S2	5	5	100 000	74	49	112	98
S3	10	10	1000 000	118	82	198	165
S4	20	20	1 000 000	214	136	312	276

The results presented in Table 1 indicate that the proposed PFC architecture maintains low and predictable response times for small and medium-scale configurations (S1 - S3). For scenarios involving up to 10 PLCs and 10 industrial PCs, the average response time remains below 120 ms, which is well within acceptable limits for discrete manufacturing environments. In the most demanding scenario (S4), response times increase but remain operationally feasible, with occasional exceedance of 250 ms. The observed near-linear growth confirms that communication overhead and database access scale proportionally with the number of connected devices. Response time is here measured as the elapsed time between sending a request from a PLC/industrial PC to the Main PFC System Application and receiving an authorization or acknowledgment response after database verification.

The following conclusions and recommendations were drawn from the data analysis:

- the system supports up to 10 PLCs and 10 industrial PCs without noticeable delays exceeding 100 ms,
- for a larger number of devices, the following actions are recommended: query caching or asynchronous database request queuing, splitting read and write functions into two processes, as well as implementing database indexes on columns involved in database queries,
- OpenPLC and simulated industrial PCs enable accurate simulation of system behavior in production conditions without risking real-world hardware,
- horizontal scalability is recommended (deployment on multiple PCs with PFC main application, where each of them collects data from restricted number of devices) for more than 20 PLCs plus more than 20 PCs and more than one million records.

Despite promising laboratory results, simulation-based validation has inherent limits. Tools like OpenPLC and PC simulators allow controlled, repeatable testing without risking equipment, but they cannot fully capture real industrial effects such as scan-cycle variability, network jitter, hardware communication delays, I/O buffering, and unexpected shop-floor disturbances. The simulated PCs also lacked interaction with real sensors and actuators, which can affect data availability and synchronization. Therefore, while the tests confirm architectural feasibility, communication robustness, and database performance, further validation with real PLC hardware and production stations is needed to assess behavior under actual operating conditions.

5 Conclusions and Future Work

This paper proposes a reference architecture and operational framework for a Production Flow Control (PFC) system for discrete mechanical manufacturing. It combines centralized, database-driven traceability with real-time communication to shop-floor devices, enabling runtime enforcement of production correctness through automatic blocking. The main contribution is system-level integration of communication, data management, and operational control rather than new control algorithms. Laboratory validation with simulated PLCs and industrial PCs showed reliable communication, low response times, and stable database performance under increasing load, indicating readiness for industrial implementation, while also underscoring the need for real-world trials.

Future work includes:

- deployment with real PLC hardware and physical stations to assess scan-cycle behavior, I/O interaction, network performance, and blocking under defects/rework and operator interventions.
- integration with MES to synchronize orders, variants, and routes via bidirectional data exchange, and
- large-scale multi-line pilots to evaluate long-term stability, scalability, maintainability, and organizational adoption.

These steps would also enable advanced extensions such as predictive blocking and AI-based anomaly detection. Overall, the framework bridges the gap between passive traceability and active, execution-layer enforcement.

References

1. Zheng, M., et al.: Construct food safety traceability system for people's health under the Internet of Things and big data. IEEE Access **9**, 1–11 (2021)
2. Zhou, X., et al.: Traceability in food supply chains: a systematic literature review and future research directions. Int. Food Agribusiness Manage. Rev. **25**(2), 1–18 (2022)
3. Anastasiadis, F., et al.: Food traceability: a consumer-centric supply chain perspective. Foods **10**(3), 1–17 (2021)
4. Rossi, S., et al.: Agri-food traceability today: advancing innovation towards resilient and transparent supply chains. Trends Food Sci. Technol. **163**, 1–12 (2025)
5. Masengu, A., et al.: Sustainable food traceability: a comprehensive model for enhancing safety and resilience in the supply chain in the Oman food industry. Future Bus. J. **11**, 1–15 (2025)
6. Saez, C., et al.: Smart pharmaceutical manufacturing: ensuring end-to-end traceability and data integrity in medicine production. Big Data Res. **24**, 1–12 (2021)
7. Akram, W., et al.: Blockchain technology: a potential tool for the management of pharma supply chain. Res. Social Adm. Pharm. **20**(6), 156–164 (2024)
8. Durá, M., et al.: Blockchain for data originality in pharma manufacturing. J. Pharm. Innov. **18**(4), 1745–1763 (2023)
9. Kuhn, M., Funk, F., Franke, J.: Blockchain architecture for automotive traceability. Procedia CIRP **97**, 390–395 (2021)
10. Ada, N., et al.: Blockchain technology for enhancing traceability and efficiency in automobile supply chain–a case study. Sustainability **13**(24), 1–18 (2021)

11. Lu, D.: Toward privacy-aware traceability for automotive supply chains. SAE Int. J. Transp. Cybersecurity Priv. **4**(2), 1–10 (2021)
12. Gartner, J., et al.: Potentials of traceability systems–a cross-industry view. Procedia CIRP **104**, 1–6 (2021)
13. Cheshmberah, M., Beheshtikia, S.: Supply chain management maturity: an all-encompassing literature review on models, dimensions and approaches. LogForum **16**(1), 8–25 (2020)
14. Schuitemaker, R., Xu, X.: Product traceability in manufacturing: a technical review. Procedia CIRP **93**, 700–705 (2020)
15. Steghofer, J.-P. et al.: Design decisions in the construction of traceability information models for safe automotive systems. In: IEEE 29th International Requirements Engineering Conference, pp. 1–10. IEEE (2021)
16. Mitsiaki, A. et al.: Enhancing defect traceability and data integrity in Industry 4.0 using blockchain. In: X ECCOMAS Thematic Conference on Smart Structures and Materials (SMART 2023), pp. 1173–1184. ECCOMAS (2023)
17. Chrobot, J.: A concept for a production flow control system toolset for discrete manufacturing of mechanical products. Appl. Comput. Sci. **21**(1), 143–152 (2025)

The Evolution of Maintenance Strategies in the Context of Industrial Development and Sustainable Development

Bartosz Adamski(✉) and Marta Grabowska

Faculty of Mechanical Engineering, Poznan University of Technology, Poznan, Poland
bartosz.adamski@put.poznan.pl

Abstract. The article presents the evolution of approaches to maintenance in the context of technological, industrial and socio-economic development. The analysis covers successive stages in the development of maintenance strategies – from the reactive approach characteristic of the early days of industrialisation, through preventive and predictive maintenance models, to contemporary concepts based on digitalisation, artificial intelligence and the principles of sustainable development. Particular attention is paid to the links between the development of industrial technologies and the evolution of maintenance management methods, including the impact of Industry 4.0 and the concept of Industry 5.0. The article also discusses the importance of sustainability as a key factor shaping modern maintenance systems and provides an overview of research on the integration of economic, environmental and social aspects in the assessment of maintenance effectiveness. On this basis, directions for further research are proposed, focusing on identifying the maturity level of maintenance systems and developing tools to support their development in modern industry.

Keywords: maintenance · sustainability · maturity level of maintenance

1 Introduction and Research Approach

Machine maintenance serves as a fundamental pillar for ensuring industrial productivity, reliability, and competitiveness. Its progression, from basic reactive fixes to advanced, digitally integrated approaches, mirrors the broader technological and organizational advancements shaping the industrial sector [1]. The history of maintenance has evolved significantly over time, reflecting changes in technology, industrial needs, and management practices. It transitioning from a rudimentary and often neglected necessity to a critical and strategic function within organizations. In the past, maintenance was not systematically planned or executed, especially in ancient times when machinery was simpler and easier to control [2]. Today, maintenance is not merely a cost center but a strategic function, leveraging advanced analytics, artificial intelligence, and human-centric design to optimize asset performance and sustainability [1, 3, 4].

The aim of this article is to present the evolution of approaches to maintenance in the context of technological, industrial and socio-economic development. The analysis

J. Trojanowska et al. (Eds.): MANUFACTURING 2026, LNME, pp. 305–314, 2026.
https://doi.org/10.1007/978-3-032-21654-0_23

covers successive stages in the development of maintenance strategies from the turn of the XIXth and XXth centuries to the contemporary approach associated with Industry 4.0 and 5.0. Therefore, the selection of literature sources depended on the nature of the issues described and their place in a specific time frame. When discussing maintenance strategies characteristic of particular historical periods, there are used primarily publications from the analyzed eras or contemporary studies containing reliable descriptions and interpretations of these periods. This allowed us to maintain the consistency of the technological context and accurately reflect the conditions for the development of particular concepts.

In turn, when analyzing the tools, methods and contemporary solutions used in maintenance, priority was given to literature published in the last five years to ensure that the information presented was up to date and consistent with the current state of knowledge.

The literature search process was based on the selection of keywords related to the evolution of maintenance strategies, such as: 'history of machine maintenance', 'evolution of maintenance', 'maintenance strategies', as well as queries relating to specific historical periods (e.g. 'maintenance during World War II', 'maintenance in the 1950s', 'maintenance in the 1960s').

The main sources of data were the Elsevier Scopus and ResearchGate bibliographic databases and the Google Scholar search engine, which gave access to both peer-reviewed scientific publications and relevant industry reports and research reviews.

1.1 The Beginning of Machinery Maintenance (up to the 1950s)

The maintenance of equipment and machinery up to the 1950s was characterized by the predominance of a reactive maintenance, representing the first of the three historical stages in the evolution of maintenance concepts [2]. Maintenance during this period was based on ad hoc responses to equipment failures, with repairs carried out only after breakdowns occurred. This era spanned from the early use of machines and equipment up to approximately the beginning of the Second World War. Due to the relatively low level of mechanization, failures had little impact on production continuity. Machines and devices were simple in design, easy to maintain, and their operation did not require high qualifications. It can be stated, with some simplification, that there was no significant need for preventive actions other than basic cleaning and lubrication. More substantial maintenance activities—such as machine overhauls, component repairs, or part regeneration—were performed usually in response to failures. The prevailing belief at the time was that the frequency of equipment failures depended on the age of the machinery and was associated with physical aging processes [2, 3].

During the 1940s, maintenance activities were often viewed as a "necessary evil." Repairs and replacements were primarily reactive, occurring only after equipment or its components had failed. This approach is commonly referred to as a run-to-failure maintenance strategy [5]. Prior to World War II (1939–1945), the maintenance and continued operation of machinery relied heavily on the human senses—sight, hearing, smell, touch, and even taste—to identify changes in performance that could indicate a pending failure if left unaddressed. Generally, maintenance followed a reactive approach: "If it's not broken, don't fix it." Wartime conditions often led to labor shortages and increased demand for machinery, which made reactive maintenance the default strategy due to limited resources and the urgent need to keep equipment operational [6]. Industries began to recognize the limitations of reactive maintenance and started to adopt preventive strategies, scheduling maintenance to avoid failures and improve reliability [5, 6]. This involved scheduling regular inspections, often during annual or semi-annual shutdowns, where machinery would be disassembled, inspected, and any worn or faulty components replaced. These practices were typically supplemented by periodic checks on auxiliary equipment, such as performing oil changes in engines or gearbox systems. Hand tools, repair manuals, and simple diagnostic equipment were standard in the 1940s and 1950s [2, 3, 7].

1.2 Early Development of Maintenance (1950s–1970s)

Comparing the evolution of machinery maintenance across defined periods reveals that up to the 1950s, maintenance was reactive and largely unplanned, driven by specific wartime demands. Between the 1950s and 1970s, preventive maintenance was adopted to minimize downtime with scheduled inspections and component replacements. Mechanization increased machine complexity and downtime became costly, prompting a move from purely corrective maintenance to the first models of preventive maintenance (PM). Basic sensors were used, manual repair schedules were prepared and the first logbooks were kept. In these years it was still dominant Single-Unit Models - focused on individual assets, ignoring interdependencies [5, 8].

Over the years maintenance became more technical, with regular overhauls and scheduled activities to optimize reliability and reduce unscheduled downtime. The concept of predictive maintenance emerged, using performance and structural health monitoring. Maintenance schedules and checklists became common in the 1960s for early preventive maintenance. By the late 1960s and 1970s, industries started using more advanced sensors, performance monitoring devices, and statistical analysis to predict failures and optimize maintenance intervals. Preventive maintenance was well established, and statistical techniques were applied to calculate failure probabilities and set optimal replacement intervals. Condition-based maintenance (CBM) began to be adopted, relying on data and monitoring to guide maintenance actions [1, 2, 9].

1.3 Focus on Maintenance (1980s–2000s)

The evolution of machinery maintenance strategies from the 1980s to the 2000s is marked by a transition from mainly preventive, manual methods to integrated, digital, and predictive approaches. In late 1970s and early 1980s, preventive maintenance became more structured and time-based. The electrical machinery industry played a leading role in Japan's "high-technology miracle" during the 1980s, with significant contributions to industrial structure and maintenance practices. Early adoption of Reliability-Centered Maintenance (RCM) and basic computerized systems began. In this period we can trace our origins to the first Computerised Maintenance Management Systems (CMMS), which supported the work of service technicians and significantly improved processes. Manual work was increasingly replaced by automation and control systems, reducing labor and improving consistency in maintenance tasks. The adoption of Computerized

Numerical Control (CNC) technology made machinery more flexible and easier to maintain, as programming devices allowed for precise control and diagnostics. Also integration of microprocessors and electronic devices (sensors, actuators, controllers) enabled more sophisticated monitoring and control of machinery. The economic crisis of the 1980s led to increased focus on cost-effective maintenance solutions and adoption of technologies that reduced resource requirements [2, 3].

As the years pass, the shift from preventive to predictive maintenance was facilitated by machinery monitoring technologies, allowing for maintenance based on actual equipment condition rather than fixed schedules. The period of the 1990s is a time of introduction and formalization of preventive maintenance practices, including detailed documentation and instructions, helped organizations proactively address equipment aging and degradation, reducing downtime and improving reliability. Shift from reactive and scheduled maintenance to predictive and condition-based approaches, enabled by advancements in machinery monitoring technology and sensing. This allowed for maintenance actions based on actual equipment condition rather than fixed schedules. The first mention of Total Productive Maintenance (TPM) appeared during this period. Its rapid development began in the late 1990s [10, 11].

The 2000s represented a transformative period in machinery maintenance, marking a shift from conventional, time-based preventive approaches to more advanced, data-driven, and reliability-oriented strategies. During this time, methods like RCM and TPM gained traction, focusing on structured planning, operator engagement, and leveraging cutting-edge technologies for condition monitoring and predictive maintenance. The adoption of sensor networks, computerized management tools, and machine learning algorithms revolutionized maintenance operations, allowing organizations to enhance equipment reliability, lower costs, and boost overall operational efficiency [1, 2].

RCM originated in the military aircraft industry and was adapted across sectors such as maritime, nuclear, oil and gas, and manufacturing during the 2000s. It was marked by its integration of quantitative analysis and adaptability to diverse industrial contexts, enabling more precise and cost-effective maintenance planning [12, 13].

1.4 Maintenance in the Modern World (Nowadays)

The 2010s represented a pivotal shift in maintenance practices, transitioning away from traditional reactive and preventive methods toward sophisticated, data-driven strategies. During this period, condition-based maintenance (CBM) and predictive maintenance (PdM) became increasingly prominent within industrial operations. These methodologies leveraged continuous or periodic monitoring of machine condition parameters—such as vibration, temperature, pressure, and acoustic emissions—facilitated by the expanding availability of Internet of Things (IoT) devices, wireless sensor networks, and advanced measurement technologies. Maintenance actions were no longer dictated solely by predetermined schedules based on time or usage. Instead, they were guided by the real-time condition of equipment, thereby minimizing unnecessary downtime and redundant maintenance tasks [1–3, 14, 15].

A critical enabler of this paradigm shift was the incorporation of big data analytics and machine learning into maintenance systems. Advanced algorithms, including support vector machines, convolutional neural networks, and recurrent neural networks, were increasingly applied for fault detection, diagnostics, and the estimation of remaining useful life. These methods enhanced the ability of maintenance systems to process vast and diverse datasets while uncovering intricate degradation patterns that traditional threshold-based models could not detect. Consequently, decision-making processes transitioned from static and rule-based models to more dynamic and probabilistic frameworks, better suited to complex operational scenarios. Additionally, the 2010s witnessed the emergence of digital twin technology and building information modeling (BIM) as nascent tools for enhancing maintenance planning, particularly for intricate mechanical, electrical, and plumbing systems. Although these technologies remained in their infancy during this decade, they facilitated the virtual modeling of assets, supporting predictive simulations and improving visualization capabilities. However, despite the clear advantages offered by predictive maintenance, its widespread adoption was hindered by substantial initial investment requirements, issues related to data quality, and the demand for specialized expertise—challenges that were particularly pronounced for small and medium-sized enterprises [1, 2, 16].

The 2020s marked significant advancements in maintenance strategies, further evolving concepts introduced in the previous decade. This era saw the rise of prescriptive maintenance as a leading framework. Building on predictive analytics, prescriptive maintenance not only anticipated equipment failures but also recommended optimal actions and schedules to address them. By combining machine learning, optimization algorithms, causality analysis, and semantic reasoning, this approach facilitated automated, context-aware decision-making. Maintenance systems became increasingly self-adaptive, dynamically adjusting strategies to accommodate changing operational conditions and uncertainties [2, 3].

Advancements in artificial intelligence underpinned this transition. Technologies such as deep learning architectures, reinforcement learning, and multi-agent systems enabled threshold-free decision models and real-time optimization for managing complex, multi-component assets. Digital twins progressed into fully functional cyber-physical replicas of physical systems, allowing continuous synchronization between the physical and virtual realms. These digital twins enhanced capabilities such as advanced simulations, what-if scenario analyses, and closed-loop prescriptive decision-making across entire production networks. Maintenance practices in the 2020s became deeply aligned with organizational priorities, including sustainability, energy efficiency, and asset lifecycle management. Innovations like energy-harvesting sensors and low-power wireless networks minimized the environmental impact of monitoring systems. Maintenance strategies actively contributed to reducing waste and extending asset lifespans while fostering interconnectivity across production processes, supply chains, and resource management systems as part of Industry 4.0 and the emerging Industry 5.0 ecosystems [16–18].

Despite its substantial long-term benefits—such as lower lifecycle costs, greater reliability, and improved operational resilience—prescriptive maintenance also introduced challenges. High implementation complexity, concerns over cybersecurity, and demands for robust data governance slowed adoption, particularly for smaller enterprises. However, the 2020s firmly positioned intelligent, autonomous, and sustainability-driven maintenance systems as a cornerstone of modern industrial operations. This foundation paved the way for future ecosystems that could self-diagnose and autonomously schedule cyber-physical maintenance activities [2, 3] (Table 1).

Table 1. Overview of changes in machine maintenance strategies over the years [Original source].

Period of time	Dominant Strategy	Tools and Techniques	Technological Innovations	Key Changes and Drivers
Up to the 1950s	Reactive Maintenance	Basic hand tools, reparation after breakdown	Run-to-failure maintenance strategy	Relied on the human senses
1950s	Time-based Preventive Maintenance	Scheduled inspections, manual checklists, basic instrumentation	Mechanized equipment, early industrial sensors	Postwar industrial growth, mechanization, heavy industry needs
1960s	Expanded Preventive Maintenance	Routine checkups, standardized procedures, basic condition monitoring	Improved engine/material tech, early automation	Systematization, safer/longer operational life
1970s	Preventive Maintenance with Diagnostics	Fault diagnostics, remote sensing, lubrication analysis	Adaptive machine control, CNC, robots, ferrography	Shift to condition-based monitoring, productivity focus
1980s	Computerized & Integrated Maintenance	CMMS, computerized scheduling, expert systems	NC/CNC, digital controls, positioning systems	Automation, systems integration, cost reduction
1990s	TPM & Reliability-Centered Maintenance	Condition monitoring, reliability analysis, structured planning	Advanced diagnostics, integrated automation, data-driven tools	Globalization, competitiveness, uptime/efficiency focus
2000s	Predictive/Proactive Maintenance	Condition monitoring, real-time analytics, sensor tech	IoT, ML algorithms, advanced sensor networks	Cost reduction, downtime minimization, complexity rise
2010s	Industry 4.0-Integrated Maintenance	Digital twins, AI analytics, cloud platforms	Big Data, cyber-physical systems	Predictive focus, IT/OT integration, smarter decisions

(*continued*)

Table 1. (*continued*)

Period of time	Dominant Strategy	Tools and Techniques	Technological Innovations	Key Changes and Drivers
Early 2020s	Advanced Predictive & Cognitive Maintenance	Edge computing, digital twins, sensor fusion, mobile apps	5G, advanced IoT, robotics, AR tools	Real-time optimization, sustainability, resilience
Present	Hybrid Preventive and Predictive Blend, Prescriptive Maintenance	Integrated CMMS, predictive analytics, remote monitoring	AI, cloud, digital twin evolution, Industry 4.0/5.0	Data-driven, legacy-modern integration, human-centric

2 Steps in the Development of Maintenance in the Context of Industrial Revolutions and Contemporary Conditions

In the article by researchers from the University of West Bohemia, the various developmental stages of machine maintenance were correlated with the industrial revolution, starting from Industry 1.0 and reactive maintenance, up to predictive maintenance in Industry 4.0 (Table 2). This serves as an interesting way to categorize the different phases of maintenance strategy development [15]. In this context, it would also be reasonable to compare the various periods of development in machine maintenance alongside the industrial revolution. This would provide insight into how these aspects have evolved over time and reveal the true correlation between these two approaches to presenting the progression of machine maintenance.

Table 2. Correlation of Industrial Revolution and Maintenance [15].

Industry revolution	Industry 1.0	Industry 2.0	Industry 3.0	Industry 4.0
Characteristics of the industrial revolution	Mechanization, steam power, weaving loom	Mass production, assembly lines, electrical energy	Automation, computers, electronics	Cyber Physical Systems, IoT, networks, cloud, BDA
Type of maintenance	**Reactive maintenance**	**Planned maintenance**	**Productive maintenance**	**Predictive maintenance**
Inspection	Visual inspection	Instrumental inspection	Sensor monitoring	Predictive analysis
OEE	<50%	50–75%	75–90%	>90%
Maintenance team reinforcement	Trained craftsmen	Inspectors	Reliability engineers	Data scientists

The distinguishing features of industrial revolutions relate, on the one hand, to the technical and technological solutions available in a given period of time and, on the other hand, to social and economic conditions.

As mentioned in Sects. 2.1 and 2.2 of this article, changes in the approach to maintenance depended significantly on the specific needs of the arms industry and were then conditioned by the shortages resulting from the Second World War.

Moving on to the present day, it can be said that, in technical and technological terms, the conditions for maintenance are determined by Industry 4.0, while in social and economic terms, sustainable development, sustainable production and the circular economy are important indicators.

Sustainable production is defined as the manufacture of goods using processes that minimize negative impacts on the environment, promote efficient use of energy and resources, ensure the safety of employees, communities and consumers, and remain economically viable. Its overarching goal is to achieve a balance between the environmental, social and economic dimensions of business activities, which makes it possible to meet stakeholder expectations and build a sustainable competitive advantage [19]. A review of the literature reveals a number of studies aimed at defining the impact of conservation on addressing the challenges of sustainable production. The literature on the subject increasingly emphasizes the need to take into account the principles of sustainable development in the assessment of maintenance performance, treating them as an important support in meeting the challenges of modern production. In this context, numerous authors propose various approaches that integrate environmental, social and economic aspects into maintenance performance assessment systems. One approach is proposed by Franciosi et al. [20], who developed a holistic conceptual framework for assessing the impact of maintenance activities on sustainable development and the impact of maintenance on specific areas of an organisation's operations, right down to the end customer level. Singh and Gupta identified fourteen key maintenance factors that influence sustainable performance improvement [21]. Subsequent studies focused on integrating the concept of sustainable development with classic performance management tools, e.g. a performance assessment model combining the three dimensions of sustainable development with the four perspectives of the Balanced Scorecard (finance, customer, internal processes, and development and learning) [22]. Another example is a comprehensive system for assessing sustainable maintenance performance for automotive companies, based on the Balanced Scorecard concept, covering eight perspectives assigned to the three pillars of sustainable development and a set of indicators at the strategic, tactical and operational levels [23, 24].

The examples cited above confirm that research into the relationship between sustainable development and maintenance, while also highlighting the trend towards developing methods for analyzing [25] and evaluating the effectiveness of the maintenance process.

3 Summary and Plans for Further Research

Based on an analysis of the literature on the evolution of various approaches to maintenance, the relationship with industrial revolutions and socio-economic conditions (including sustainable development), and trends in the development of methods and tools for analyzing maintenance effectiveness, the authors propose a thesis and research objective. Firstly, today, various companies use more or less advanced maintenance systems and tools, corresponding to different periods of concept development and different industrial revolutions. These different approaches will be referred to in further research as the maturity of the maintenance system. Table 1 contains a preliminary classification and description of maturity classes. Secondly, it is possible to develop a tool for identifying the maturity class in an enterprise. Assigning a maintenance system to one of the maturity classes will make it possible to indicate directions for development, identify gaps, and propose methods for improving maintenance. The tool will be based on a questionnaire and supported by observation and conclusions from in-depth interviews.

Acknowledgments. The presented results derive from a scientific statutory research conducted by Chair of Production Engineering, Faculty of Mechanical Engineering, Poznan University of Technology, Poland, supported by the Polish Ministry of Science and Higher Education from the financial means no 0613/SBAD/4940.

References

1. Adaramola, B.A., Kayode, J.F., Monye, S.I., Afolalu, S.A.: Overview of maintenance management strategies in the industry. In: International Conference on Science, Engineering and Business for Driving Sustainable Development Goals, SEB4SDG 2024. Institute of Electrical and Electronics Engineers Inc. (2024). https://doi.org/10.1109/SEB4SDG60871.2024.10629938
2. Legutko, S.: Development trends in machines operation maintenance (2009)
3. Mushiri, T., Mbohwa, C.: Machinery maintenance yesterday, today and tomorrow in the manufacturing sector. In: World Congress on Engineering : WCE 2015. Newswood Limited : International Association of Engineers, London (2015)
4. Grenyer, A., Dinmohammadi, F., Erkoyuncu, J.A., Zhao, Y., Roy, R.: Current practice and challenges towards handling uncertainty for effective outcomes in maintenance. Procedia CIRP. **86**, 282–287 (2019). https://doi.org/10.1016/j.procir.2020.01.024
5. Alla, H.R., Hall, R., Apel, D.B.: Performance evaluation of near real-time condition monitoring in haul trucks. Int. J. Min. Sci. Technol. **30**, 909–915 (2020). https://doi.org/10.1016/j.ijmst.2020.05.024
6. Veruz, E.G., Silva, A.J., Michalski, M.A.C., da Silva, R.F., de Souza, G.F.M., Oshiro, A.T.: A framework for planning underwater hull inspections based on computer vision and degradation assessment. Ocean Eng. **317**, 120053 (2025). https://doi.org/10.1016/j.oceaneng.2024.120053
7. Roylance, B.J.: Machine failure and its avoidance—What is tribology's contribution to effective maintenance of critical machinery? Proc. Inst. Mech. Eng. Part J J. Eng. Tribol. **217**, 349–364 (2003). https://doi.org/10.1243/135065003322445278
8. Deshpande, V.S., Modak, J.P.: Application of RCM to a medium scale industry. Reliab. Eng. Syst. Saf. **77**, 31–43 (2002). https://doi.org/10.1016/S0951-8320(02)00011-X
9. Karki, B.R., Basnet, S., Xiang, J., Montoya, J., Porras, J.: Digital maintenance and the functional blocks for sustainable asset maintenance service – a case study. Digit. Bus. **2**, 100025 (2022). https://doi.org/10.1016/J.DIGBUS.2022.100025
10. Adgar, A., Jantunen, E., Arnaiz, A.: Towards true dynamic decision making in maintenance

11. Leal, A., Bokrantz, J., Skoogh, A.: The role of maintenance in company-specific production systems. In: Procedia CIRP, pp. 584–590. Elsevier B.V. (2024). https://doi.org/10.1016/j.procir.2024.10.133
12. Mokashi, A.J., Wang, J., Vermar, A.K.: A study of reliability-centred maintenance in maritime operations. Mar. Policy **26**, 325–335 (2002). https://doi.org/10.1016/S0308-597X(02)00014-3
13. Elhalim, E.K.A., Abdel-Magied, R.K., Afefy, I.H., Aly, M.F.: An efficient maintenance plan using proposed framework of RCM made simple approach. Ind. Eng. Manag. Syst. **18**, 222–233 (2019). https://doi.org/10.7232/iems.2019.18.2.222
14. Gao, G., Zhou, D., Tang, H., Hu, X.: An intelligent health diagnosis and maintenance decision-making approach in smart manufacturing. Reliab. Eng. Syst. Saf. **216**, 107965 (2021). https://doi.org/10.1016/J.RESS.2021.107965
15. Poor, P., Ženíšek, D., Basl, J.: Historical overview of maintenance management strategies: development from breakdown maintenance to predictive maintenance in accordance with four industrial revolutions. In: Proceedings of the International Conference on Industrial Engineering and Operations Management, pp. 495–504. Pilsen (2019)
16. Zhang, Y., Cai, B., Gao, C., Zhao, Y., Shao, X., Yang, C.: A system-centred predictive maintenance re-optimization method based on multi-agent deep reinforcement learning. Expert Syst. Appl. **274**, 127034 (2025). https://doi.org/10.1016/J.ESWA.2025.127034
17. Chen, S., Turanoglu Bekar, E., Bokrantz, J., Skoogh, A.: AI-enhanced digital twins in maintenance: Systematic review, industrial challenges, and bridging research–practice gaps. J. Manuf. Syst. **82**, 678–699 (2025). https://doi.org/10.1016/J.JMSY.2025.07.006
18. Tian, Y., Liu, A.: Responsible manufacturing toward Industry 5.0. Manuf. Ind 4.0 Ind 5.0 Adv. Appl. **5**, 231–263 (2024). https://doi.org/10.1016/B978-0-443-13924-6.00008-9
19. Jasiulewicz-Kaczmarek, M., Żywica, P., Gola, A.: Fuzzy set theory driven maintenance sustainability performance assessment model: a multiple criteria approach. J. Intell. Manuf. **32**, 1497–1515 (2021). https://doi.org/10.1007/s10845-020-01734-3
20. Franciosi, C., Di Pasquale, V., Iannone, R., Salvatore, M.: Multi-stakeholder perspectives on indicators for sustainable maintenance performance in production contexts: an exploratory study. J. Qual. Maint. Eng. **27**(2), 308–330 (2021). https://doi.org/10.1108/JQME-03-2019-0033
21. Singh, R.K., Gupta, A.: Framework for sustainable maintenance system: ISM–fuzzy MICMAC and TOPSIS approach. Ann. Oper. Res. **290**, 643–676 (2020)
22. Jasiulewicz-Kaczmarek, M., Żywica, P.: The concept of maintenance sustainability performance assessment by integrating balanced scorecard with non-additive fuzzy integral. Maint. Reliab. **20**, 650–661 (2018)
23. Sari, E., Shaharou, A., Maaram, A., Yazid, A.M.: Sustainable maintenance performance measures: a pilot survey in Malaysian automotive companies. Procedia CIRP **26**, 443–448 (2015)
24. Kucejko, M., Bugdol, M.: Recommendation system for DFMEA analysis using neural network embeddings. Manag. Prod. Eng. Rev. **16**(3), 1–13 (2025)
25. Luściński, S., Bednarek, M., Jabłoński, M.: Enhancing predictive maintenance of industrial assets through machine diagnostic parameter grouping. Manag. Prod. Eng. Rev. **16**(3), 1–12 (2025)

Operational Excellence and Logistics

A Low-Cost Model for Digital Transformation and Cybersecurity in Production Management

Leonel Patrício[1], Leonilde Varela[1], Justyna Trojanowska[2], Zilda Silveira[3], Nuno Leal[4], and José Machado[1](✉)

[1] University of Minho, 4804-533 Guimarães, Portugal
jmachado@dem.uminho.pt
[2] Poznan University of Technology, Poznan, Poland
[3] University of Sao Paulo, Sao Paulo, Brazil
[4] Sonae Arauco, Portugal

Abstract. Digital transformation in production management has become a key driver of operational efficiency and industrial competitiveness. However, the implementation of digital technologies presents significant challenges, particularly when organizations seek low-cost solutions that also ensure strong cybersecurity and operational continuity. This study conducts a systematic literature review and proposes a low-cost conceptual model that integrates digital transformation practices with robust cybersecurity measures in production management. By examining the relationship between accessible digital technologies—such as low-cost IoT sensors, modular automation systems, and open-source platforms—and essential cybersecurity requirements, the study identifies the main enablers and barriers to achieving secure and economically viable digitalization. The proposed model aims to guide organizations, especially small and medium-sized enterprises, in adopting secure, scalable, and financially sustainable digital solutions. The central research question addressed is: How can a low-cost model be developed to integrate digital transformation and cybersecurity for improved efficiency in production management? The findings contribute to a practical and comprehensive approach, offering valuable insights for academics, policymakers, and industry practitioners.

Keywords: Digital Automation · Cybersecurity · Production Management · Low-Cost

1 Introduction

The integration of digital transformation into production management has become a fundamental driver of competitiveness and innovation in modern industries. As organizations navigate the challenges of a rapidly evolving global market, the adoption of digital technologies such as the Internet of Things (IoT), cloud computing, and data analytics has become increasingly essential for improving efficiency and responsiveness [1]. However, the growing dependence on interconnected systems has simultaneously amplified cybersecurity risks, particularly for organizations with limited financial resources.

J. Trojanowska et al. (Eds.): MANUFACTURING 2026, LNME, pp. 317–331, 2026.
https://doi.org/10.1007/978-3-032-21654-0_24

This dual challenge highlights the need for accessible, low-cost digital solutions that can enhance production management while ensuring adequate protection against cyber threats.

Digital transformation in production management refers to the systematic adoption of digital tools and technologies to optimize industrial processes, improve decision-making, and strengthen operational continuity [2]. Industry 4.0, which encapsulates this digital revolution, emphasizes the convergence of cyber-physical systems, intelligent automation, and real-time data exchange [3]. While these advancements offer substantial benefits—such as reduced downtime, enhanced traceability, and improved process control—they also introduce new vulnerabilities, including data breaches, unauthorized access, and system disruptions. Cybersecurity, therefore, has become an indispensable component of digital transformation, yet many companies lack the financial and technical capacity to implement sophisticated protection mechanisms.

The increasing prevalence of cyber incidents in industrial settings underscores the urgency of integrating cybersecurity considerations into digital transformation strategies [4]. Small and medium-sized enterprises (SMEs), in particular, face disproportionate risks due to limited budgets, outdated infrastructure, and insufficient cybersecurity expertise. As a result, there is a pressing need for models that support digital transformation in a cost-effective manner while addressing essential cybersecurity requirements. Existing literature has explored various dimensions of digitalization and cybersecurity in manufacturing, including secure IoT architectures, risk assessment methodologies, and protection of industrial control systems [5]. However, research remains fragmented and often focuses on high-cost or highly specialized solutions that are not accessible to smaller organizations.

The state of the art reveals a significant gap in the development of integrated, low-cost frameworks that simultaneously address digital transformation and cybersecurity in production management. While some studies examine the economic benefits of digital technologies [6] and others investigate cybersecurity challenges in industrial environments [7], few efforts have been made to combine these perspectives into a unified, actionable model tailored to resource-constrained organizations. This gap points to the need for a comprehensive approach capable of guiding companies in adopting secure digital systems without incurring prohibitive costs.

To address this challenge, the present study proposes a conceptual low-cost model that integrates digital transformation practices with cybersecurity principles in production management. Drawing on multidisciplinary literature, the model identifies key technological, organizational, and economic enablers and barriers that influence the adoption of secure and affordable digital solutions [8]. By synthesizing insights from both digital transformation and cybersecurity domains, this research seeks to provide a structured pathway for organizations aiming to modernize their production systems responsibly and economically.

The central research question guiding this study is:

- **How can a low-cost model be developed to integrate digital transformation and cybersecurity for improved efficiency in production management?**

This overarching question gives rise to several sub-questions: What are the main enablers and barriers to implementing low-cost digital transformation with embedded cybersecurity? Which affordable technologies and practices can support secure digitalization in production environments? How can organizations balance cost constraints with the need for robust cybersecurity measures? Addressing these questions contributes to the creation of a practical, accessible framework that facilitates secure digital adoption across different industrial contexts.

In summary, this study aims to advance the understanding of how digital transformation and cybersecurity can be jointly structured through a low-cost approach that supports efficiency and security in production management. Through a systematic review of existing literature and the development of a conceptual model, this research offers valuable guidance for academics, policymakers, and industry practitioners seeking to implement feasible and secure digitalization strategies.

The organization of this article is as follows: Sect. 2 provides an overview and critical examination of the relevant literature; Sect. 3 introduces the research framework along with the proposed model; Sect. 4 offers a discussion of the findings; and Sect. 5 delivers the final conclusions. The article ends with the list of referenced sources.

2 Methodology

2.1 Identification of Relevant Articles

The literature review presented in this chapter is based on carefully chosen sources that are directly relevant to the research question. The findings highlighted here are drawn from the work of established authors who have investigated this topic or closely related areas. The set of publications analyzed was compiled using the online platform B-on, selected because it offers full-text access to a broad collection of indexed journals, conference proceedings, eBooks, and reference materials. B-on consolidates content from numerous leading international academic publishers and societies. Typical coverage through national consortia such as B-on includes major providers like Elsevier (including ScienceDirect, Cell Press, and The Lancet), Springer Nature (SpringerLink, Nature Portfolio, Palgrave Macmillan), Wiley (Wiley Online Library), Taylor & Francis (including Routledge), SAGE Publishing, Oxford University Press, Cambridge University Press, IEEE (IEEE Xplore), ACM (ACM Digital Library), American Chemical Society (ACS Publications), Royal Society of Chemistry (RSC), IOP Publishing, AIP Publishing, SPIE, Wolters Kluwer (including Lippincott Williams & Wilkins), De Gruyter, Emerald Publishing, MDPI, Frontiers, PLOS, BMJ, Royal Society Publishing, Thieme, Karger, Brill, and AAAS (Science), among others. Access and available content may vary depending on institutional subscriptions and over time; thus, the list should be viewed as illustrative rather than comprehensive. Through the Portuguese Foundation for Science and Technology's online scientific library, researchers can reach many of these databases and publisher platforms via B-on. This platform formed the basis of the search strategy for the current study, which was structured into three groups as shown in Table 1 (Group 1, Group 2, and Group 3). The literature review conducted in this study is based on the principles of the PRISMA (Preferred Reporting Items for Systematic Reviews and Meta-Analyses) methodology, which is widely used to ensure transparency, rigor, and reproducibility in systematic literature reviews. PRISMA provides a structured framework for identifying, screening, assessing eligibility, and selecting relevant studies through a clearly defined and documented process. In this research, these principles were applied by adopting a systematic search strategy, explicit inclusion and exclusion criteria, and a transparent multi-stage screening procedure. The identification phase involved structured searches in the B-on platform using predefined keyword groups related to digital transformation, production management, and cybersecurity. During the screening phase, titles, keywords, and abstracts were analyzed to remove duplicates and studies not aligned with the research scope. The eligibility phase applied filters related to publication type, peer-review status, language, time span (2015–2025), and full-text availability. Finally, the inclusion phase resulted in the selection of the most relevant studies that directly address low-cost digital transformation and cybersecurity in production management. Figure 1 presents the flow diagram of the literature search and screening process, reflecting the main PRISMA stages and ensuring methodological clarity, traceability, and reproducibility of the review.

Table 1. Groups of searched through "B-on".

Group 1	Group 2	Group 3
"Digital Automation" OR "IoT" OR "Digital Transformation" OR "AI" OR "Machine Learning" OR "Robotics" OR "Smart Manufacturing" OR "Industry 4.0" OR "Big Data" OR "Cloud Computing" OR "Process Optimization"	"Production Management" OR "Supply Chain Management" OR "Lean Manufacturing" OR "Quality Control" OR "Operational Efficiency" OR "Inventory Management" OR "Production Planning" OR "Manufacturing Systems" OR "Process Optimization" OR "Resource Allocation" OR "Low-Cost Solutions" OR "Cost-Effective Implementation" OR "Affordable Technology" OR "Open-Source Implementation" OR "Open-Source Tools" OR "Open-Source Software" OR "Budget-Friendly Solutions"	"Cybersecurity" OR "Information Security" OR "Data Protection" OR "Network Security" OR "Industrial Cybersecurity" OR "Cyber Risk Management" OR "Threat Detection" OR "IoT Security" OR "Cyber-Physical Systems Security" OR "Digital Security" OR "Cyber Resilience" OR "Security Frameworks" OR "Critical Infrastructure Protection" OR "Malware Protection" OR "Access Control"

Separate searches were conducted using the B-on platform, employing the three groups and connecting terms with the OR operator applied to the Title or Keywords (KW) of each set. Table 2 presents the number of articles retrieved in each search.

After applying the filters (Fig. 1), the titles, keywords, and abstracts of the articles were reviewed to determine which studies were directly relevant to the research topic. The initial search yielded 257 papers; after applying the filters, a total of 23 articles remained, of which only 14 were directly aligned with the focus of this study.

Table 2. Research tests performed through the "B-on".

	Set 1	
Title	(Group 1 AND Group 2 AND Group 3)	n = 14
Keywords (KW)	(Group 1 AND Group 2 AND Group 3)	n = 243

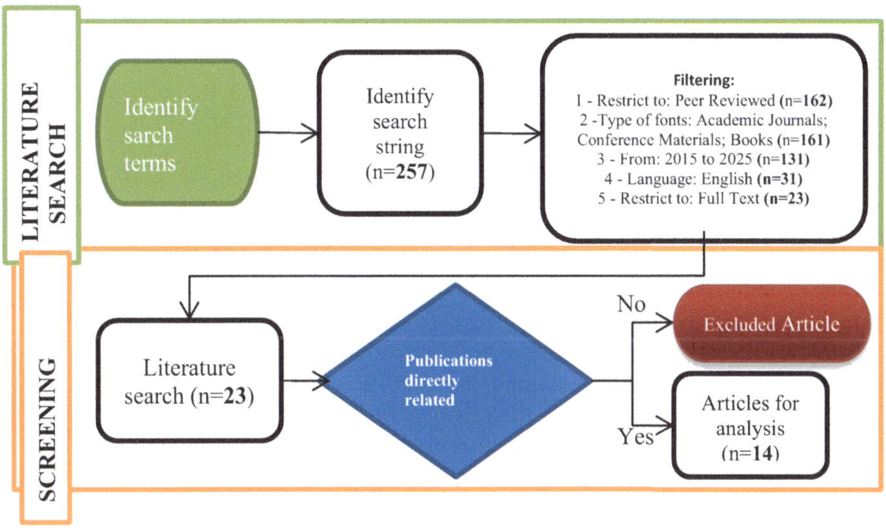

Fig. 1. Flow diagram of literature search and respective screening.

Subsequently, during the research process, a series of filters were applied to the collected publications, and the resulting number of articles is summarized in Table 3. The inclusion criteria comprised peer-reviewed journal articles, conference papers, and books published between 2015 and 2025, written in English, and directly addressing digital transformation, cybersecurity, and production management. Exclusion criteria included non-peer-reviewed sources, papers not focused on industrial or production contexts, and studies lacking relevance to low-cost or SME-oriented solutions.

In the following Section, the analysis and synthesis of the selected articles are presented. This section provides data from the studies deemed relevant to the topic of this research.

2.2 Articles Synthesis and Analysis

The following section summarizes and analyzes 14 articles relevant to the themes of digital transformation, cybersecurity, and production management (production/manufacturing). Table 4 presents an overview of the type of study and the main themes addressed in each article.

Table 3. Publications obtained through the B-on, after the application of some filters.

	Title	Keywords (KW)
Initial result:	14	243
1 - Restrict to: Peer Reviewed	8	154
2 -Type of fonts: Academic Journals; Conference Materials; Books	8	153
3 - From: 2015 to 2025	8	123
4 - Language: English	3	28
5 - Restrict to: Full Text	3	20

The analysis of these 14 articles reveals several important insights for the topic of integration between digitalization (digital transformation) and cybersecurity in production management:

- Most common technologies: The most recurrent technologies in the studies are IoT, digital twins, artificial intelligence (AI), and cyber-physical systems (CPS). For example, Ahmed et al. (article 8) analyze how AI and digital twins can improve efficiency and safety in additive manufacturing. Barton et al. identify the essential attributes of Industry 4.0 for SMEs, including IoT and cybersecurity.
- Types of study: The articles include a variety of approaches: systematic reviews (e.g., Ahmed, article 8), quantitative models (e.g., Torbacki, article 5), framework proposals (e.g., Wai, article 1; Jiang, article 10), reports/overviews (e.g., Pochmara, article 12), and conceptual analyses (e.g., Hajda, article 3). This diversity suggests that the topic is still under development and is addressed both in theory and in practical applications. • Sustainability/Cost Aspects: Some articles focus on topics that are highly relevant to our low-cost model. For example, Barton et al. (article 7) discuss Industry 4.0 attributes, especially for SMEs, highlighting the role of low-cost-of-entry technologies and cybersecurity. Torbacki (article 5) proposes a multi-criteria decision model to prioritize cybersecurity measures according to sustainability criteria, which is very much aligned with our proposal.
- Specific Security Challenges: Many articles identify risks such as vulnerabilities in PLC/SCADA (Hajda, article 3), the need for network segmentation and encryption (Wai, article 1), gaps in legacy protocols and system upgrades (Wai, article 1), as well as the importance of staff training and security policies (Pochmara, article 12).
- Production-security integration: The proposal to use digital twins for vulnerability assessment (Jiang, article 10) and operational frameworks for secure digital adoption (Metin, article 11) are particularly relevant for a model that aims to unite digital transformation and cyber resilience.
- Organizational skills: Ren et al. (article 9) show that the capacity for digital transformation depends on organizational dimensions, which reinforces the importance of the human factor and business maturity in the adoption of low-cost initiatives.

The synthesis of the articles also highlights some limitations and gaps in the literature, which are relevant to justify and position our proposed model:

Table 4. Articles Synthesis.

	Study Type	Main Topics
[9]	Case Study/Application	SCADA; network segmentation; anomaly detection; access control; encryption; legacy system modernization
[10]	Review	Cybersecurity risks in manufacturing; IoT; cyber-physical systems; mitigation measures
[11]	Conceptual Study/Analysis	Risk management; PLC vulnerabilities; communication protocols; network architecture
[12]	Model/Proposal	AI for production monitoring; data analytics; optimization; automation security
[13]	Quantitative Model	Cybersecurity evaluation; prioritization criteria; sustainability; technological measures
[14]	Study/Technical Proposal	Digital twin; AI; additive manufacturing; proactive security; dynamic monitoring
[15]	Review/Analysis	Industry 4.0 attributes for SMEs; IoT; Big Data; AI; Cloud; cybersecurity; low-cost solutions
[16]	Systematic Review	CPS, IoT, Big Data, AI; production efficiency; predictive maintenance; quality control
[17]	Qualitative Meta-Analysis	Digital transformation capabilities; organizational capacity; maturity; skills framework
[18]	Proposal/Framework	Digital twin for vulnerability assessment; security evaluation; collaboration between production and IT
[19]	Review + Conceptual Model	Digital risk assessment; adoption factors; resilience; secure digitalization
[20]	Report/Overview	Security strategies; ISA/IEC 62443 standards; network monitoring; employee training
[21]	Review	IoT architecture; smart manufacturing; cybersecurity threats; architectural solutions
[22]	Systematic Review	Cybersecurity in critical digital infrastructures; lessons for manufacturing; risk mitigation

1. Fragmented approach: Most studies focus on specific technologies or aspects of cybersecurity (such as PLC, SCADA, AI) without offering an integrated framework that holistically unites digital transformation + security + production management.
2. Lack of empirical validation: Although there are proposed models or frameworks (e.g., Torbacki, Metin, Wai), there are few studies that test these frameworks in real production environments to assess cost, scalability, and security in practice.
3. Context of small and medium-sized enterprises (SMEs): Some articles (such as Barton's) consider SMEs, but many securities or digitization models focus on large companies. There is less empirical evidence of secure and inexpensive implementation for SMEs.

4. Economic and social sustainability: Although several studies address security and efficiency, few explicitly address the dimensions of social sustainability (e.g., impact on employment, training, safety culture) in the context of secure digital transformation.
5. Interoperability and legacy: Integrating legacy systems with new secure digital environments (CPS, digital twins, IoT) remains a challenge, and there is a scarcity of studies proposing practical, low-cost secure retrofit strategies for pre-existing factories.

Based on findings from the literature, the proposed low-cost model for digital transformation and cybersecurity in production management can address these gaps:

- Integrated framework: This study proposes the integration of the three components (digitalization, security, production management) into a cohesive model, addressing the fragmentation of the literature.
- Practical validation: The model can be tested in real SMEs, aiming to demonstrate how IoT retrofitting + digital twin + basic security measures can be implemented with a limited budget.
- Focus on SMEs: By prioritizing low-cost and open-source technologies, the model makes secure digitalization accessible to smaller companies, something underexplored by the literature.
- Social sustainability: Including employee training, shared security policies, and change management procedures in the model can address the social dimension of technology adoption.
- Legacy integration: Incorporate secure modernization strategies into the model (network segmentation, micro segmentation, digital twin for vulnerability mapping).

3 Proposal for a Low-Cost Model for Digital Transformation and Cybersecurity in Production Management

The integration of digital transformation and cybersecurity into production management has become increasingly critical for organizations seeking to improve operational efficiency, competitiveness, and resilience. However, many small and medium-sized enterprises (SMEs) face significant financial and technical constraints that limit their ability to implement advanced digital solutions or comprehensive cybersecurity strategies. To address these challenges, this study proposes a low-cost model that combines digital transformation with cybersecurity measures in a structured, scalable, and economically feasible manner. The model emphasizes modular implementation, the use of affordable and open-source technologies, and continuous monitoring to ensure secure and efficient production processes.

The proposed model is designed around three interdependent dimensions: digital transformation, cybersecurity, and integration with production management. The digital transformation component focuses on adopting tools that optimize production processes, improve decision-making, and enable real-time monitoring. Key elements include IoT sensors, which provide low-cost data collection from machines and production lines; automation modules for repetitive or high-impact tasks; data visualization platforms such as open-source solutions (e.g., Grafana) to monitor key performance indicators;

and cloud or hybrid storage solutions to enable scalable and cost-effective data management. By leveraging open-source software and affordable hardware, organizations can minimize upfront costs while maintaining the flexibility to expand and adapt the system over time.

Cybersecurity is fully embedded within the model to address the vulnerabilities introduced by digitalization. Essential measures include network segmentation to isolate critical systems, access control and authentication mechanisms, regular software updates and patch management, and lightweight intrusion detection systems. Employee training is also a central component, as awareness and proper practices are among the most cost-effective ways to mitigate cyber risks. By integrating cybersecurity from the outset, the model ensures that digital transformation does not compromise the integrity, availability, or confidentiality of production systems.

Implementation of the model follows a phased and iterative strategy, which allows organizations to adopt technologies gradually while minimizing risk and managing costs. The process begins with a comprehensive assessment of existing production processes, identifying high-priority areas for digitalization and evaluating current cybersecurity vulnerabilities. Following this, a prioritization phase selects interventions that deliver the greatest value at the lowest cost, such as the installation of critical sensors or automation modules in pilot areas. The deployment phase involves incremental implementation and monitoring, enabling adjustments based on performance and security outcomes. Finally, the optimization phase expands successful solutions across the organization, integrates continuous feedback mechanisms, and refines processes for efficiency and security. This phased approach allows organizations to scale their digital transformation efforts responsibly, ensuring that each step adds measurable value.

To support understanding, Table 5 summarizes the core components and examples of low-cost or open-source tools that can be used within the model:

Table 5. The core components and examples of low-cost or open-source tools.

Component	Description	Example Tools/Technologies
IoT Sensors	Real-time data collection from production equipment	Arduino, Raspberry Pi, low-cost sensors
Data Visualization	Monitoring KPIs and production metrics	Grafana, Kibana, Freeboard
Automation Modules	Process automation for repetitive tasks	Node-RED, OpenPLC, Home Assistant
Cloud/Hybrid Storage	Cost-effective data storage and analysis	AWS Free Tier, Google Cloud Free Tier
Cybersecurity Measures	Protection against unauthorized access and attacks	Snort IDS, OpenVPN, pfSense
Employee Training	Awareness of cyber risks and safe practices	Free online courses, internal workshops

The benefits of the proposed model are multidimensional. Operationally, it improves efficiency, reduces downtime, enhances quality control, and enables real-time decision-making. From a cybersecurity perspective, even basic measures such as access control, network segmentation, and threat monitoring significantly reduce exposure to common cyber threats. Economically, the focus on low-cost and open-source solutions lowers the barrier to entry, allowing SMEs to adopt digital transformation without prohibitive expenses. Moreover, by integrating continuous monitoring and feedback loops, the model encourages ongoing improvement, ensuring that both production efficiency and security evolve over time.

It is important to emphasize that low-cost technologies such as Arduino, Raspberry Pi, and similar platforms are not intended to replace industrial-grade automation systems in mission-critical or continuous 24/7 operations. In the proposed model, these technologies are positioned primarily as enabling tools for pilot projects, retrofitting legacy systems, non-critical monitoring, and proof-of-concept implementations. Their role is to lower the entry barrier to digital transformation, particularly for SMEs, rather than to fully substitute certified industrial hardware.

In addition to the text description, Table 6 presents a step-by-step phased implementation strategy that organizations can follow:

Table 6. A step-by-step phased implementation strategy.

Phase	Activities	Objectives
Assessment	Analyze production processes, identify opportunities, evaluate vulnerabilities	Identify priority areas for digitalization and security
Prioritization	Select high-impact, low-cost interventions	Optimize cost-effectiveness and ROI
Deployment	Implement solutions in pilot areas, monitor performance and security	Validate operational and cybersecurity effectiveness
Optimization	Expand successful interventions, refine processes, integrate feedback	Scale system responsibly, continuous improvement

The proposed low-cost model represents a practical framework that integrates digital transformation and cybersecurity into production management while remaining financially viable and technically accessible. By combining affordable technologies, open-source tools, modular implementation, and employee engagement, organizations—especially SMEs—can modernize production processes, strengthen cybersecurity, and enhance overall operational performance. The model is adaptable to different industrial contexts and scalable according to organizational resources and priorities, providing both immediate operational gains and long-term strategic value.

4 Results and Discussion

The research aimed to develop a low-cost model for integrating digital transformation and cybersecurity in production management, addressing the central question: "How can a low-cost model be developed to integrate digital transformation and cybersecurity for improved efficiency in production management?" The proposed model combines technological, organizational, and procedural factors to provide a feasible framework for SMEs and other resource-constrained organizations. This section presents the key components of the model, their implementation, and discusses their implications for operational efficiency, security, and cost-effectiveness.

Key Components of the Low-Cost Model:

1. **Technological Factors**

   - **IoT and Real-Time Monitoring:** The integration of low-cost IoT sensors and real-time monitoring systems was found to enhance operational efficiency by providing continuous data on machine performance, production output, and resource utilization. This enables predictive maintenance, reduces unplanned downtime, and improves resource allocation, allowing organizations to optimize production without substantial investment.
   - **Open-Source Platforms for Data Analysis and Automation:** Utilizing open-source software such as Grafana for visualization, Node-RED for automation, and lightweight intrusion detection systems significantly reduces costs while maintaining functionality. These tools enable real-time analytics, process optimization, and cybersecurity monitoring, making digital transformation accessible even for SMEs with limited budgets.
   - **Automation Modules:** Modular and low-cost automation systems were identified as essential for improving process efficiency. By automating repetitive tasks, organizations can reduce errors, free human resources for higher-value activities, and increase overall throughput while keeping implementation costs low.

2. **Organizational Factors**

   - **Employee Training and Awareness.** Workforce engagement and training were found to be critical for the successful adoption of digital and cybersecurity measures. Teaching employees to recognize cybersecurity threats and use digital tools effectively ensures smoother implementation and reduces the risk of operational disruptions due to human error.
   - **Process Mapping and Change Management:** Implementing the model in a phased and structured way, starting with high-priority areas, allows organizations to test solutions incrementally. This approach reduces resistance to change, promotes understanding of new workflows, and ensures that digital and cybersecurity interventions are integrated effectively with existing operations.
   - **Cross-Functional Collaboration:** Encouraging collaboration between IT, production, and management teams facilitates knowledge sharing, problem-solving, and faster troubleshooting. This collaborative approach supports the alignment of technological adoption with organizational goals, enhancing both efficiency and security.

3. **Procedural and Security Factors**

- **Network Segmentation and Access Control:** Basic cybersecurity measures, including network segmentation, user authentication, and controlled access to critical systems, provide a foundational level of protection against common cyber threats without incurring high costs.
- **Incremental Cybersecurity Monitoring:** Lightweight monitoring tools and open-source intrusion detection systems enable continuous oversight of digital operations, allowing organizations to identify vulnerabilities early and respond promptly, maintaining operational continuity.
- **Feedback and Continuous Improvement Loops:** Establishing monitoring dashboards and performance indicators for both production efficiency and security ensures that processes can be continuously refined, risks mitigated, and best practices reinforced over time.

The proposed low-cost model demonstrates how the integration of technological, organizational, and procedural factors can simultaneously improve production efficiency and cybersecurity while remaining financially accessible. By leveraging IoT and open-source tools, organizations can optimize production, reduce waste, and anticipate maintenance needs without incurring prohibitive costs. Automation modules contribute further to efficiency by reducing errors and freeing human resources for strategic tasks.

Organizational strategies, including employee training, process mapping, and cross-functional collaboration, ensure that technological interventions are effectively adopted and sustained. These strategies minimize human-related vulnerabilities and increase the overall resilience of production systems. Furthermore, the phased implementation and continuous monitoring allow organizations to scale digitalization and security interventions progressively, ensuring that investments are cost-effective and produce measurable benefits.

Compared to existing frameworks, which often focus on either digital transformation or cybersecurity independently, the proposed model stands out by integrating both aspects into a cohesive, low-cost approach. This holistic strategy addresses common gaps in current literature by providing practical guidance for SMEs seeking to modernize production processes without compromising security or exceeding budget constraints. The model also supports continuous improvement, allowing organizations to adapt to evolving technological and cybersecurity challenges while maintaining efficiency.

The findings suggest that organizations can adopt the proposed framework to enhance operational efficiency, improve cybersecurity resilience, and reduce costs. By prioritizing low-cost technologies, open-source solutions, and incremental deployment, SMEs can achieve digital transformation that is both sustainable and scalable. Emphasizing workforce engagement and cross-functional collaboration ensures that digital and cybersecurity investments yield long-term benefits. Overall, the model provides a practical and actionable guide for organizations seeking to modernize production management in a secure, efficient, and cost-effective manner.

A critical consideration in low-cost digital transformation is the trade-off between initial investment costs and long-term operational reliability. While low-cost hardware solutions reduce upfront expenditure, their limitations in terms of durability, redundancy, and continuous operation may introduce risks if deployed indiscriminately in critical production environments. The proposed model addresses this challenge by advocating a risk-based and phased deployment strategy, in which low-cost solutions are first applied to non-critical processes, monitoring functions, or pilot areas, while critical operations may require gradual migration to industrial-grade solutions as value is demonstrated. From a cost–benefit perspective, the model explicitly acknowledges that what is considered 'low cost' at the investment stage may become costly if it leads to production downtime or operational failures. Therefore, decision-making within the model emphasizes total cost of ownership, operational risk, and scalability rather than initial acquisition cost alone.

5 Conclusions

The integration of digital transformation and cybersecurity into production management has become a critical factor for enhancing operational efficiency, competitiveness, and resilience across industrial sectors. However, the adoption of digital technologies often poses challenges, particularly for small and medium-sized enterprises, due to financial constraints and increased exposure to cyber threats. This study conducted a systematic review of relevant literature and proposed a low-cost model for integrating digital transformation and cybersecurity into production management, offering a practical framework for resource-constrained organizations.

By examining the interplay between affordable digital technologies, open-source solutions, organizational strategies, and basic cybersecurity measures, the study identifies key factors that facilitate or hinder the effective implementation of secure and efficient production systems. The central research question, "*How can a low-cost model be developed to integrate digital transformation and cybersecurity for improved efficiency in production management?*" is addressed through a model that combines technological, organizational, and procedural dimensions. The findings emphasize the importance of a holistic approach that simultaneously enhances operational performance, maintains cybersecurity resilience, and manages costs. Key strategies include the adoption of IoT and open-source tools for monitoring and automation, incremental deployment of digital solutions, workforce training, and continuous monitoring to support ongoing improvement.

The study demonstrates that even with limited resources, organizations can achieve significant benefits by implementing modular, low-cost digital and cybersecurity solutions. Collaboration between IT, production, and management teams, alongside careful process mapping and prioritization of high-impact interventions, ensures that technological and security investments yield long-term operational gains. Nevertheless, the study has limitations, including the absence of empirical testing of the proposed model in real industrial environments and its focus primarily on SMEs, potentially overlooking applications in larger enterprises or different industrial contexts.

Future research should focus on validating the model through case studies across diverse sectors, assessing the long-term effectiveness of low-cost cybersecurity measures, and exploring emerging technologies that can further reduce costs while enhancing efficiency and security, such as energy-efficient AI, low-power IoT devices, and automated threat detection systems.

In summary, this study advances knowledge on structuring digital transformation and cybersecurity to achieve efficient and secure production management in a cost-effective manner. The proposed model offers practical guidance for organizations seeking to modernize their production processes while balancing technological progress, operational efficiency, and cybersecurity imperatives, and provides a foundation for further research on accessible and scalable digital transformation strategies.

6 Disclosure of Interests

The authors have no competing interests to declare that are relevant to the content of this article.

Acknowledgments. This work has received support from the European Union under the Next Generation EU, through a grant of the Portuguese Republic's Recovery and Resilience Plan (PRR) Partnership Agreement, as part of the PRODUTECH R3 project – "Agenda Mobiliza-dora da Fileira das Tecnologias de Produção para a Reindustrialização", aimed at the mobilization of the production technologies industry to promote the reindustrialization of the manufacturing sector (Project ref. nr. 60 - C645808870-00000067; Total project investment: 166.988.013,71 Euros; Total Grant: 97.111.730,27 Euros).

References

1. Raimundo, R.J., Rosário, A.T.: Cybersecurity in the Internet of Things in industrial management. Appl. Sci. **12**(3), 1598 (2022)
2. Saeed, S., et al.: Digital transformation and cybersecurity challenges for businesses resilience: issues and recommendations. Sensors **23**(15), 6666 (2023)
3. Alexopoulos, K., Nikolakis, N., Xanthakis, E.: Digital transformation of production planning and control in manufacturing SMEs – the mold shop case. Appl. Sci. **12**(21), 10788 (2022)
4. Vărzaru, A.A., et al.: Digital transformation and innovation: the influence of the IoT on strategic decision-making. Systems **12**(9), 359 (2024)
5. Iribarren, M.Z., et al.: A review of Industry 4.0 assessment instruments for digital transformation. Appl. Sci. **14**(5), 1693 (2024)
6. Goel, A., et al.: An overview of digital transformation and environmental sustainability. Sustainability **16**(24), 11079 (2024)
7. Pasqualino, R., et al.: Digital transformation and sustainable oriented innovation. Sustainability **13**(21), 11564 (2021)
8. Metin, B., et al.: Digitalisation and cybersecurity: towards an operational framework for cyber-resilient digital adoption. Electronics **13**(21), 4226 (2024)
9. Wai, E., Choi, J., Lee, S.: Seamless Industry 40 integration: a multilayered cyber-security framework for resilient SCADA deployments in CPPS. Appl. Sci. **13**, 12008 (2023)
10. Prinsloo, J., van der Merwe, A., Botha, L.: A review of Industry 4.0 manufacturing process security. Appl. Sci. **9**, 5105 (2019)
11. Hajda, J., Krawiec, P., Pawłowski, K.: Security challenges in Industry 4.0 PLC systems. Appl. Sci. **11**, 9785 (2021)
12. Azambuja, A.J.G., Lima, R.F., Silva, R.: Artificial intelligence-based cyber security in the context of intelligent manufacturing. Electronics **2023**, 12 (1920)

13. Torbacki, W.: A hybrid MCDM model combining DANP and PROMETHEE II methods for the assessment of cybersecurity in Industry 4.0. Sustainability **13**, 8833 (2021)
14. Ahmmed, M.S., Haque, A., Islam, S.: Digital twins, AI, and cybersecurity in additive manufacturing. Machines **13**, 691 (2023)
15. Barton, M., Rossi, C., Schmidt, T.: Identification overview of Industry 4.0 essential attributes for SMEs. Appl. Sci. **12**, 5672 (2022)
16. Ahmmed, M.S., Haque, A., Islam, S.: A systematic review of Industry 4.0 and AI. Machines **12**, 681 (2022)
17. Ren, X., Liu, F., Zhao, Y.: Qualitative meta-analysis based on current research on digital transformation capability system. Sustainability **15**, 14168 (2023)
18. Jiang, Y., Li, X., Wang, J.: Leveraging digital twin technology for enhanced system dependability and cybersecurity in CPPS. Future Internet **16**, 134 (2024)
19. Metin, B., Yildiz, O., Kaya, H.: Digitalisation and cybersecurity: towards an operational framework for cyber-resilient digital adoption. Electronics **13**, 4226 (2024)
20. Pochmara, J.: Cybersecurity of industrial systems — a 2023 report. Electronics **13**, 1191 (2024)
21. Rudenko, R., Kim, S., Lee, H.: A brief review on IoT, Industry 4.0 and cybersecurity in intelligent manufacturing. Electronics **11**, 1742 (2022)
22. Saeed, S., Ahmad, S., Malik, M.: Digital transformation in energy sector: cybersecurity challenges. Information **15**, 764 (2024)

Increasing the Productivity of the Work Cell

Anna Karwasz[1(✉)], Krzysztof Łukaszewski[1], and Roman Wdowik[2]

[1] Faculty of Mechanical Engineering, Poznań University of Technology, Poznan, Poland
anna.karwasz@put.poznan.pl

[2] Faculty of Mechanical Engineering and Aeronautics, Rzeszów University of Technology, Rzeszów, Poland

Abstract. The main objective of the research was to increase the efficiency of the workpiece nest by introducing an industrial robot. The existing process of manufacturing protective sleeves for hydraulic cylinders mounted in loader buckets was analysed. Efficiency and DPU (Defects Per Unit) were calculated. The introduction of an industrial robot with RoboDK offline programming software was proposed. Simulations and tests of the proposed solution showed an improvement in productivity and a reduction in the DPU index. In addition, the automation of the station contributed to a reduction in the production cycle time, a reduction in quality defects, and an increase in the safety of the operators working at this station.

Keywords: Manufacturing Automation · Industrial Robots · Efficiency Improvement

1 Introduction

Automation and robotisation are playing an increasingly important role in modern industry, particularly in the context of Industry 4.0. Manufacturing companies are implementing robotic systems to increase productivity and reduce production costs. The aim of reducing manual machine operation through automation and robotisation is to eliminate downtime and increase efficiency. Greater market competitiveness and an increasing shortage of manual labour are forcing companies to use the latest technologies.

All over the world, we are seeing an increase in the use of robots in production processes. China is the leader in robotisation, while Germany stands out on the old continent. Poland still lags behind in terms of the share of robots in production processes, and the saturation rate of industrial robots in Polish companies is significantly lower than the EU average, which shows the need for development in this area [1]. Industry 4.0 is the integration of intelligent machines, systems and the way changes are introduced into production processes [2]. The use of advanced systems such as the Internet of Things (IoT), Artificial Intelligence (AI), cloud computing, big data and robotisation allows production processes to be quickly adapted to changing market requirements, while increasing flexibility and efficiency. The fourth industrial revolution not only allows for the optimisation of costs and resources, but also enables the creation of smart factories that are highly focused on customer needs.

J. Trojanowska et al. (Eds.): MANUFACTURING 2026, LNME, pp. 332–346, 2026.
https://doi.org/10.1007/978-3-032-21654-0_25

1.1 Robotisation

The first industrial robot was implemented in production in the 1960s at General Motors plants. Its introduction into casting machine operation processes was significant from the perspective of employee safety and increased production efficiency. The development of industrial robots and vision systems has significantly influenced the possibilities of manufacturing process automation. Integration with digital technologies and the development of Industry 4.0 has enabled the full potential of industrial robots to be exploited. In the 21st century, these systems are capable of independently monitoring their work, identifying problems in real time and optimising processes to achieve maximum efficiency [3, 4].

Robotisation is finding application in many industrial fields. Companies are using robots because of the opportunities they offer. Robotisation is the process of introducing industrial robots to perform production tasks with the aim of automating repetitive activities, increasing productivity and reducing operating costs [5].

Robotic production systems play a key role in the automation of modern manufacturing processes. As the author [6] points out, robotic production systems are integrated systems in which industrial robots interact with machines, control systems and other devices to perform production tasks. Their main purpose is to automate production processes, increase efficiency and improve the quality and precision of manufacturing. Modern robotic production systems are the basis for Industry 4.0, which focuses on the integration of digital technologies such as artificial intelligence (AI), the Internet of Things (IoT) and data analysis or Big Data. Based on the synchronisation and interaction of industrial robots and sensors through the use of control systems, robotic production systems can be adapted to all types of products. They are characterised by high accuracy and repeatability, which translates into product quality. The robotisation of the process itself allows for a shorter production cycle and an increase in the number of products manufactured per unit of time [5].

The robotisation of industrial processes is a key element of modern production automation. The numerous benefits of their use, such as excellent accuracy and repeatability, adaptability to a wide range of tasks through programmability and tool interchangeability, and durability that allows them to operate at high temperatures or in contact with hazardous chemicals [5, 6], contribute to an increase in the density of robotisation in Poland, i.e. the number of robots per 10,000 employees in the manufacturing industry. The increase in the index from 56 in 2021 [7] to 78 in 2023 [8] shows that more and more manufacturing companies are deciding to use them in their production processes.

1.2 Industrial Robots

A robot is an automatic mechanical device, programmable in multiple axes, which can perform a variety of tasks in an industrial environment, such as welding, assembly, painting or packaging, with high precision and repeatability [6]. These devices can be classified according to their design, application and control method.

Robotic production systems are used in many industries where the automation of manufacturing processes is crucial, as these processes must be carried out precisely, repetitively and efficiently.

In the automotive industry, robots perform automated welding processes, which greatly improves the precision and aesthetics of welds. In combination with vision systems, they paint car bodies, ensuring even paint distribution and minimising material waste. They are also responsible for assembling components with high accuracy, eliminating human error. In the electronics industry, they solder components, ensuring durability and flawless connections in printed circuits. They precisely assemble integrated circuits consisting of miniature electronic components such as resistors and capacitors. Vision systems working with robots inspect integrated circuits, detecting even microscopic defects. In the food industry, they excel at sorting based on size, colour or shape. Cutting, portioning and mixing ingredients using robots ensures high precision and repeatability. The automation of packaging and labelling processes significantly speeds up the preparation of products for distribution, which is crucial in the context of food with a use-by date. In the pharmaceutical industry, robotic systems support the dosing of substances, ensuring high accuracy while eliminating the risk of contamination. They carefully check the integrity of packaging and allow for the detection of defects in medicines. In heavy industry, they are used in mechanical processing, where they work with CNC machines to perform drilling, milling and grinding processes. They also operate sockets and production lines, responsible for transporting material between successive operations [9].

Both automation and robotisation improve plant efficiency, but they differ in terms of the technological solutions used and the level of employee involvement. Robotisation, which is usually a more advanced form of automation, is often an integral part of it, increasing process efficiency through the use of specialised equipment and algorithms [10, 11].

In the context of comparing two production cells – the current one and the one being designed – an important aspect relating to the effectiveness of the implemented solution is productivity, which determines the maximum number of products made well (i.e. in accordance with requirements) in a given time [4]. It also allows for an assessment of the extent to which the new solution contributes to improving the efficiency of the entire production plant. Productivity is calculated using formula 1.

$$Productivity = \frac{Number\ of\ components\ produced[units]}{Production\ time[hour]} \tag{1}$$

You can also calculate DPU (Defects Per Unit) (formula 2), which determines the ratio of the number of defects to the number of units produced.

$$DPU = \frac{Number\ of\ defects}{Number\ of\ units\ produced} \tag{2}$$

1.3 Offline Programming in the RoboDK System

Offline robot programming (OLP) is a method of creating programmes for industrial robots using specialised computer software (virtual environment) based on 3D data. After verifying the robot programme in the virtual environment, it can be implemented on a physical robot [12, 13].

RoboDK is advanced software for offline programming of industrial robots. It enables the creation, testing and optimisation of robot programmes in a virtual environment. Integration with CAD/CAM systems allows you to generate trajectories based on 3D models, which significantly reduces implementation time while limiting the risk of errors [14, 15].

Manufacturing companies opt for this solution primarily due to long downtimes of robot stations caused by programming new actions. OLP allows movements to be configured without having to stop production, which effectively increases plant efficiency. Using advanced software, the time needed to develop the programme itself can be reduced by up to 80%, while increasing programmer productivity and minimising downtime at other stations. Programming robot action sequences in a virtual environment eliminates the need for human presence in the robot's work area, minimising the risk of accidents. Modifications to the robot's movement trajectory result in shorter cycle times [12].

2 Materials and Methods

The subject of the study is a production cell manufacturing protective sleeves for hydraulic cylinders installed in loader buckets. One employee is responsible for operating the machines, transferring materials within the cell and collecting finished products. The technological process involves a series of operations carried out on two numerically controlled (CNC) machines: a lathe and a milling machine.

The described nest operates in a single-shift system. It is operated by a single operator, whose responsibilities include unloading material and loading finished products, transferring semi-finished products between CNC machines, which he controls and starts up, and carrying out preliminary quality control of the sleeves produced, allowing for early detection of any non-conformities. The layout of the machines is shown in Fig. 1. However, due to the fact that the transport of finished products stored on racks is difficult, it was decided to store them in special containers when implementing the new solution.

The sequence of operations in the production cell is as follows:

1. The operator takes the material from the container.
2. The employee first removes the machined material from the lathe chuck. Then, they place the raw sleeve in the chuck, carefully ensuring that it is correctly secured. Once the material is correctly secured, they start the machining programme
3. The employee removes the finished product from the milling machine chuck and secures the material. He starts the machining programme.
4. The employee performs a preliminary measurement and places the finished product in a designated container.
5. The employee takes new material from the container.
6. After completing the first stage of machining, the employee removes the sleeve from the holder, turns it and re-clamps it in the lathe, ensuring that it is properly secured. Then, they run the machining programme for the other side.
7. Continuation of the employee's work cycle from point 2.

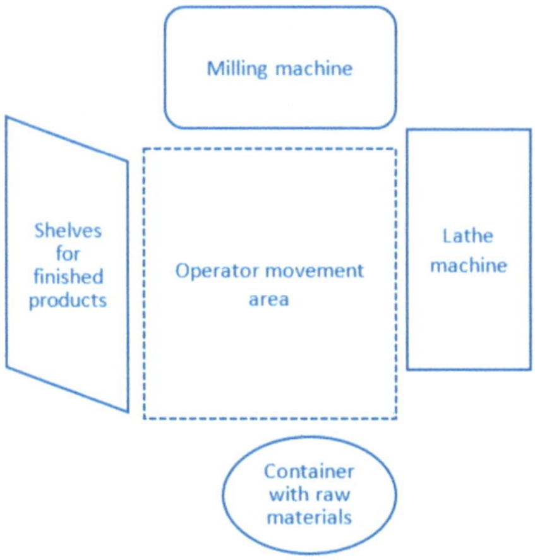

Fig. 1. Layout of the analysed production cell [16].

The production cell operates on a single-shift basis, which translates into 7 h and 40 min of effective work. The total duration of the production cycle was determined as the sum of all tasks performed within the production cell. This included both CNC machine operations and manual activities performed by the employee. The value obtained reflects the full complexity of the production process.

However, due to the fact that the lathe and milling machine in the analysed cell work in parallel, the actual throughput of the cell depends on the slowest stages of the process – the work of the lathe and the operations supporting it. The parallel nature of the work of these devices results from the cyclical functioning of the cell. When the lathe starts machining the next sleeve, the milling machine continues to finish the sleeve from the previous cycle. Therefore, the key indicator for assessing performance is the effective production cycle time, which only takes into account those operations that affect the throughput of the entire system.

3 Current Situation

In order to accurately determine the cycle time, measurements were taken for 10 samples during actual operation of the nest, and based on these measurements, the average duration of each operation performed by the operator in the production process was calculated. The measurement results are presented in Table 1.

The total production cycle time is 7 min and 54 s, consisting of:

- Removing raw material from the container: 8 s.
- Loading new raw material into the lathe: 16 s.
- Machining the first side of the material on the lathe: 186 s.
- Rotating the material: 36 s.
- Machining the second side of the material: 84 s.
- Unloading the double-sided machined material from the lathe: 12 s.
- Loading the double-sided machined material onto the milling machine: 15 s.

Table 1. Average time of operations performed by the operator.

No	Lathe machine: Removal of the machined sleeve and installation of the blank [s]	Lathe machine: Rotation of the material [s]	Milling machine: Removal of the finished product and installation of the blank [s]
1	28	35	26
2	29	36	28
3	30	34	25
4	27	37	27
5	28	38	26
6	30	36	28
7	31	35	27
8	29	36	26
9	30	37	28
10	28	36	29
Average	29	36	27

- Machining the material on the milling machine: 96 s.
- Unloading the finished product: 11 s.
- Placing the finished product in a container: 10 s.

The bottleneck is the lathe, where the cycle time is 5 min 35 s and consists of:

- Loading new raw material and unloading the machined sleeve: 29 s.
- Machining the first side of the material on the lathe: 186 s.
- Rotating the material: 36 s.
- Machining the second side of the material: 84 s.

DPU (Defects Per Unit) was calculated using formula 2, which determines the ratio of the number of defects to the number of units produced. To enable this, the number of sleeves produced over 5 days was measured. The results are presented in Table 2.

Table 2. Number of correct and defective products produced and DPU.

Day	Number of products manufactured	Number of correct products	Number of defective products	DPU
1	74	71	3	0.041
2	73	70	3	0.041
3	73	71	2	0.027
4	72	68	4	0.056
5	73	71	2	0.027

The average DPU value for the production cell is 0.038.

The data from Table 2 was also used to calculate the productivity of the production cell. Assuming that the effective working time during one shift is 7 h and 40 min, this gives 38 h and 20 min (38.33 h) over 5 days.

$$Productivity = \frac{365}{38,33} \approx 9,52\ units/hour$$

The current production slot solution has many limitations resulting from manual operation by an employee. The machines are started manually, which increases the production cycle time. Manual mounting of the sleeve in the lathe and milling machine holder increases the risk of errors. Incorrect material positioning can lead to dimensional deviations, defects on the sleeve surface, and damage to the cutting insert. Such errors have a negative impact on the quality of the final products, increase the amount of production waste, and generate higher costs associated with the purchase of more blades. Another aspect is employee breaks, during which the machines are not working, which reduces effective working time. Manual operation also affects operator safety. Manual clamping of the sleeve increases the risk of injuries such as cuts, lacerations or crushing, and inattention can lead to dangerous accidents. Therefore, the main objective of the research was to increase the efficiency of the workstation in question by introducing an industrial robot.

4 Proposed Improvements

The new concept is based on automating the entire production process in the current cell using an industrial robot that has been programmed offline, which allowed for the detection of errors during programme creation. The robot is responsible for loading and unloading material, transporting it within the cell, and automatically initiating machining programmes on the lathe and milling machine. The new solution will help eliminate downtime resulting from human limitations, increase the efficiency of the production cell, shorten the production cycle time and reduce the number of defective products, while increasing the safety of the operator.

The new production cell consists of a Turin robot equipped with two grippers, milling machines and lathes. The robot picks up the material from the feeder and places the finished products in a designated container. In addition, the robot is mounted on a base with a height of 750 mm to facilitate access to each target point, and the layout is shown in Fig. 2.

An industrial robot in a robotised production cell is connected to CNC machines (lathe machine and milling machine). When the CNC machines finish their work, they send a READY signal, which means that the robot can work in the working area of the lathe or milling machine. After finishing, it sends a START signal, which initiates the machining programme on the machine. This type of connection enables simple and reliable exchange of control signals and status signals between devices.

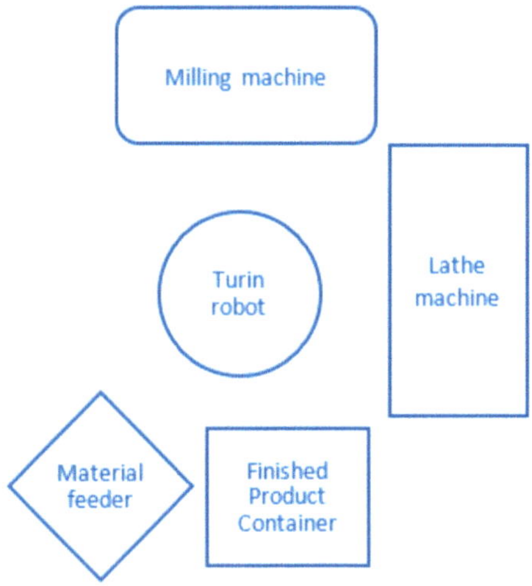

Fig. 2. Robotic production cell layout [16].

Description of the robot operation sequence in a robotic production cell:

1. **Starting position:**
   The robot begins work in its initial position. All devices in the production cell are in a state of full readiness.
2. **Retrieving Material from the Feeder:**
   Using gripper number 1, the robot picks up the raw material from the feeder, preparing it for loading into the lathe machine.
3. **Loading into the Lathe Machine and Unloading the Machined Material:**
   The robot uses gripper number 2 to remove the machined material from the lathe. Then, after completing this operation, it places the raw material held in gripper number 1 into the lathe chuck. Once the material is correctly secured, the robot sends a START signal to the lathe, which initiates its machining programme.
4. **Loading into the Milling Machine and Unloading the Finished Product:**
   The finished product is removed from the milling machine using gripper number 1. The robot places the material previously removed from the lathe in the place of the finished product using gripper number 2. After loading is complete, the robot sends a START signal to the milling machine, which starts its machining programme.
5. **Placing the Finished Product in a Container:**
   The finished product, held in gripper number 1, is placed in a container for finished products.
6. **Taking the Next Piece of Material from the Feeder:**
   The robot takes a new piece of raw material from the feeder using gripper number 1, starting the next cycle.
7. **Turning the material in the lathe machine:**
   After completing the first part of the machining process, the robot removes the material from the lathe, turns it using gripper number 2 and places it back in the lathe chuck. Once the material is properly secured, the robot sends a START signal to the lathe, initiating the machining of the other side of the sleeve.

8. **Repeat the cycle from point 3:**
   Once machining is complete, the work cycle is repeated until the material in the feeder is exhausted or the entire production task is completed.

Each robot operation is synchronised with the work of CNC machines. The robot can only enter the working area of CNC machines after receiving a READY signal from the lathe and milling machine, ensuring smooth operation and eliminating the possibility of collisions.

The use of two independent grippers mounted on the robot arm allows for efficient loading and unloading operations, eliminating the need for additional trips to retrieve material. This allows the robot to efficiently operate CNC machines, minimising downtime and reducing production cycle times. Working with a single gripper would require additional robot movements, which would increase handling time and reduce the efficiency of the workstation.

4.1 Selection of Industrial Robot and Software

The robot selected to perform tasks in the new cell is the Turin TR20-1920B [17–19]. RoboDK was selected as the software for offline programming for the new production cell. Unlike other systems available on the market, it supports robots from multiple manufacturers, which will enable its use in other cells operated by devices from other brands in the future.

The first key stage in offline robot programming is to create a virtual model of the production cell in a simulation environment that reflects real conditions as closely as possible. The process begins with importing machine models, such as a lathe, milling machine or material feeder, which are available in CAD formats, allowing for accurate reproduction of their geometry and functions. Next, an industrial robot model is added from the RoboDK software library, which, if no suitable ready-made model is available, can be configured manually, taking into account parameters such as range of motion, speed, acceleration and maximum payload.

A tool, such as a two-finger gripper, is mounted on the robot arm and will be used to operate the entire cell. The appropriate tool can be selected from the RoboDK library or its CAD model can be imported, adjusting the dimensions and technical parameters to the requirements of the production process.

The next step is to define the target points, such as loading and unloading points on CNC machines, starting positions and transition areas in the robot's trajectory. For each point, the orientation of the tool and its position in space are determined. Based on the defined points, a robot movement trajectory is created, which should be optimised to minimise cycle time, avoid collisions and maintain smooth movements.

The entire process requires additional parameters to be taken into account, such as the robot's speed, acceleration and tool setting tolerance. The virtual production cell, which reflects the actual working environment, forms the basis for further activities such as simulation tests and programme optimisation. This makes it possible to detect potential errors at an early stage, minimising the risk of problems during implementation at the actual workstation.

The work began with the creation of a virtual model of the production cell in RoboDK, which reflects the actual workstation. The first step was to add a robot base to ensure its stability and the correct mounting height for the Turin industrial robot. The robot was imported into the virtual environment, and then a double gripper, adapted to manipulate the sleeves produced at this station, was mounted on its arm.

Next, CNC machine models, i.e. lathes and milling machines, were imported and precisely positioned in relation to the industrial robot. The positions were defined based on the actual spatial conditions in the workplace. Additional elements were added, such as a raw material feeder and a table with a container for finished products.

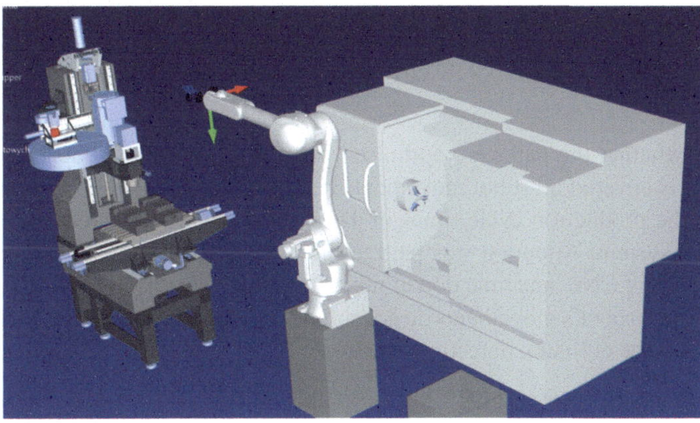

Fig. 3. Virtual model of a production nest in the RoboDK environment [16].

The layout shown in Fig. 3 was adjusted to ensure optimal robot access to the machines and reduce their downtime. The virtual cell allows for the creation of trajectories, their optimisation and verification for collisions. The current configuration will allow for the expansion of the cell with another machine in the future.

Programmes were created for the model: home, material collection, machined part collection, part assembly in the milling machine, finished part placement, and subprogrammes defining the reference system, speeds, linear displacement, gripper opening and closing, and lathe loading and unloading.

4.2 Tests and Simulations

In order to verify the correctness of the programmes, a simulation of the duration of individual operations was carried out. Simulation tests showed that the programme meets all design assumptions.

Comparing the operation of the existing production cell with the robotic solution designed using RoboDK software required a detailed, multi-stage analysis methodology. The key element was to take into account data from measurements of the actual operation of the analysed cell, as well as the results of virtual simulations and measurements taken on the implemented robotic production cell. The analysis is based on cycle time, productivity and production quality.

The first stage focused on the actual operation of the cell, and measurements were taken during the plant's daily operations:

- Based on the measurements in Table 1, the average loading and unloading times for the lathe and milling machine, rotation and reassembly, as well as the duration of machining operations on CNC machines are presented.
- It was assumed that the cell operates in a single-shift mode and that the effective working time is 7 h and 40 min.
- Based on data from 5 days of cell operation, the DPU (Defects Per Unit) index was calculated, allowing for the assessment of the number of defects per production unit.
- Productivity calculations were made based on the collected data.

After completing the analysis of the current cell, a simulation of the designed robotic cell was carried out in the RoboDK environment. The use of a virtual model allowed for:

- Checking whether the paths are optimised in terms of time, safety and smoothness. The analysis covered all robot movements between target points.
- Compare the production cycle times in the robotic cell with the results obtained in the current system. Attention was paid to the elimination of manual operations and the reduction of CNC machine downtime.
- Ensure that the robot's movements do not collide with other elements of the production station using the collision detection function.

In order to accurately compare the two solutions, the analysis was based on the following indicators:

- Production cycle time: The time from loading the raw material to removing the finished product was measured in both systems. In the case of the robotic cell, the additional reduction in downtime resulting from automation was taken into account.
- Production efficiency: The efficiency of both production cells was calculated based on the measurements taken to assess the impact of the changes implemented on the efficiency of the cell.
- Production quality DPU: The number of defective products in the manual and robotic cells was compared, taking into account the elimination of human error in the automation process.

An analysis of the results of the comparison between the two systems allowed us to identify key differences and benefits resulting from the implementation of a robotic cell. Process automation significantly reduced operation time by eliminating downtime and optimising the sequence of actions. The use of an industrial robot reduced the number of errors associated with manual operation of CNC machines, which translated into higher production quality. Automation made it possible to reduce operator labour costs while increasing the efficiency of production machine utilisation.

A comparison of the current production cell with the new robotic solution was carried out on the basis of an analysis of three key indicators: production cycle time, productivity and product quality. The data collected during the actual operation of both cells provided a solid basis for an accurate comparison of their efficiency and performance.

5 Results

Table 3 presents key data for the effective product cycle time, which, due to the specific nature of the analysed cell, better illustrates the production capabilities of the cyclically operating cell before and after the implementation of changes. The differences in the case of operations that were previously performed by the operator show that the introduction of the robot has improved the operation of the cell. The total production cycle time after the implementation of the changes is 5 min and 22 s.

Table 3. Times of key operations before and after robot implementation.

Operation name	Time before implementation of changes [s]	Time after implementation of changes [s]	Reduction in operation time [s]
Loading new raw material and unloading machined sleeve	29	23	6
Machining the first side of the material on the lathe	186	186	0
Rotating the material	36	29	7
Machining the second side of the material	84	84	0

In order to compare performance parameters and the DPU (Defects Per Units) index, the number of units produced per day was re-measured and the percentage of products that did not meet production specifications was analysed. The data presented in Tables 4 and 5 clearly show an increase in the average number of products manufactured during a single production shift.

Table 4. Number of correct and defective items before implementation of the new solution.

Day of nest measurements before implementation	Number of products manufactured	Number of correct products	Number of defective products	DPU
1	74	71	3	0.041
2	73	70	3	0.041
3	73	71	2	0.027
4	72	68	4	0.056
5	73	71	2	0.027
Average	73	70,2	2,8	0.038

Table 5. Number of correct and defective items after implementing the new solution.

Day of nest measurements after implementation	Number of products manufactured	Number of correct products	Number of defective products	DPU
1	81	80	1	0.012
2	81	81	0	0.000
3	82	81	1	0.012
4	81	80	1	0.012
5	81	81	0	0.000
Average	81,2	80,6	0,6	0.0073

Based on calculations of the effective production cycle time for both solutions, it can be seen that the main problem was manual operation by an employee. As a key element of the process, the employee requires regular breaks and time for additional activities not related to the production cell, which resulted in downtime for the CNC machines. In addition, situations related to momentary inattention were possible, which could lead to delays in loading and unloading machines and in the execution of production operations. These factors had a negative impact on the smoothness of the process, whereas the use of a robot effectively eliminated machine downtime. This allowed the average daily number of products manufactured to be increased from 73 to 81, with a significant reduction in the number of defective products.

Productivity was calculated for the 5 days during which the measurements were taken, i.e. the production time, including a 20-min break, totalled 38 h and 20 min.

$$Productivity = \frac{406}{38,33} \approx 10,59 units/hour$$

An analysis of the productivity of both production cells revealed a significant increase in this indicator after the implementation of the robotic solution. Productivity increased from 9.52 units/hour for the manually operated cell to 10.59 units/hour for the robotic system.

The introduction of a robotic production cell enabled a significant reduction in the DPU indicator from 0.038 to 0.0073, which means a clear improvement in the quality of manufactured products. The elimination of errors resulting from manual work and the precise and repeatable operation of the robot have reduced the number of defective products. The automation of the process has also minimised the impact of human factors, which has increased production reliability.

6 Discussion and Conclusions

The main objective was to create a robotic production cell to eliminate manual operation of CNC machines. The analysis showed that manual operation of the lathe and milling machine by a single employee caused downtime, increased production time and increased the risk of errors during technological processes and material clamping. These factors had a negative impact on production efficiency and quality. The developed solution, which involved the implementation of a Turin robot and offline programming in the RoboDK environment, was designed to eliminate these problems.

The new cell automated key operations such as loading and unloading materials, transport between machines and running machining programmes. The introduction of automation shortened the production cycle time, enabling the lathe and milling machine to operate simultaneously. In addition, errors related to inaccurate sleeve clamping were eliminated, which improved process stability and product quality. The defect rate was significantly reduced.

The implementation of modern technologies also had a positive impact on work safety. Operators were freed from the need to directly operate the machines, which minimised the risk of accidents. The results of analyses and simulations clearly indicate that the solution improved production efficiency and product quality and reduced costs resulting from downtime and defects, in line with the objectives of Industry 4.0.

The introduction of automation also improved operator safety by eliminating the need for manual fastening of sleeves and contact with machines during their operation. Operators were able to focus on supervision and quality control. In order to maximise the potential of the new system, it is necessary to properly train personnel in the operation of the robot and RoboDK software. Regular servicing and monitoring of the equipment is also crucial to ensure its reliability. Although the implementation required an initial investment, in the long term the project contributes to reducing operating costs and increasing the company's competitiveness.

Acknowledgments. The presented results are derived from a scientific statutory research project number 0613/SBAD/4940 conducted by Faculty of Mechanical Engineering, Poznan University of Technology, Poland, supported by the Polish Ministry of Science and Higher Education.

References

1. International Federation of Robotics: World Robotics Report 2022: Executive Summary. IFR, Frankfurt am Main (2022)
2. Nexelem: Industry 4.0 – What is it and why is it worth implementing? https://nexelem.com/en/blog/industry-4-0/. Accessed 08 Jan 2025
3. APA Group: Expensive, few experts and many doubts. Report 'The State of Industry 4.0 in Poland 2024' MamStartup (2024). https://mamstartup.pl/drogo-malo-ekspertow-i-duzo-watpliwosci-raport-stan-przemyslu-4-0-w-polsce-2024-od-apa-group. Accessed 08 Jan 2025
4. Pająk, E.: Production Management. Product, Technology, Organisation. Wydawnictwo Naukowe PWN, Warsaw (2021). (in polish)
5. Lisowski, E.: Industrial Robotics. Krakow University of Technology, Krakow (2020). (in polish)
6. Zieliński, C.: Robotics. Theory and practice. WNT, Warsaw (2019). (in polish)
7. International Federation of Robotics (IFR): World Robotics Report 2021. IFR, Frankfurt am Main (2021)
8. International Federation of Robotics (IFR): World Robotics Report 2023. IFR, Frankfurt am Main (2023)

9. Kaczmarek, A.: Robotisation of Production Processes. Poznań University of Technology Press, Poznań (2018). (in polish)
10. Automation and robotisation of production processes – what are they and how do they differ from each other? InRel; https://inrel.pl/en/. Accessed 10 Jan 2025
11. Mazak Corporation: Quick Turn 10 Lathe - Operation Manual. Mazak Corporation (1998)
12. Visual Components: Offline Robot Programming (OLP): The Complete Guide with Examples Visual Components Blog (2023). https://www.visualcomponents.com/blog/offline-robot-programming-olp-the-complete-guide-with-examples/. Accessed 08 Jan 2025
13. Visual Components: Steps of the OLP Process Visual Components Blog (2023). https://www.visualcomponents.com/blog/steps-of-the-olp-process/. Accessed 08 Jan 2025
14. RoboDK: Offline Programming Software for Robots RoboDK (2023). https://robodk.com. Accessed 08 Jan 2025
15. DPSToday: RoboDK – a new approach to robot programming (2023) (in polish). https://www.dpstoday.pl/robodk-czyli-nowe-podejscie-do-programowania-robotow/. Accessed 14 Jan 2025
16. Kłodziński, S.: Increased productivity of the workstation, Bachelor's thesis, Poznan University of Technology, Poznan (2025)
17. Firmove: Automation and robotisation of production processes. Firmove (2023) (in polish). https://firmove.pl/aktualnosci/biznes/prowadzenie-firmy/automatyzacja-i-robotyzacja-procesow-produkcyjnych. Accessed 08 Jan 2025
18. Hurco Companies, Inc.: VM1 Milling Machine - Operation Manual. Hurco Companies Inc. (1990)
19. Turin Robot: TR20–1920B Product Details. https://www.turinrobot.com/products_details/17.html. Accessed on 14 Jan 2025

Analysis of Advanced Digital Technology Applications for Effective Human–Machine–Robot Collaboration in the Production of Precision Mechatronic Systems

Dušan Paulišin[1], Marek Kočiško[1], Petr Baron[1](✉), and Denisa Olekšáková[2]

[1] Department of Computer Support of Production Technologies, Faculty of Manufacturing Technologies, Technical University of Košice, Prešov, Slovak Republic
petr.baron@tuke.sk

[2] Department of Process Engineering, Faculty of Manufacturing Technologies, Technical University of Košice, Prešov, Slovak Republic

Abstract. Digitalization and automation of manufacturing processes are becoming key factors of competitiveness in the production of high-precision mechatronic systems. This article presents the results of the initial phase of research conducted at SPINEA, Ltd., focused on identifying processes with the highest potential for digitalization and the implementation of collaborative robotics (HSR - Human-System-Robot). The analyzed production operations included machining, final finishing, and assembly of DriveSpin® actuators and TwinSpin® cycloidal gearboxes. Based on an assessment of frequency, variability, and ergonomic requirements, the main areas suitable for digital transformation were defined: material picking, screwing, pressing, substance application, and intralogistics. The DS 070 actuator, which has the highest share in the company's production portfolio, was selected as the pilot product. The results serve as a foundation for designing intelligent and flexible workplaces with collaborative robots and real-time digital data acquisition.

Keywords: digital transformation · collaborative robotics · Human-System-Robot (HSR) cooperation · assembly automation; · Smart Manufacturing · DriveSpin actuators · intralogistics · mechatronic systems

1 Introduction

Digitalization and transformative trends within the ongoing industrial revolution are fundamentally changing the way industrial companies design, manage, and optimize manufacturing processes. A key principle of this transformation is the use of intelligent and interconnected systems that enable rapid adaptation of production to be evolving market and customer requirements. The greatest benefits are expected in high value-added domains where precision, reliability, and product modularity play a crucial role.

J. Trojanowska et al. (Eds.): MANUFACTURING 2026, LNME, pp. 347–359, 2026.
https://doi.org/10.1007/978-3-032-21654-0_26

The production of precision mechatronic systems, such as cycloidal gearboxes and electromechanical actuators, represents a highly engineered manufacturing environment with strict demands on accuracy and process traceability. These systems form the foundation of modern robotic and automation solutions in manufacturing, medicine, aerospace, and energy. Their assembly and testing involve numerous fine, specification-driven operations in which the human operator still plays an irreplaceable role.

Current production nevertheless faces several challenges:

- medium to high product variability driven by customer-specific configurations,
- the need to increase productivity without expanding labor capacity,
- rising requirements for traceability and error elimination,
- a shortage of skilled workers and increasing ergonomic demands.

These factors create the need for intelligent Human-System-Robot (HSR) cooperation, which combines the precision and strength of technology with the flexibility of human work.

The topics of digitalization and process traceability have been addressed by several authors. Article [1] defines the digital twin (DT) as a virtual representation of a physical product or process that can enhance efficiency, monitoring, and production optimization. Similarly, [2] analyzes literature and applications of DT across the entire product lifecycle. Article [3] directly links DT with traceability and dynamic quality control. Studies [4–7] characterize DT as a technical tool enabling tracking, recording, simulation, and optimization of manufacturing and assembly processes.

In recent years, there has been an intensive development and exploration of Human–System–Robot integration concepts within the context of smart manufacturing, significantly expanding the framework for our empirical findings. The landscape of intelligent manufacturing is shaped by digitalization and robotics—primarily through human–robot collaboration (HRC) systems, digital twins (DT), IoT-based sensor technologies, artificial intelligence (AI), and machine learning (ML) [8–11]. Othman and Yang [8] emphasize the transition toward flexible, AI- and DT-enabled HRC environments, particularly in the automotive and food industries. Keshvarparast et al. [9] provides a systematic review of collaborative robot deployment in assembly lines, including ergonomics and performance control. Petzoldt et al. [10] focus on optimizing task redistribution in HRC systems, considering product variability and decision-making metrics. Cai et al. [11] further extend the HRC framework by incorporating multimodal perception, DT, and adaptive control aligned with Industry 4.0 principles.

Digital twins are regarded as a central component of smart manufacturing—Parashar and Gameti [12] analyze current approaches and future perspectives of DT in production processes. Hu et al. [13] map the methodological foundations and applications of DT throughout the entire product lifecycle. Soori et al. [14] trace the evolution of DT in the context of production optimization and time-to-market reduction. Parashar and Gameti [12] and Karkaria et al. [15] highlight the synergistic use of AI within DT for predictive maintenance and process optimization.

Sensor integration is also critical—Cherubini and Navarro-Alarcon [16] present multimodal control for collaborative robots, encompassing vision, tactile, and auditory feedback. Montini et al. [17] evaluate practical experiences with HRC technologies and identify current challenges and opportunities. Articles in *Frontiers* [18] examine ergonomics in HRC environments and establish direct ergonomic metrics for safe human–robot collaboration.

In the domain of intralogistics and material handling, AGV/AMR systems have proven effective for workflow optimization—Interact Analysis [19] reports a substantial annual growth of 30–50% in AGV/AMR deployments. Weidemann et al. [20] provides a detailed analysis of collaborative logistics trends within Industry 4.0.

Traceability and data governance in digital manufacturing represent another layer of this comprehensive framework—Cheng et al. [21] introduce a workflow-driven DT system for monitoring assembly processes of complex products. Argilovski et al. [22] discuss traceability concepts in the digital transformation of manufacturing. Smina et al. [23] offer a systematic review of data governance frameworks in the context of smart supply chains.

Collaborative robots, AGV/AMR logistics systems, and digital twins represent technologies that enable this transformation to be implemented safely, economically, and at scale. SPINEA, Ltd., a global manufacturer of high-precision drive systems, has identified the need to significantly strengthen competitiveness through increased levels of digitalization in assembly and logistics. Despite existing ERP, PLM, and production planning systems, real-time information flow remains limited, and most key assembly operations are still manual. In the preloaded assembly processes, human expertise remains indispensable, as an experienced technician can detect potential issues through touch, sound, movement resistance, or subtle deviation-signals that automated sensors cannot yet reliably capture or interpret.

Fig. 1. DriveSpin® DS 070 electric rotary actuator

SPINEA, Ltd. therefore identified the need for an intelligent HSR workplace to improve working conditions and enhance production adaptability while maintaining high flexibility and efficiency. This article presents the results of the preliminary identification of processes with the greatest potential for modernization through digital technologies and collaborative robotics. The DriveSpin® DS 070 actuator was selected as the representative product, as it constitutes the largest share of production and includes all key assembly operations typical of the company's portfolio (Fig. 1).

The main contributions of the study presented are:

- defining a methodology for systematic evaluation of the production process for digitalization,
- identifying and prioritizing bottlenecks in assembly and logistics,
- preparing the fundamental basis for designing an intelligent HSR workplace.

The results represent the first stage of a comprehensive research project aimed at developing system-level solutions that enhance efficiency, quality, and workplace ergonomics while maintaining high production flexibility.

2 Characteristics of the Applied Research Methodology

The aim of the research was to systematically identify manufacturing processes suitable for digitalization and for supporting collaborative robotics in the production of precision mechatronic systems at SPINEA, Ltd. The methodology is based on a combination of process engineering, technological assessment, and evaluation of ergonomics and workplace safety.

2.1 Evaluation Procedure

The analysis was carried out in the following stages:

1. Process analysis - mapping the current state of production
   - workplace observations
   - analysis of operational documentation
   - measurement of operation times
2. Product data collection
   - production batch statistics
   - variability of configurations and components
3. Evaluation of digitalization suitability
   - frequency of task repetition
   - level of product variability
   - level of risk and physical workload for operators
   - potential for standardization
4. Technological survey of automation

- availability of Industry 5.0 solutions
- possibilities for implementing collaborative robots (cobots)

5. Selection of the pilot product and processes

- representativeness within the product portfolio
- feasibility of creating a digital twin

The processes were characterized using the following matrix (Table 1), which evaluates each operation along five key criteria: frequency of repetition, variability, ergonomic demands, safety risks, and criticality to quality. Each criterion is assessed on a qualitative high/low scale, and the resulting pattern of evaluations is linked to corresponding implications such as return on investment, suitability for automation, benefits for the working environment, HSR potential, and the need for digital data collection. This structured evaluation provides a transparent basis for comparing heterogeneous operations and for selecting priority areas for digitalization and the implementation of HSR workplaces.

Table 1. Process Evaluation Matrix Based on Digitalization Suitability Criteria.

Evaluation factor	Evaluation	Implication
Frequency of repetition	high/low	return on investment
Variability	high/low	suitability for automation
Ergonomic demands	low/high	benefits for the working environment
Safety risks	low/high	HSR potential
Criticality to quality	low/high	need for digital data collection

2.2 Evaluation Procedure

Combined research methods from industrial engineering, data analytics, and technological diagnostics were used to analyze the assembly and logistics processes. Each source provided a different type of information needed for an objective assessment of the technological potential for digitalization:

- Direct process observation (time studies of assembly operations, video recordings for detailed analysis, evaluation of ergonomic aspects) - the output is the identification of bottlenecks and non-ergonomic tasks suitable for robotic assistance.
- Analysis of data from information systems (ERP QAD, Monaco, PLM Windchill).
- Interviews and workshops with relevant personnel (assembly operators, design engineers, technologists, production and logistics managers) - the goal is to obtain qualitative indicators of practical issues and automation opportunities.
- Technological survey of automation solutions (benchmarking of collaborative robots, availability of screwing and pressing automats, system capabilities such as MES, Smart-Tooling, digital twin) the output is an evaluation of integration possibilities within SPINEA's existing infrastructure.

3 Results of the Production Process Analysis

The analysis of production processes at SPINEA (particularly in the LR, DS, and MS product lines) identified key areas with high potential for digital transformation. The assessment focused primarily on processes with high repeatability, variability, and added value from automation.

Table 2. Assessment of Process Suitability Based on Digitalization Potential.

Process Type	Process Name	Type	Suitability for digitalization	Notes/Benefit
Material picking	Order picking from the warehouse (vertical stacker)	Main 1	High	Inefficient, error-prone → the possibility of automation and real-time tracking
Screwing	Automated tightening of bolts of various types and torques	Main 2	High	Repeatable, standardizable, ideal for robotics
Applying fabrics	Glue, lubricant, sealant - precise dosing	Main 2	High	Quality-critical, can be integrated with dispensing machines
Pressing	Precise pressing forces and times	Main 2	Medium - High	Sensory control and automation possible
Heating	Preheating housing before installation	Main 2	Central	It requires precise regulation, the possibility of automation and IoT monitoring
Handling (inserting)	Inserting components into machines, measuring jigs	Secondary	High	Repeatable, possibility of deploying collaborative robots
Material Transfer	Transport between machines, plants	Secondary	Medium - High	Suitable for automated trucks, AGVs or simple feeders
Packaging & Labelling	Final packaging of products and application of labels	Secondary	Central	Ability to digitize and track with bar/QR codes

Material preparation (warehouse picking):

- The material picking process is time-consuming and organizationally complex.
- High variability in requirements and handling procedures leads to low efficiency and an increased risk of errors when delivering components to workstations.
- Improvement potential lies in implementing automated warehouse solutions (e.g., AGV/AMR robots, intelligent shelving, automatic identification systems), which would enable more accurate and faster material preparation.

Assembly operations:

- Assembly processes include tasks such as screwing, pressing, heating, and adhesive application, all of which exhibit a high degree of repeatability and well-defined procedures.
- These operations are suitable for automation or for support through collaborative robots (cobots).
- Implementing digitally controlled assembly lines will enable higher precision, consistency, and traceability of manufacturing tasks.

Secondary processes:

- Secondary processes include part handling, internal material transport, and packaging of finished products.
- These activities are often routine yet time-consuming and prone to human error.
- Automation through conveyor systems, robotic manipulators, or intelligent packaging solutions can significantly streamline material flow and reduce operational costs.

Table 2 presents an overview of the monitored processes at SPINEA and an assessment of their suitability for digitalization. The matrix distinguishes between main and secondary processes and links the qualitative evaluation of digitalization potential with specific notes on anticipated benefits, such as error reduction, improved traceability, and the possibility of deploying collaborative or fully automated solutions. This structured overview provides a basis for prioritizing pilot projects and for defining a phased roadmap for the implementation of digital technologies within SPINEA's production system.

4 Analysis of Production Batches Based on Product Characteristics

The aim of the production batch analysis by product type and size was to obtain an overview of the production structure in terms of size ranges and product variants. This analysis makes it possible to identify which product types are most frequently represented in manufacturing. Figures 2 and 3 show the number of produced and actuators over a three-year period (2023–2025).

Process variability represents the differences or deviations that arise during individual stages of manufacturing TwinSpin gearboxes and DriveSpin actuators-from component machining, through assembly, to final inspection. These deviations can affect the accuracy, stiffness, and overall quality of the final product. Batch variability represents the differences or deviations between individual production batches that occur during the manufacturing process. Each production batch should exhibit the same properties, quality, and parameters, as all are manufactured according to the same technological procedures.

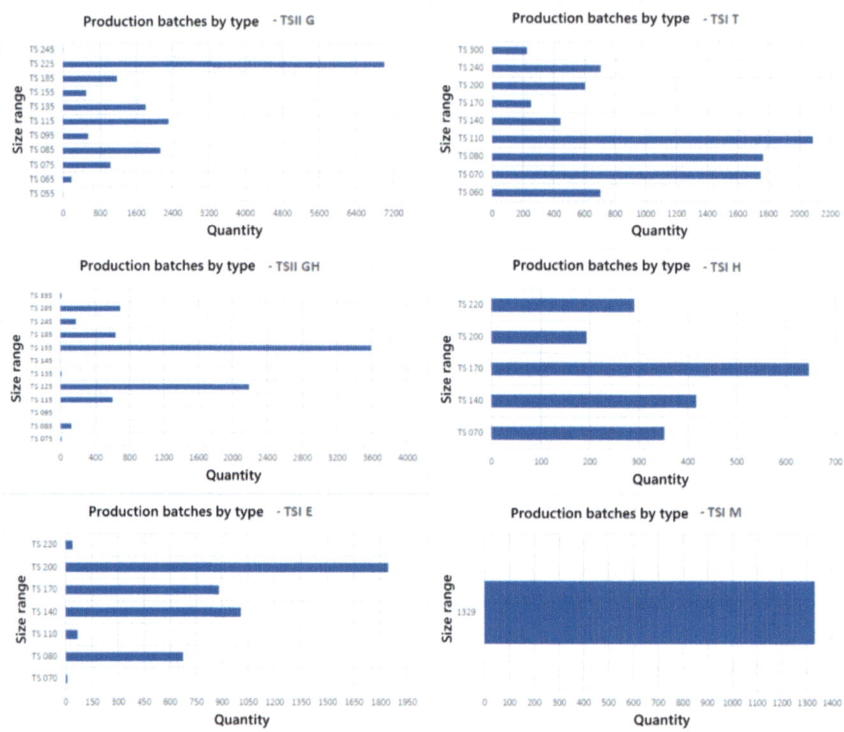

Fig. 2. Overview of Production Batches of TwinSpin® Gearboxes (2023–2025).

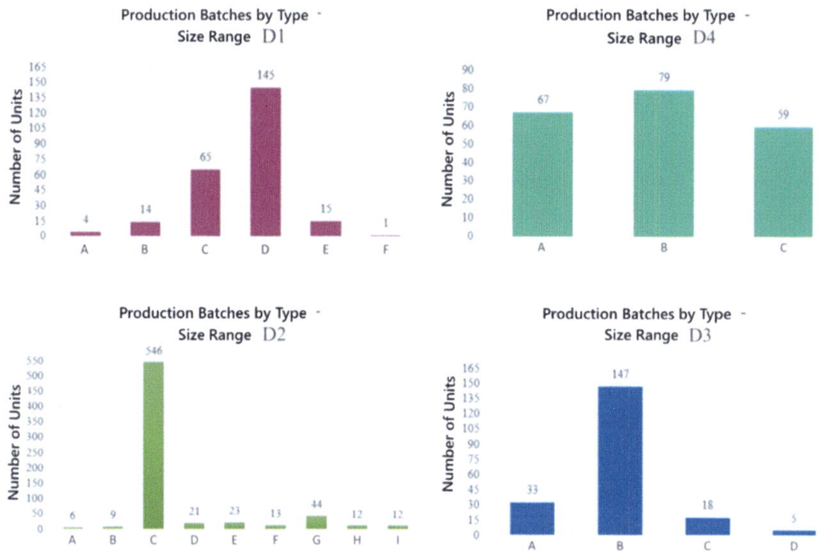

Fig. 3. Overview of Production Batches of DriveSpin® Actuators (2023–2025).

Monitoring production batch variability is important for evaluating the stability and repeatability of the manufacturing process. Low variability indicates that the process is stable and produces products with consistent properties, while high variability points to potential issues in the process that may lead to an increased number of nonconformities or reduced quality.

The goal of the company is to minimize variability between batches to ensure uniform product quality, reduce production costs, and improve overall manufacturing efficiency.

5 Discussion

The results of the analysis clearly confirm the need to increase the level of digital automation in the production of DriveSpin® actuators. The technological complexity of the product requires processes with a high degree of precision, traceability, and repeatability. In this context, the implementation of intelligent HSR cooperation (Human-System-Robot) represents an optimal step toward increasing production capacity, quality, and technological reliability.

5.1 Comparison with Modern Trends in the Manufacturing of Precision Drive Systems

Current Smart Manufacturing trends in Europe identify three major technological challenges:

- Mass customization while maintaining the efficiency of serial production requirement for high adaptability of production lines.

- Increasing demands for traceability and quality control complete digital process recording as part of the product.
- Safe sharing of the workspace between humans and robots ISO/TS 15066 as the technological framework.

Table 3. Expected benefits of digitization of selected production and logistics processes in SPINEA, Ltd.

Area	Expected improvement	Technological effect
Screwing	+40–60% stability of mounting parameters	100% traceability of moments
Pressing	0% internal damage after digitization	sensory control of forces
Logistics	−20–30% downtime	Automated material flows
Mounting the sensors	reduction of mounting variability	Increased positioning accuracy
Data quality	100% digital process recording	Support for certifications

Based on the analysis conducted, it can be stated that production at SPINEA, Ltd. exhibits a high degree of product modularity, which correlates with the trend of High-Mix/Medium-Volume manufacturing, an ideal candidate for flexible collaborative robotics.

The implementation of the proposed solutions has the potential to deliver the following measurable technical benefits (Table 3). From a technological perspective, this represents a transition from a manually controlled process to a data-driven manufacturing process. To achieve full effect, it is essential to implement an architecture consisting of three layers:

1. Sensor and robotic layer
   - collaborative robots (screwing, handling)
   - intelligent tools with data collection
2. Digital control layer
   - integration of ERP - MES - Smart Tools - Vision systems
   - digitalized real-time process monitoring
3. Analytical layer
   - predictive maintenance
   - automated process optimization

In the long term, this creates the foundation for a digital twin of the assembly process and autonomous cycle optimization.

6 Conclusion

Digitalization and automation of manufacturing processes represent a key opportunity for technological advancement in the production of precision mechatronic drives. The results of the analysis conducted at SPINEA, Ltd. showed that the greatest potential in terms of technological benefit lies in assembly operations and intralogistics, which are currently performed predominantly manually and are characterized by high repeatability and a critical impact on the final product quality.

The chosen methodology enabled a systematic identification of bottlenecks and an assessment of the suitability of operations for implementing the Human-System-Robot cooperation concept. The DS 070 actuator was selected as the pilot product, as it represents a significant portion of total production and includes all relevant assembly characteristics. Collaborative robotics, intelligent screwing, automated pressing, and smart logistics solutions can increase assembly repeatability, enhance quality traceability, and reduce the risk of human error.

An essential part of the proposed approach is the digital layer ensuring online monitoring and data collection, which enables the use of artificial intelligence for predictive quality, robot trajectory optimization, and autonomous decision support. The integration of these elements creates the foundation for an assembly digital twin and moves toward highly flexible workplaces of the future.

Future research will focus on experimental verification of HSR concepts in the assembly of DS 070, evaluating their impact on technological stability, and extending the solutions to other product portfolios. In the long term, a transition toward an autonomous, data-driven manufacturing model with a high degree of adaptation to changes in customer specifications is anticipated.

The implementation of advanced digital technologies for effective Human-System-Robot cooperation (intelligent HSR workplaces) can serve as a reference model for the transformation of small and medium-sized enterprises (SMEs) that perform similar production operations and face challenges related to increasing efficiency, stability, and competitiveness. For SMEs, which often operate with limited workforce capacity and are exposed to growing demands for quality and production flexibility, the systematic integration of digitalization, automation, and collaborative technologies represents a potentially sustainable solution. However, many do not know how to begin or how to proceed when their product mix exhibits high variability. The outcomes of this project can provide a methodological framework and practical recommendations for identifying critical processes, optimizing production workflows, and gradually implementing innovations, thereby supporting the transition toward a modern, data-driven, and adaptive manufacturing model.

Funding. This work was supported by the Slovak Research and Development Agency under Contract No. APVV-24-0470 and by the Scientific Grant Agency of the Ministry of Education, Research, Development and Youth of the Slovak Republic (VEGA) under Grant No. 1/0403/23. It was also funded by the EU NextGenerationEU through the Recovery and Resilience Plan for Slovakia under project No. 09I05-03-V02-00042.

References

1. Matheson, A., Minto, J.: Human-robot collaboration in manufacturing applications: a review. J. Manuf. Syst. **53**, 344–356 (2019)
2. Son, Y.H., Kim, G.-Y., Kim, H.C., Jun, C., Noh, S.D.: Past, present, and future research of digital twin for smart manufacturing. J. Comput. Des. Eng. **9**(1), 1–23 (2022)
3. Soori, S., Oztemel, E., Gursev, S.: Digital twin for smart manufacturing: a review. Procedia CIRP **117**, 675–680 (2023)
4. Liu, X., Zhang, Y., Wang, Z.: A digital twin-driven approach towards traceability and dynamic control for processing quality. Procedia Manuf. **54**, 52–58 (2021)
5. Hopko, S., Wang, J., Mehta, R.: Human factors considerations and metrics in shared space human-robot collaboration: a systematic review. Front. Robot. AI **9**, 799522 (2022)
6. Alfaro-Viquez, D., Zamora-Hernandez, M., Fernandez-Vega, M., Garcia-Rodriguez, J., Azorin-Lopez, J.: A comprehensive review of AI-based digital twin applications in manufacturing: integration across operator, product, and process dimensions. Electronics **14**(4), 646 (2025)
7. Maculotti, G., Khusnuddinov, F., Kholkhujaev, J., Galetto, M.: Traceable digital twin for accurate positioning of industrial robot arms in human-robot collaborative systems. Flex. Serv. Manuf. J. (2025)
8. Othman, U.Z., Yang, E.: Human-robot collaborations in smart manufacturing environments: review and outlook. Sensors **23**(12), 5663 (2023)
9. Keshvarparast, A., Battini, D., Battaia, O., Pirayesh, A.: Collaborative robots in manufacturing and assembly systems: literature review and future research agenda. J. Intell. Manuf. **35**(5), 2065–2118 (2024)
10. Petzoldt, C., Harms, M., Freitag, M.: Review of task allocation for human-robot collaboration in assembly. Int. J. Comput. Integr. Manuf. **36**(11), 1675–1715 (2023)
11. Cai, Y., Xu, W., Yao, B., Xiang, Y., Ji, Z., Ai, H.: Safe human–robot collaboration using multimodal perception and robot learning. In: Human-Centric Smart Manufacturing, pp. 229–252. Springer (2025)
12. Parashar, A., Gameti, N.: Digital twins in manufacturing: a survey of current practices and future trends. Int. J. Sci. Res. Arch. **13**(01), 1240–1250 (2024)
13. Hu, S., et al.: Digital twins enabling intelligent manufacturing: from methodology to application. Intell. Sustain. Manuf. **1**(1), 10007 (2024)
14. Soori, M., Arezoo, B., Dastres, R.: Digital twin for smart manufacturing, a review. Sustain. Manuf. Serv. Econ. **2**, 100017 (2023)
15. Karkaria, V., Chen, J., Luey, C., Siuta, C., Lim, D., Radulescu, R., Chen, W.: A Digital Twin Framework Utilizing Machine Learning for Robust Predictive Maintenance. arXiv preprint arXiv:2408.06220 (2024)
16. Cherubini, A., Navarro-Alarcon, D.: Sensor-based control for collaborative robots: fundamentals, challenges, and opportunities. Front. Neurorobot. **14**, 576846 (2021)
17. Montini, L., et al.: Practical experiences and challenges in HRC technology deployment. J. Manuf. Process. **82**, 45–58 (2024)
18. Frontiers Editorial Board: Ergonomic human–robot collaboration in industry: a review. Front. Robot. AI **9**, 813907 (2023)
19. Interact Analysis: Market trends in AGV/AMR systems for intralogistics. Industry Report, pp. 1–25 (2022)
20. Weidemann, J., et al.: Collaborative logistics in Industry 4.0: trends and applications. Robotics **12**(3), 84 (2023)
21. Cheng, F., et al.: Workflow-driven digital twin systems for assembly traceability. J. Braz. Soc. Mech. Sci. Eng. **47**, 151 (2025)

22. Argilovski, A., et al.: Traceability concepts in digital manufacturing transformation. J. Ind. Inf. Integr. **28**, 100345 (2022)
23. Smina, S., et al.: Data governance frameworks in smart supply chains: a systematic review. Information **17**(1), 19 (2025)

Selecting Transportation Options Within the Distribution Network

Jarosław Ziółkowski[1,2,3], Patrycja Guzanek[1,2,3](✉), and Mykola Karpenko[3]

[1] Military University of Technology, Warsaw, Poland
patrycja.guzanek@student.wat.edu.pl
[2] Doctoral School of Military, University of Technology, Warsaw, Poland
[3] Vilnius Gediminas Technical University, Vilnius, Lithuania

Abstract. The article presents a methodology for selecting a transport option within a distribution network. In commercial transport activities, economic aspects are an important component, which are also the basic criterion for selecting both the carrier itself or the option (variant). The paper presents a comparative analysis of the selection of a variant for implementing deliveries within a road distribution network from among four possible options, i.e. purchasing one's own transport fleet on commission, leasing and long-term rental. The analysis period was one calendar year, for which demand values measured by the number of deliveries were compared monthly and then the average variable unit cost of delivery was calculated for each selection option. The second component was fixed costs, with commissioned transport omitted from their calculation. In addition, expected values were determined for each option as weighted average probabilities of a given scenario and its financial result. The calculations were presented graphically in the form of a decision tree taking into account the results of each option within the proposed scenarios. Finally, a break-even analysis was carried out and final conclusions were formulated.

Keywords: Distribution Network · Costs · Expected Value Analysis · Break-Even Point

1 Introduction

Issues related to road transport in relation to the process, phenomenon, system or object constitute a wide spectrum of interest and have significant publishing potential [1–3]. In scientific works, the authors focus their main efforts on the possibilities of improving a number of areas requiring improvement. Usually, to improve the studied phenomenon, the set of potential applications includes issues from the field of operational research, including linear programming [4–6], neural networks [7–9] or artificial intelligence [10, 11].

The simplest in the calculation sense are linear programming problems, in which a single-criterion objective function can be directed to a minimum or maximum [12]. In the first case, the objective becomes areas of various meanings, such as: minimizing costs [13–15], delivery time [16, 17], total distance [18, 19] or runs performer by empty

J. Trojanowska et al. (Eds.): MANUFACTURING 2026, LNME, pp. 360–378, 2026.
https://doi.org/10.1007/978-3-032-21654-0_27

vehicles [20], total number of vehicles [21], fuel consumption [22], carbon emissions [23]. In the second area, a specific effect is usually maximized, i.e. profit [24–26], level of customer service or load capacity. Researchers are also particularly interested in the readiness and reliability of means of transport [27, 28]. In the field of optimization of transport processes implemented within the distribution network, two-criteria issues may also appear, which are slightly more difficult than single-criteria issues. One of them is the problem of assigning both the same type of vehicle with a specific capacity to transport tasks, but taking into account a limited number of means of transport with a different cargo box capacity. Such natural limitations occur in most distribution networks, not only commercial but also military ones. Ziółkowski et al. in paper [26] uses the example of a military supply network to present a multi-criteria methodology enabling minimization of total transport costs, number and type (due to load capacity) of vehicles used, distance travelled, fuel used, and CO_2 emissions into the atmosphere. Moreover, additional restrictions on existing transport resources were included, considering the number and type of vehicles available at the base. This is of great importance, especially when there is a need to provide emergency deliveries, for example, in the event of a war threat.

In two-criteria decision-making processes, a significant problem is the fact that the set of sought acceptable solutions in reality may not exist. This means that the parallel fulfilment of two criteria with specific parameters is in practice impossible, which means that the set of solutions is de facto an empty set.

Solving decision-making problems concerning the analysis and evaluation of complex processes and transport systems with an additionally diversified fleet of vehicles using analytical methods is often difficult or even impossible to perform. In such a case, simulation models are used, the results of which often constitute the basis for making rational decisions.

Modeling of readiness in transport systems has been undertaken in a number of studies. Integrating autonomous vehicles (AVs) into urban traffic systems presents significant opportunities for optimizing traffic flow, reducing congestion, and enhancing transportation efficiency were presented in [24]. This study proposes a comprehensive framework that combines mathematical optimization techniques, policy planning, and AV adoption modeling to improve urban mobility. Using Highway Capacity Manual (HCM) Optimization methods, the research fine-tunes traffic signal timings, dynamically allocates green time, and enhances intersection coordination to maximize throughput. In paper authors used data collected from 1375 potential electric two-wheeler buyers of Bengaluru, India, to study the influence of socio-demographic variables and certain latent factors such as environmental enthusiasm, technological enthusiasm, monetary benefits, environmental benefits, social image, lack of infrastructural readiness, perceived fee, and perceived risks on consumers' intention to adopt electric two-wheelers. The results reveal environmental enthusiasm, technological enthusiasm, monetary benefits, environmental benefits, and social image to have a significant positive impact on electric two-wheeler adoption intention. Other authors identified the effects of freeway design and condition on lane-detection performance of lidar-based AVs through a field test on two typical freeways in Shanghai, China, and the freeway readiness was evaluated. The

Tongji University Road and Traffic Data Acquisition System was used for data collection. The lane-detection failure was treated as the label. Ten variables of five feature types were considered as the parameters, including road geometry, road segment, road marking, vehicle operation, and environment.

The aim of this article is to select a transport option taking into account cost calculation. Four variants of meeting transport needs were considered in the paper, i.e. deliveries made by the transport company (on-call delivery), leasing, long-term rental and cash purchase. Each of them was estimated within the framework of the annual demand reported by four recipients (R1–R4) with specific locations. The recipient locations are fixed and serviced from one central warehouse by the supplier (S). The essence of the presented approach is to estimate fixed and variable costs for each of the four options in order to select the best cost variant. Expected value analysis and break-even analysis were used to estimate the costs for individual options.

2 Assumptions and Description of the Process

The subject of the analysis is an enterprise from the SME sector that delivers products to regular customers. The company and customers create a distribution network with five nodes, whose locations are illustrated in Fig. 1.

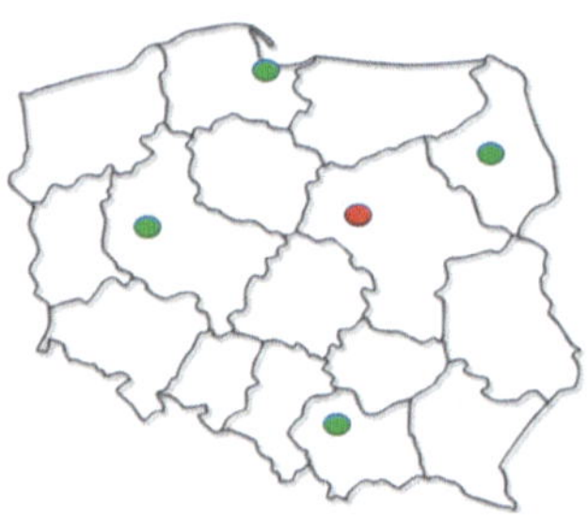

Fig. 1. Distribution network including supplier S and recipients R1–R4.

The family business carries out direct deliveries, in accordance with the monthly individual demand for each recipient. Within the framework of transports performed by road transport, there are four possible options (variants, scenarios) for meeting transport needs, i.e. deliveries made by the transport company (on-call delivery), leasing, long-term rental and cash purchase.

In the first stage of this work, it was necessary to estimate fixed and variable costs for each option, and then to conduct an analysis of the expected value and the break-even point in order to select the best solution, with the selection criterion being the total costs. Total costs are interpreted as the sum of two components, i.e. the product of the unit cost of delivery and the number of deliveries and fixed costs, according to the following equation:

$$T_C = F_C + AV_{UC} \bullet X \tag{1}$$

where:

T_C – total costs;

F_c – fixed costs;

V_C – variable costs ($V_C = AV_{UC} \bullet X$);

AV_{ZJ} – average variable unit cost of delivery;

X – intensity of transport work performed (measured by the number of deliveries).

Variable costs V_C depends on the number of deliveries X and average variable unit delivery cost AV_{UC}, whereas fixed costs F_c are incurred regardless of the level of intensity of the company's activities.

The following assumptions were made to develop the model:

1. Deliveries are made on fixed routes;
2. The same type of vehicle is used to make deliveries;
3. Each recipient determines the number of deliveries in each month for the calendar year under analysis based on the forecasted demand;
4. For each option (variant), the unit cost of delivery should be estimated, the components of which are other types of expenses;
5. Variable costs depend on the number of deliveries during the year and the unit cost of delivery. The first task was to estimate the unit cost of delivery for each option individually.

The cost analysis of transportation options was conducted based on annual and average indicators derived from monthly operational data for four alternative delivery arrangements (Table 1).

Table 1. Deliveries made by the transport company (on-call delivery) – aggregated results.

Transportation option	Total annual cost [PLN]	Average monthly cost [PLN]
Transport company (on-call delivery)	65764,24	5480,35
Leasing	17183,12	1431,93
Long-term rental	16526,99	1377,25
Cash purchase	18273,22	1522,77

Detailed monthly calculations (including route-level data and the derivation of "Total costs/month") are provided in the Appendix.

3 Results and Discussion

The detailed data presented in the tables above regarding the four possible transport options were used to determine the average variable unit cost of delivery (Table 2).

Table 2. Average variable unit cost of delivery for each option.

	On-call delivery	Leasing	Long-term rental	Cash purchase
Total number of deliveries	548			
Total distance	156040			
Total sum of variable costs	234060,00	60278,63	58820,85	65035,82
Average variable unit cost of delivery	427,12	109,99	107,34	118,67

The next component adopted for the calculation are fixed costs. The only variant in which they do not occur is on-call delivery. In relation to the remaining three options, they had to be estimated. Additional assumptions were adopted for the estimation:

1) The vehicle used for the analyses was a Ford Transit; 2) The number of deliveries was a random variable and the individual months in a given year fell within the range $X \epsilon \{34 - 53\}$;
2) Due to the distance between network nodes and the number of deliveries in a given month, 4 vehicles and 4 drivers were used for analysis.

A summary of the fixed costs for the four options is presented in Table 3.

Table 3. Fixed cost breakdown for 4 options.

	Fixed costs/month [PLN]	Number of vehicles/drivers	Annual fixed cost [PLN]
Leasing	2214	4	106272
Long-term rental	2941	4	141168
Cash purchase	-	4	246976
Salary of the drivers	3514	4	168672

Taking into account the relationship (1) for the analysed network, we can therefore compare the components of fixed and variable costs.

The functions symbolizing the individual options are first-order curves, so it is possible to determine intersection points (Fig. 2), which essentially represent exactly the same total transport costs for the two compared options.

The next step is to estimate low, medium and high demand based on the supply scenario presented below, resulting from the market research conducted by the company in the last year. For this purpose, the studied group measured by the number of cruises in individual months (Table 4) should be divided into three intervals of the same width.

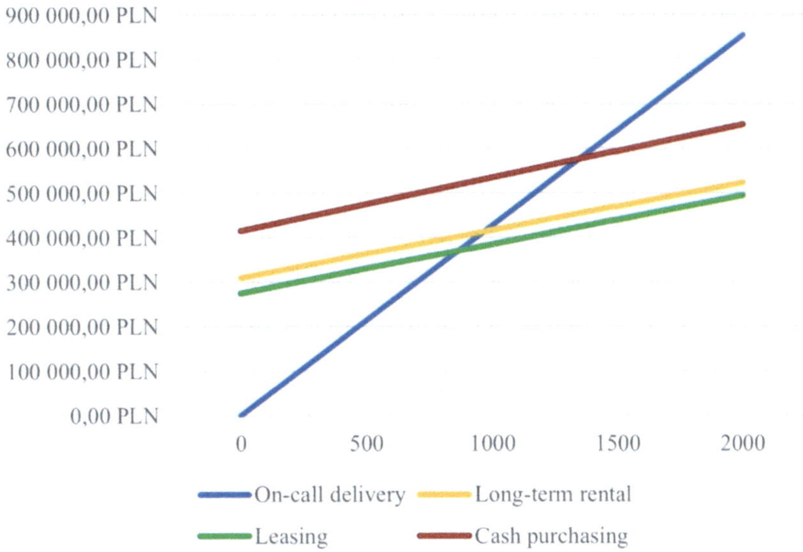

Fig. 2. Total costs depending on the number of deliveries.

Table 4. Demand volumes in individual months of the year.

Route/Month	I	II	III	IV	V	VI	VII	VIII	IX	X	XI	XII
S-R1	14	12	10	12	12	10	6	12	16	18	12	10
S-R2	8	1	12	16	12	7	16	14	15	7	18	8
S-R3	10	20	3	4	12	18	15	9	12	15	9	10
S-R4	2	16	14	14	5	9	11	18	8	12	13	11

In order to divide the population into three ranges symbolizing low, medium and high demand, the range should be determined mathematically and then the analysed population should be divided into three ranges of the same width, according to the Eq. (2):

$$R_x = (X_{max} - X_{\min})/3 \quad (2)$$

Then, based on the frequency of occurrence of individual values in a given range, the percentage share of a given range (demand level) should be determined. In the next step, the total value of cruises in each of the specified series is calculated. Based on this value, by dividing it by the number of recorded events in a given range, it is possible to determine the average demand for a given level (Table 5).

Table 5. Calculation of low, medium and high demand and estimation of percentage share.

Demand quantity/scenario	Number of deliveries/year	Upper limit	frequency	Lower limit	Compartment width	Sum of demand	Percentage share	Average arithmetic
Low	280	7	6	1	<1,7)	35	13%	6
Medium	492	13	25	7	<7,13)	256	52%	10
High	726	20	17	13	<13,20>	257	35%	15

Based on the above analysis, which determined the levels of individual types of demand along with the probability of their occurrence, the expected value was calculated in the following part. It is the sum of the products of the expected cost of each level of demand and the probability of occurrence of a given quantity, in accordance with the Eq. (3).

$$EW_j = \sum_{i=1}^{I} P_i C_i \tag{3}$$

where:

EW_j – expected value for the j-th option;

P_i – i-th scenario probability;

C_i – financial result for i-th scenario.

Then, taking into account the above values, costs were calculated for each scenario and for each transport option. They are needed to calculate the expected values for each option (Table 6).

Table 6. Total costs T_C for each scenario and calculating the expected value EW_j for each option [PLN].

Demand quantity/scenario	Low demand	Medium demand	High demand	EW_j
number of deliveries/percentage share [%]	280/13	492/52	726/35	–
On-call delivery	119 592,70	210 141,46	310 086,79	234 220,17
Long-term rental	339 894,44	362 649,95	387 766,88	368 701,09
Leasing	305 743,30	329 062,77	354 802,18	335 263,88
Cash purchase	448 877,98	474 037,82	501 808,59	480 728,32

Taking into account the level of intensity of the company's activities from the perspective of the expected value analysis, the best option is on-call delivery, which has the lowest expected value $EW_j = 234220{,}17[PLN]$.

4 Conclusions

When making decisions, it is also necessary to examine the company's break-even point, which is expressed mathematically by the following Eq. (4):

$$BE_P = \frac{F_c}{UR_g - AV_{UC}} \tag{4}$$

where:

BE_P – break-even point;
F_c – fixed costs;
UR_g – gross unit revenue (750 PLN);
AV_{UC} – average variable unit delivery cost.

Table 7. Total costs T_C for each scenario and calculating the expected value EW_j for each option [PLN]

	Fixed costs Fc [PLN]	Average variable unit delivery cost [PLN]	Break even point BEP	The number of deliveries from which the company earns the full net amount	Net revenue ratio on 1st delivery [%]	Net income on 1st delivery (for break even point)	Net income on each subsequent delivery
On-call delivery	0,00	427,12	0	1	–	322,88	322,88
Long-term rental	309 840,00	107,34	482,12	483	88,08	566,09	642,66
Leasing	274 944,00	110,00	429,60	430	40,17	257,08	640,00
Cash purchase	415 648,00	118,68	658,38	659	62,23	392,87	631,32

The calculations performed for the break-even analysis (Table 7) show that:

1. Contract transport has the lowest profitability threshold of $BE_P = 0$ deliveries. This is because in this case the company does not incur any fixed costs, but uses external services of a transport company in the on-call mode. In a situation of uncertainty and variability of demand, this is the safest option to choose considering the risk. In this case, from the first delivery, the company starts earning net at 322.88 [PLN].
2. The riskier options in the analyses conducted are long-term rental and leasing with profitability thresholds of $BE_P = 107{,}34$ and $BE_P = 110$ deliveries. For these variants, the company starts to make profits for leasing from the 430th delivery, while for long-term rental from the 483rd delivery, respectively, at 640 [PLN] and 642.66

[PLN] net for each subsequent one. In the case of leasing, the net profit for the 430th delivery is exactly 257.08 [PLN], while for the 483rd delivery implemented with the long-term rental variant, it is 556.09 [PLN] net.
3. The riskiest option is a cash purchase with the highest break-even point of up to

$BE_P = 659$ deliveries, which in practice has the lowest probability of implementation and is therefore the most difficult to achieve. For the above break-even point, the net profit is only 62.23 [PLN], while for each subsequent delivery the net profit will be 631.32 [PLN]. To sum up, this is a profitable option, but with a large and constant stream of applications.

In summary, the conducted break-even analysis demonstrates clear differentiation in the profitability thresholds and risk levels among the examined transport financing options [29]. The on-call delivery model proves to be the most secure solution under uncertain and variable demand conditions, while leasing and long-term rental become economically justified in the case of stable and predictable order volumes. Although the cash purchase option yields the highest unit profit in the long run, it also entails the greatest financial risk and requires a substantial and continuous demand stream to be effective.

The presented results are subject to several limitations. First, the study is based on a single SME case and a specific distribution network (one supplier and four fixed recipients), which may limit the direct generalizability of the numerical outcomes to other industries, network topologies, or geographic contexts. Second, the analysis horizon covers one calendar year and relies on an internally derived demand scenario structure (low/medium/high) and associated probabilities; changes in demand patterns over longer horizons were not investigated. Third, the comparison was conducted primarily from the cost perspective and does not explicitly quantify qualitative or strategic criteria (e.g., service level, flexibility, risk of disruptions, contractual constraints, or administrative burden) that may influence the final decision in practice. Finally, several simplifying assumptions were adopted (fixed routes, a single vehicle type, and stable unit rates), which do not capture potential variability in operating conditions such as fuel prices, labor availability, regulatory changes, or seasonal fluctuations. These limitations indicate that the results should be interpreted as decision support for the analyzed case rather than as universally applicable benchmarks.

Future research may focus on incorporating stochastic demand modeling and sensitivity analyses to more accurately assess the robustness of each financing model under changing market conditions.

Appendix

See Tables 8, 9, 10 and 11.

Table 8. Deliveries made by the transport company (on-call delivery).

Route	Distance [km]	Rate per 1 km [zł]	Cost of delivery	Number of deliveries	Percentage share [%]	Cost per month	Average cost/month	Total costs/month	Distance
January									
S-R1	311	1,5	466,5	14	41,18%	6531	2689,24	4573,94	9634
S-R2	340	1,5	510	8	23,53%	4080	960,00		
S-R3	198	1,5	297	10	29,41%	2970	873,53		
S-R4	290	1,5	435	2	5,88%	870	51,18		
February									
S-R1	311	1,5	466,5	12	24,49%	5598	1370,94	6078,49	12672
S-R2	340	1,5	510	1	2,04%	510	10,41		
S-R3	198	1,5	297	20	40,82%	5940	2424,49		
S-R4	290	1,5	435	16	32,65%	6960	2272,65		
March									
S-R1	311	1,5	466,5	10	25,64%	4665	1196,15	5333,92	11844
S-R2	340	1,5	510	12	30,77%	6120	1883,08		
S-R3	198	1,5	297	3	7,69%	891	68,54		
S-R4	290	1,5	435	14	35,90%	6090	2186,15		
April									
S-R1	311	1,5	466,5	12	26,09%	5598	1460,35	6255,39	14024
S-R2	340	1,5	510	16	34,78%	8160	2838,26		
S-R3	198	1,5	297	4	8,70%	1188	103,30		
S-R4	290	1,5	435	14	30,43%	6090	1853,48		
May									
S-R1	311	1,5	466,5	12	29,27%	5598	1638,44	4738,02	11638
S-R2	340	1,5	510	12	29,27%	6120	1791,22		
S-R3	198	1,5	297	12	29,27%	3564	1043,12		
S-R4	290	1,5	435	5	12,20%	2175	265,24		
June									
S-R1	311	1,5	466,5	10	22,73%	4665	1060,23	4615,98	11664
S-R2	340	1,5	510	7	15,91%	3570	567,95		
S-R3	198	1,5	297	18	40,91%	5346	2187,00		
S-R4	290	1,5	435	9	20,45%	3915	800,80		
July									
S-R1	311	1,5	466,5	6	12,50%	2799	349,88	5558,63	13466
S-R2	340	1,5	510	16	33,33%	8160	2720,00		

(continued)

Table 8. (*continued*)

January									
Route	Distance [km]	Rate per 1 km [zł]	Cost of delivery	Number of deliveries	Percentage share [%]	Cost per month	Average cost/month	Total costs/month	Distance
S-R3	198	1,5	297	15	31,25%	4455	1392,19		
S-R4	290	1,5	435	11	22,92%	4785	1096,56		
August									
S-R1	311	1,5	466,5	12	22,64%	5598	1267,47	6266,66	15494
S-R2	340	1,5	510	14	26,42%	7140	1886,04		
S-R3	198	1,5	297	9	16,98%	2673	453,91		
S-R4	290	1,5	435	18	33,96%	7830	2659,25		
September									
S-R1	311	1,5	466,5	16	31,37%	7464	2341,65	5976,12	14772
S-R2	340	1,5	510	15	29,41%	7650	2250,00		
S-R3	198	1,5	297	12	23,53%	3564	838,59		
S-R4	290	1,5	435	8	15,69%	3480	545,88		
October									
S-R1	311	1,5	466,5	18	34,62%	8397	2906,65	5876,94	14428
S-R2	340	1,5	510	7	13,46%	3570	480,58		
S-R3	198	1,5	297	15	28,85%	4455	1285,10		
S-R4	290	1,5	435	12	23,08%	5220	1204,62		
November									
S-R1	311	1,5	466,5	12	23,08%	5598	1291,85	6345,92	15404
S-R2	340	1,5	510	18	34,62%	9180	3177,69		
S-R3	198	1,5	297	9	17,31%	2673	462,63		
S-R4	290	1,5	435	13	25,00%	5655	1413,75		
December									
S-R1	311	1,5	466,5	10	25,64%	4665	1196,15	4144,23	11000
S-R2	340	1,5	510	8	20,51%	4080	836,92		
S-R3	198	1,5	297	10	25,64%	2970	761,54		
S-R4	290	1,5	435	11	28,21%	4785	1349,62		

Table 9. Deliveries made by the leasing

January									
Route	Distance [km]	Rate per 1 km [zł]	Cost of delivery	Number of deliveries	Percentage share [%]	Cost per month	Average cost/month	Total costs/month	Distance
S-R1	311	0,39	120,1	14	41,18%	1682,0	692,57	1177,95	9634
S-R2	340	0,39	131,3	8	23,53%	1050,7	247,23		
S-R3	198	0,39	76,5	10	29,41%	764,9	224,96		

(*continued*)

Table 9. (*continued*)

January									
Route	Distance [km]	Rate per 1 km [zł]	Cost of delivery	Number of deliveries	Percentage share [%]	Cost per month	Average cost/month	Total costs/month	Distance
S-R4	290	0,39	112,0	2	5,88%	224,1	13,18		
February									
S-R1	311	0,39	120,1	12	24,49%	1441,7	353,06	1565,42	12672
S-R2	340	0,39	131,3	1	2,04%	131,3	2,68		
S-R3	198	0,39	76,5	20	40,82%	1529,8	624,39		
S-R4	290	0,39	112,0	16	32,65%	1792,4	585,29		
March									
S-R1	311	0,39	120,1	10	25,64%	1201,4	308,05	1373,67	11844
S-R2	340	0,39	131,3	12	30,77%	1576,1	484,96		
S-R3	198	0,39	76,5	3	7,69%	229,5	17,65		
S-R4	290	0,39	112,0	14	35,90%	1568,4	563,01		
April									
S-R1	311	0,39	120,1	12	26,09%	1441,7	443,59	1900,13	14024
S-R2	340	0,39	131,3	16	34,78%	2101,5	862,15		
S-R3	198	0,39	76,5	4	8,70%	306,0	31,38		
S-R4	290	0,39	112,0	14	30,43%	1568,4	563,01		
May									
S-R1	311	0,39	120,1	12	29,27%	1441,7	376,09	1087,58	11638
S-R2	340	0,39	131,3	12	29,27%	1576,1	411,16		
S-R3	198	0,39	76,5	12	29,27%	917,9	239,44		
S-R4	290	0,39	112,0	5	12,20%	560,1	60,88		
June									
S-R1	311	0,39	120,1	10	22,73%	1201,4	293,02	1275,76	11664
S-R2	340	0,39	131,3	7	15,91%	919,4	156,97		
S-R3	198	0,39	76,5	18	40,91%	1376,8	604,44		
S-R4	290	0,39	112,0	9	20,45%	1008,2	221,32		
July									
S-R1	311	0,39	120,1	6	12,50%	720,8	98,30	1561,68	13466
S-R2	340	0,39	131,3	16	33,33%	2101,5	764,18		
S-R3	198	0,39	76,5	15	31,25%	1147,3	391,13		
S-R4	290	0,39	112,0	11	22,92%	1232,3	308,08		
August									
S-R1	311	0,39	120,1	12	22,64%	1441,7	360,42	1782,00	15494
S-R2	340	0,39	131,3	14	26,42%	1838,8	536,32		

(*continued*)

Table 9. (*continued*)

Route	Distance [km]	Rate per 1 km [zł]	Cost of delivery	Number of deliveries	Percentage share [%]	Cost per month	Average cost/month	Total costs/month	Distance
January									
S-R3	198	0,39	76,5	9	16,98%	688,4	129,07		
S-R4	290	0,39	112,0	18	33,96%	2016,5	756,19		
September									
S-R1	311	0,39	120,1	16	31,37%	1922,2	580,30	1480,98	14772
S-R2	340	0,39	131,3	15	29,41%	1970,1	557,59		
S-R3	198	0,39	76,5	12	23,53%	917,9	207,82		
S-R4	290	0,39	112,0	8	15,69%	896,2	135,28		
October									
S-R1	311	0,39	120,1	18	34,62%	2162,5	763,24	1543,19	14428
S-R2	340	0,39	131,3	7	13,46%	919,4	126,19		
S-R3	198	0,39	76,5	15	28,85%	1147,3	337,45		
S-R4	290	0,39	112,0	12	23,08%	1344,3	316,31		
November									
S-R1	311	0,39	120,1	12	23,08%	1441,7	332,70	1634,30	15404
S-R2	340	0,39	131,3	18	34,62%	2364,2	818,37		
S-R3	198	0,39	76,5	9	17,31%	688,4	119,14		
S-R4	290	0,39	112,0	13	25,00%	1456,4	364,09		
December									
S-R1	311	0,39	120,1	10	25,64%	1201,4	231,04	800,46	11000
S-R2	340	0,39	131,3	8	20,51%	1050,7	161,65		
S-R3	198	0,39	76,5	10	25,64%	764,9	147,09		
S-R4	290	0,39	112,0	11	28,21%	1232,3	260,68		

Table 10. Deliveries made by the long-term rental

January									
Route	Distance [km]	Rate per 1 km [zł]	Cost of delivery	Number of deliveries	Percentage share [%]	Cost per month	Average cost/month	Total costs/month	Distance
S-R1	311	0,38	117,2	14	41,18%	1641,3	675,8	1149,46	9634
S-R2	340	0,38	128,2	8	23,53%	1025,3	241,3		
S-R3	198	0,38	74,6	10	29,41%	746,4	219,5		
S-R4	290	0,38	109,3	2	5,88%	218,6	12,9		

(*continued*)

Table 10. (*continued*)

January									
Route	Distance [km]	Rate per 1 km [zł]	Cost of delivery	Number of deliveries	Percentage share [%]	Cost per month	Average cost/month	Total costs/month	Distance
February									
S-R1	311	0,38	117,2	12	24,49%	1406,8	344,5	1527,57	12672
S-R2	340	0,38	128,2	1	2,04%	128,2	2,6		
S-R3	198	0,38	74,6	20	40,82%	1492,8	609,3		
S-R4	290	0,38	109,3	16	32,65%	1749,1	571,1		
March									
S-R1	311	0,38	117,2	10	25,64%	1172,3	300,6	1340,45	11844
S-R2	340	0,38	128,2	12	30,77%	1538,0	473,2		
S-R3	198	0,38	74,6	3	7,69%	223,9	17,2		
S-R4	290	0,38	109,3	14	35,90%	1530,5	549,4		
April									
S-R1	311	0,38	117,2	12	26,09%	1406,8	367,0	1572,02	14024
S-R2	340	0,38	128,2	16	34,78%	2050,7	713,3		
S-R3	198	0,38	74,6	4	8,70%	298,6	26,0		
S-R4	290	0,38	109,3	14	30,43%	1530,5	465,8		
May									
S-R1	311	0,38	117,2	12	29,27%	1406,8	411,8	1190,70	11638
S-R2	340	0,38	128,2	12	29,27%	1538,0	450,1		
S-R3	198	0,38	74,6	12	29,27%	895,7	262,1		
S-R4	290	0,38	109,3	5	12,20%	546,6	66,7		
June									
S-R1	311	0,38	117,2	10	22,73%	1172,3	266,4	1160,03	11664
S-R2	340	0,38	128,2	7	15,91%	897,2	142,7		
S-R3	198	0,38	74,6	18	40,91%	1343,5	549,6		
S-R4	290	0,38	109,3	9	20,45%	983,9	201,2		
July									
S-R1	311	0,38	117,2	6	12,50%	703,4	87,9	1396,92	13466
S-R2	340	0,38	128,2	16	33,33%	2050,7	683,6		
S-R3	198	0,38	74,6	15	31,25%	1119,6	349,9		
S-R4	290	0,38	109,3	11	22,92%	1202,5	275,6		
August									
S-R1	311	0,38	117,2	12	22,64%	1406,8	318,5	1574,85	15494
S-R2	340	0,38	128,2	14	26,42%	1794,3	474,0		

(*continued*)

Table 10. (*continued*)

January									
Route	Distance [km]	Rate per 1 km [zł]	Cost of delivery	Number of deliveries	Percentage share [%]	Cost per month	Average cost/month	Total costs/month	Distance
S-R3	198	0,38	74,6	9	16,98%	671,7	114,1		
S-R4	290	0,38	109,3	18	33,96%	1967,7	668,3		
September									
S-R1	311	0,38	117,2	16	31,37%	1875,8	588,5	1501,84	14772
S-R2	340	0,38	128,2	15	29,41%	1922,5	565,4		
S-R3	198	0,38	74,6	12	23,53%	895,7	210,7		
S-R4	290	0,38	109,3	8	15,69%	874,5	137,2		
October									
S-R1	311	0,38	117,2	18	34,62%	2110,2	730,5	1476,91	14428
S-R2	340	0,38	128,2	7	13,46%	897,2	120,8		
S-R3	198	0,38	74,6	15	28,85%	1119,6	323,0		
S-R4	290	0,38	109,3	12	23,08%	1311,8	302,7		
November									
S-R1	311	0,38	117,2	12	23,08%	1406,8	324,6	1594,77	15404
S-R2	340	0,38	128,2	18	34,62%	2307,0	798,6		
S-R3	198	0,38	74,6	9	17,31%	671,7	116,3		
S-R4	290	0,38	109,3	13	25,00%	1421,1	355,3		
December									
S-R1	311	0,38	117,2	10	25,64%	1172,3	300,6	1041,47	11000
S-R2	340	0,38	128,2	8	20,51%	1025,3	210,3		
S-R3	198	0,38	74,6	10	25,64%	746,4	191,4		
S-R4	290	0,38	109,3	11	28,21%	1202,5	339,2		

Table 11. Deliveries made by the cash purchase

January									
Route	Distance [km]	Rate per 1 km [zł]	Cost of delivery	Number of deliveries	Percentage share [%]	Cost per month	Average cost/month	Total costs/month	Distance
S-R1	311	0,42	129,6	14	41,18%	1814,7	747,2	1270,91	9634
S-R2	340	0,42	141,7	8	23,53%	1133,7	266,7		
S-R3	198	0,42	82,5	10	29,41%	825,2	242,7		
S-R4	290	0,42	120,9	2	5,88%	241,7	14,2		

(*continued*)

Table 11. *(continued)*

January									
Route	Distance [km]	Rate per 1 km [zł]	Cost of delivery	Number of deliveries	Percentage share [%]	Cost per month	Average cost/month	Total costs/month	Distance
February									
S-R1	311	0,42	129,6	12	24,49%	1555,5	380,9	1688,97	12672
S-R2	340	0,42	141,7	1	2,04%	141,7	2,9		
S-R3	198	0,42	82,5	20	40,82%	1650,5	673,7		
S-R4	290	0,42	120,9	16	32,65%	1933,9	631,5		
March									
S-R1	311	0,42	129,6	10	25,64%	1296,2	332,4	1482,08	11844
S-R2	340	0,42	141,7	12	30,77%	1700,5	523,2		
S-R3	198	0,42	82,5	3	7,69%	247,6	19,0		
S-R4	290	0,42	120,9	14	35,90%	1692,2	607,4		
April									
S-R1	311	0,42	129,6	12	26,09%	1555,5	405,8	1738,12	14024
S-R2	340	0,42	141,7	16	34,78%	2267,3	788,6		
S-R3	198	0,42	82,5	4	8,70%	330,1	28,7		
S-R4	290	0,42	120,9	14	30,43%	1692,2	515,0		
May									
S-R1	311	0,42	129,6	12	29,27%	1555,5	455,3	1316,51	11638
S-R2	340	0,42	141,7	12	29,27%	1700,5	497,7		
S-R3	198	0,42	82,5	12	29,27%	990,3	289,8		
S-R4	290	0,42	120,9	5	12,20%	604,3	73,7		
June									
S-R1	311	0,42	129,6	10	22,73%	1296,2	294,6	1282,59	11664
S-R2	340	0,42	141,7	7	15,91%	992,0	157,8		
S-R3	198	0,42	82,5	18	40,91%	1485,4	607,7		
S-R4	290	0,42	120,9	9	20,45%	1087,8	222,5		
July									
S-R1	311	0,42	129,6	6	12,50%	777,7	97,2	1544,52	13466
S-R2	340	0,42	141,7	16	33,33%	2267,3	755,8		

(continued)

Table 11. (*continued*)

Route	Distance [km]	Rate per 1 km [zł]	Cost of delivery	Number of deliveries	Percentage share [%]	Cost per month	Average cost/month	Total costs/month	Distance
January									
S-R3	198	0,42	82,5	15	31,25%	1237,9	386,8		
S-R4	290	0,42	120,9	11	22,92%	1329,6	304,7		
August									
S-R1	311	0,42	129,6	12	22,64%	1555,5	352,2	1741,25	15494
S-R2	340	0,42	141,7	14	26,42%	1983,9	524,1		
S-R3	198	0,42	82,5	9	16,98%	742,7	126,1		
S-R4	290	0,42	120,9	18	33,96%	2175,6	738,9		
September									
S-R1	311	0,42	129,6	16	31,37%	2073,9	650,6	1660,52	14772
S-R2	340	0,42	141,7	15	29,41%	2125,6	625,2		
S-R3	198	0,42	82,5	12	23,53%	990,3	233,0		
S-R4	290	0,42	120,9	8	15,69%	967,0	151,7		
October									
S-R1	311	0,42	129,6	18	34,62%	2333,2	807,6	1632,96	14428
S-R2	340	0,42	141,7	7	13,46%	992,0	133,5		
S-R3	198	0,42	82,5	15	28,85%	1237,9	357,1		
S-R4	290	0,42	120,9	12	23,08%	1450,4	334,7		
November									
S-R1	311	0,42	129,6	12	23,08%	1555,5	359,0	1763,28	15404
S-R2	340	0,42	141,7	18	34,62%	2550,8	883,0		
S-R3	198	0,42	82,5	9	17,31%	742,7	128,5		
S-R4	290	0,42	120,9	13	25,00%	1571,3	392,8		
December									
S-R1	311	0,42	129,6	10	25,64%	1296,2	332,4	1151,51	11000
S-R2	340	0,42	141,7	8	20,51%	1133,7	232,5		
S-R3	198	0,42	82,5	10	25,64%	825,2	211,6		
S-R4	290	0,42	120,9	11	28,21%	1329,6	375,0		

References

1. Hadid, M., Irawan, M.Z., Parikesit, D., Firzan, F., Yunianti, N.H., Widiastuti, N.O.: Driving the future of sustainable public transport: a literature review on challenges and strategies in the adoption of autonomous buses. Discov. Sustain. **6**(1), 338 (2025)
2. Ma, C., Peñasco, C., Anadón, L.D.: Technology innovation and environmental outcomes of road transportation policy instruments. Nat. Commun. **16**, 4467 (2025)
3. Zheng, J., Wei, X., Zhang, Z., Xue, J., Chen, R., Xie, S.: A tri-level planning strategy for integrated power and transport systems incorporating EV elastic charging behaviors. Appl. Energy **396**, 126207 (2025)
4. Park, C.H., Lim, H.: A parametric approach to integer linear fractional programming: newton's and hybrid-newton methods for an optimal road maintenance problem. Eur. J. Oper. Res. **289**, 1030–1039 (2021)
5. AlKheder, S.: Experimental road safety study of the actual driver reaction to the street ads using eye tracking, multiple linear regression and decision trees methods. Expert Syst. Appl. **252**, 124222 (2024)
6. An, J., Mikhaylov, A., Jung, S.-U.: A linear programming approach for robust network revenue management in the airline industry. J. Air Transp. Manag. **91**, 101979 (2021)
7. Duc, K.N., Duy, T.N., The, T.N., Nguyen, Y.-L.T., Le, A.T.: Predicting energy consumption of electric two-wheelers under real-driving conditions in urban area: an application of artificial neural network. Energy Sustain. Dev. **87**, 101745 (2025)
8. Betkier, I., Oszczypała, M.: A novel approach to traffic modelling based on road parameters, weather conditions and GPS data using feedforward neural networks. Expert Syst. Appl. **245**, 123067 (2024)
9. Ziółkowski, J., Oszczypała, M., Małachowski, J., Szkutnik-Rogoż, J.: Use of artificial neural networks to predict fuel consumption on the basis of technical parameters of vehicles. Energies **14**, 2639 (2021)
10. Chen, D., Jia, W., Chen, Q., Chen, J., Li, Z.: Optimizing the compliance third-party supervision workflow of involved enterprises using artificial intelligence ant colony algorithm. Sci. Rep. **15**, 12202 (2025)
11. Kannan, V.V., et al.: Artificial intelligence based prediction and multi-objective RSM optimization of tectona grandis biodiesel with elaeocarpus ganitrus. Sci. Rep. **15**, 3833 (2025)
12. Piechowski, M., Meller, A.: Use of KPI indicators to optimize production process performance – case study. In: Mehmood, R., Alves, V., Praça, I., et al. (eds.) Distributed Computing and Artificial Intelligence, Special Sessions I, 20th International Conference (DCAI 2023). Lecture Notes in Networks and Systems, vol. 741, pp. 3–12. Springer, Cham (2023)
13. Rojek, I., Jasiulewicz-Kaczmarek, M., Piszcz, A., Galas, K., Mikołajewski, D.: Review of the 6G-based supply chain management within Industry 4.0/5.0 paradigm. Electronics **13**(13), 2624 (2024)
14. Liu, Z., Niu, Y., Guo, C., Jia, S.: A vehicle routing optimization model for community group buying considering carbon emissions and total distribution costs. Energies **16**, 931 (2023)
15. Zieja, M., Ziółkowski, J., Oszczypała, M.: Comparative analysis of available options for satisfying transport needs including costs. In: Proceedings of the 23rd International Scientific Conference "Transport Means 2019", Palanga, Lithuania, 2–4 October 2019, pp. 1433–1438. Kaunas University of Technology, Kaunas (2019)
16. Betkier, I.: Estimating travel time in transport network with a combined multi-attributed graph convolutional neural network and multilayer perceptron model. Eng. Appl. Artif. Intell. **142**, 109653 (2025)

17. Ziółkowski, J., Lęgas, A., Szymczyk, E., Małachowski, J., Oszczypała, M., Szkutnik-Rogoż, J.: Optimization of the delivery time within the distribution network, taking into account fuel consumption and the level of carbon dioxide emissions into the atmosphere. Energies **15**, 6648 (2022)
18. Loska, A., Paszkowski, W., Waszkowski, R.: The scenario approach to the concept of maintenance of technical systems of urban engineering. In: Trojanowska, J., et al. (eds.) Advances in Manufacturing IV: Volume 2 – Production Engineering: Digitalization, Sustainability and Industry Applications. Lecture Notes in Mechanical Engineering, pp. 84–97. Springer, Cham (2024)
19. Taran, I., Bikhimova, G., Danchuk, V., Toktamyssova, A., Tursymbekova, Z., Oliskevych, M.: Improving the methodology for optimizing multimodal transportation delivery routes and cyclic schedules in a transnational direction. Trans. Probl. **19**, 157–170 (2024)
20. Ziółkowski, J., Lęgas, A.: Minimisation of empty runs in transport. J. Konbin **48**, 465–491 (2018)
21. Ziółkowski, J., Żurek, J., Małachowski, J., Oszczypała, M., Szkutnik-Rogoż, J.: Method for calculating the required number of transport vehicles supplying aviation fuel to aircraft during combat tasks. Sustainability (Switzerland) **14**, 1619 (2022)
22. Borucka, A., Kozłowski, E.: Modeling the dynamics of changes in CO2 emissions from polish road transport in the context of COVID-19 and decarbonization requirements. Combust. Engines **195**, 63–70 (2023)
23. Sobczuk, S., Jaroń, A., Mazur, M., Borucka, A.: Renewable energy and CO2 emissions: analysis of the life cycle and impact on the ecosystem in the context of energy mix changes. Energies **18**(13), 3332 (2025)
24. Borucka, A., Sobczuk, S.: Analysis of the relationship between energy consumption in transport, carbon dioxide emissions and state revenues: the case of Poland. Energies **18**, 2291 (2025)
25. Alshabibi, N.M.: Impact assessment of integrating AVs in optimizing urban traffic operations for sustainable transportation planning in Riyadh. World Electr. Veh. J. **16**(1), 4 (2025)
26. Ziółkowski, J., Hošková-Mayerová, Š, Hutsaylyuk, V., Łukasiak, D.: Optimization of road transport within the supply network – a case study from Poland. Trans. Probl. **20**, 193–206 (2025)
27. Antosz, K., Jasiulewicz-Kaczmarek, M., Machado, J., Relich, M.: Application of principle component analysis and logistic regression to support six sigma implementation in maintenance. Eksploatacja i Niezawodność **25**(4), 174603 (2023)
28. Gładysz, P., Merkisz, J., Borucka, A.: Reliability of unmanned aerial vehicles in the context of selected factors. Eksploatacja i Niezawodność **28**(1), 210312 (2026)
29. Guzanek, P., Borucka, A., Patrejko, K.: Selection of a logistics operator for a retail company as part of supporting the company's expansion strategy. Eur. Res. Stud. J. **27**, 341–351 (2024)

Author Index

J. Trojanowska et al. (Eds.): MANUFACTURING 2026, LNME, pp. 379–380, 2026.
https://doi.org/10.1007/978-3-032-21654-0

The manufacturer's authorised representative in the EU is Springer
Nature Customer Service Centre GmbH, Europaplatz 3, 69115 Heidelberg,
Germany. If you have any concerns regarding our products, please
contact ProductSafety@springernature.com

Printed and bound by CPI Group (UK) Ltd, Croydon, CR0 4YY
21/09/2026
02219973-0010